HD
9506
.A576 Min es and mining in
M56   th icas
1986

HD   Szuprowicz, Bohdan
9506   O., 1931-
.A2
S98   How to invest in
1982   strategic metals

R00090 30840

DATE

BORROWER'S NAME

Miners and mining
in the Americas

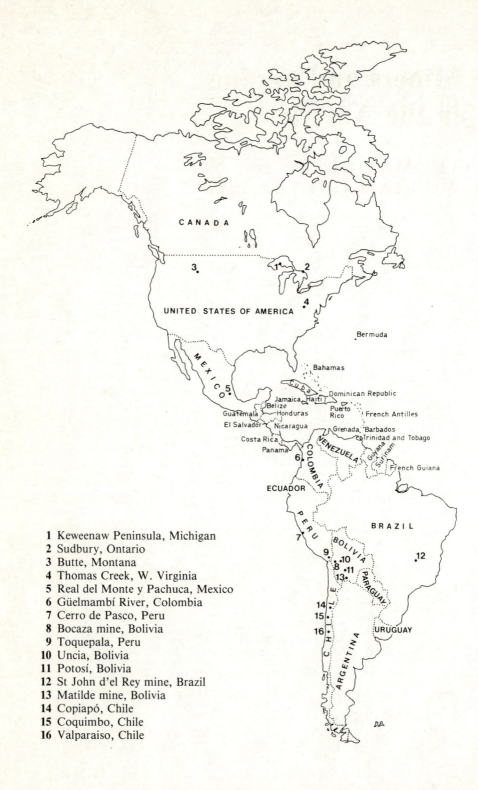

CANADA

UNITED STATES OF AMERICA

3.

1 2

4

5.

MEXICO

Bermuda

Bahamas

Cuba

Jamaica Haiti Dominican Republic
Belize Puerto French Antilles
Guatemala Honduras Rico
El Salvador Nicaragua
Costa Rica Grenada Barbados
Panama Trinidad and Tobago
6 VENEZUELA
COLOMBIA Guyana French Guiana
Surinam

ECUADOR

PERU BRAZIL

7
9 BOLIVIA .12
.10
8 .11 PARAGUAY
13

14 CHILE
15
16 URUGUAY

ARGENTINA

1 Keweenaw Peninsula, Michigan
2 Sudbury, Ontario
3 Butte, Montana
4 Thomas Creek, W. Virginia
5 Real del Monte y Pachuca, Mexico
6 Güelmambí River, Colombia
7 Cerro de Pasco, Peru
8 Bocaza mine, Bolivia
9 Toquepala, Peru
10 Uncia, Bolivia
11 Potosí, Bolivia
12 St John d'el Rey mine, Brazil
13 Matilde mine, Bolivia
14 Copiapó, Chile
15 Coquimbo, Chile
16 Valparaiso, Chile

Miners and mining
in the Americas

THOMAS GREAVES
WILLIAM CULVER *editors*

Manchester University Press

Published by Manchester University Press
Oxford Road, Manchester M13 9PL, UK
and 51 Washington Street, Dover
New Hampshire 03820, USA

British Library cataloguing in publication data
Miners and mining in the Americas.
  1. Mineral industries—North America
  2. Mineral industries—South America
  I. Greaves, Thomas   II. Culver, William
  338.2′097   HD9506.N6/

Library of Congress cataloging in publication data
Miners and mining in the Americas.
  Bibliography: p.277.
  Includes index.
  1. Mineral industries—America—History—Addresses, essays, lectures.   2. Miners—America—History—Addresses, essays, lectures.   I. Greaves, Thomas C.
II. Culver, William.
HD9506.A576M56   1985   338.2′097   85-13630

ISBN 0 7190 1761 0 *cased only*

Printed in Great Britain
by Bell and Bain Ltd., Glasgow

Contents

List of figures

For Margaret, and
For Cuca
Without whom, nothing

Miners and mining in the Americas, an introduction

WILLIAM W. CULVER and THOMAS C. GREAVES

Why publish a book about miners and their industry? Particularly, why this book, whose contributors report research spanning a hemisphere and two centuries, written from the perspectives of such varying disciplines as anthropology, economics, geography, history, political science and sociology? To answer this question we direct attention to two points.

First, although studies of mining in the western hemisphere have been a part of scholarship through much of the last century, it is our perception that more recently there has been a surge of interest from within various social science disciplines. This quickening of the research pace becomes evident about 1970. Some of the earlier contributions were, and remain, important to the study of mining topics and to theory development in our disciplines, but, with the exception of US historians, taken together they did not compose a stream of shared investigation.[1] Usually, they were isolated items. In our judgement, by the late 1960s this situation began to change. Perhaps for the first time these disciplines came to contain sufficient numbers of colleagues with an interest in mining-related research that a synergistic momentum took hold. That momentum has continued to build.

We can identify no single catalyst − a conference, a seminal publication, a major research grant − although in retrospect the 6th International Congress on Mining in Spain and Ibero-America held in Simanca, Spain, in 1973 was early and important. Instead, we suggest that the primary impetus came from a gradual accumulation of researchers in various disciplines discovering that miners offer high leverage for investigating work, technology, political conflict, worker solidarity and economic justice. As one might expect, the tendency continues for each discipline to delimit the purview of its practitioners, but there has been a growing and very salutary *leakage* across disciplinary boundaries. This book is a product of boundary perforations.

In 1982, two separate, multidisciplinary symposia were organised at the 44th International Congress of Americanists in Manchester, England, one by each of the two editors of this book. Culver's looked at the history and politics of Latin American mining; the other, organised by Greaves and Sue Turner, included contributions on North as well as Latin America, employing comparative behavioural science approaches. Most of the chapters in the present book began as papers at one or the other of these two symposia. Considerable further exchange took place at the 1983 Latin American Studies Association meeting in Mexico City, and, at this writing, we anticipate further acceleration to derive from the 1985 International Mining History Conference to be held in Melbourne.

Although the process is far from complete, studies of miners and mining have made considerable progress since 1970 toward a self-developing, multi-disciplinary scholarly field. An agenda of common research questions is taking on some reality. The cross-citation of benchmark sources is increasingly common, probably the most powerful index of, and force to deepen, pooled inquiry. The annotated bibliography in Part III of this book is an effort to accelerate that process. Especially gratifying is that our enterprise is multi-national, with active contributors from Britain, Europe, Latin America, North America and elsewhere. Although we continue to sense the effects of differing national and disciplinary scholarly agendas, we believe that at present the study of miners has made healthy progress in lowering paradigmatic and disciplinary partitions.

I A multidisciplinary study of mining

It is against this background that we and our contributor colleagues offer the present book. Its component chapters reflect the differences of format and objectives found in the various disciplines where mining is studied, as well as contrasting national scholarly traditions and priorities. But these contributors also illustrate fruitful dialogue among colleagues. There are strengthening nexuses of joint inquiry, some of which we will be listing below. There are discussions about units of analysis and the consequent constraints on explanation, and there are animated debates over the definition of terms and processes reflecting by their existence the progress of collective inquiry. The book is, at once, an index of the integrative process and perhaps a contribution to its furtherance.

The next thirteen chapters present contributions falling within a number of research areas on the current agenda of mining studies. These include: (a) the emergence of class and class consciousness (in an industry with an apparent radical labour tradition), including examination of the impact of enclave production on miner lifestyle and politics, (b) the multiple means of miners' adaptation to various production settings and political contexts, (c) the impact of export-dependent, mineral economies on local and national development or its retardation, and (d) the conjunction of politics and mine communities orchestrated by national, ideological and class-based interests. The above facets will now be elaborated and then exampled in the chapters to follow.

Emergence of class and class consciousness

The usage of the concept of 'miner' has an interesting evolution. During most of the nineteenth century the word usually referred to mine owners, their workers being described as employed labourers of various sorts. Today, 'miner' has changed from the capitalist to the person selling physical labour. Paralleling the terminological change as been our intensifying interest in the social and self-awareness of these workers, a phenomenon we know as class consciousness.

Mining has always entailed the systematic application of labour, but as technology and capital have moved the most productive mines beyond the small-scale diggings of the past, the entrepreneurial role shifted from those who toiled to others who capitalised technology, organised workers and profited from their sweat. Wherever this transition has occurred, the skills of the miner have evolved less rapidly than those of management. Mining originally called for personal acumen and the intangibles of long experience to evaluate the ore while doing a complicated series of manual tasks. Over the years, mining for most has become machinery operation. The eye for rich ores is no longer very necessary in the modern mines, which expect to rely on bulk processing and the expertise of white-collar specialists.

Miners changed and adapted, their efforts now managed by others, their toil seen as a cost to be minimised. Only collective groupings attended to the constant concerns for safety, personal support and defense of common interests. Julian Laite's research (reported in Chapter 6) analyses this change during the rise of the nickel mines at Sudbury, Ontario. In other chapters different parts of the process are singled out: Cuauhtemoc Velasco (Chapter 3) taps the archival riches of the great Mexican silver mine, Real del Monte, during the mid-nineteenth century to examine the mixture of labour arrangements occurring during the transition to capital-rich technologies. It was out of such origins that later forms of worker solidarity arose. David Becker (Chapter 13), on the other hand, looks at the period of late, high-investment capitalism in Peru and the abrasive union-management confrontations which occurred there. He finds a truncation of class consciousness development. Then Francisco Zapata (Chapter 14) examines what for many is a subsequent industrial stage, that of state capitalism in the post-nationalisation period. Here there are new players, and a different set of forces shaping worker class identity. Finally, the Greaves—Albo—Sandoval research (Chapter 10) raises perhaps a more fundamental question: to what extent is class consciousness necessarily a product of the industrial work environment? Their evidence from former peasants suggests proletarian features may be more a product of situation than any evolutionary stage.

Adapting to production and political settings

Miners adapt, just like any other group that confronts a demanding social and biotic environment. Mineral extraction, however, imposes peculiar adaptive requirements. Take, for example, the simple, often seasonal mining activities that continue today in many isolated parts of the hemisphere where ores have never attracted or no longer attract the attention of capitalists. Nina de Friedemann's material (Chapter 12), for example, demonstrates how former slaves on Colombia's western litoral use intricate rules of kinship to organise access to placer mining sites. De Friedemann's work among the Güelmambí blacks is rare ethnographic reportage of mining in an indigenous culture.

Another set of adaptations falls to the peasant-miner, found in most Latin American mining countries. In many ways these people represent a work pattern that was most common during the last century. This form of mining essentially ceased in Canada and the United States, but it continues to employ thousands of men and women in nations to the south. Ricardo Godoy's work (Chapter 8) examines peasant-based small-scale mining in highland Bolivia. Operating within a context of labour scarcity and locally enforceable concepts of usufruct, these Jukumani peasants not only interweave the fruits of mining with their agricultural economy, but are also able to demand acquiescence from Bolivian capitalists – surely a rare adaptation in twentieth-century mining. Guillermo Delgado (Chapter 9) argues that women in Bolivia's largest tin mine, more so than men, use peasant-based strategies to cope with the harsh and scarce life options at the mine, including occasional agriculture and multi-stranded exchange relationships with peasants in other areas. Mining's adaptative effects are also evident in Patricia Sach's study of retired West Virginia coal miners (Chapter 11). Long after they have ceased work, and, in fact, after the mine itself had closed, the attitudes, styles and meanings of work continued. All of these chapters affirm that mining is deeply intrusive on the lives of any who practice it. Adaptations and adjustments are as necessary for the villagers as for any hard-hatted proletarian.

The consequences of mineral export economies for national futures

Mining schools in the United States and Canada rarely question the 'staple theory of growth', that economic development in periphery countries originates in the capitalisation of export industries producing the staples of world consumption.[2] Yet, one would be hard-pressed to find any hemisphere-dwelling scholar or planner beyond US borders who would allow the theory any credence whatsoever. In the mineral-exporting economies of Middle and South America mining has severely distorted the national economies, with wealth, technology transfer and jobs being traded against heavy concessions of national autonomy (whether or not the mines are nationalised), bitterly contentious labour relations and chronic political stress.

At the same time, the mineral economies of the hemisphere are so desperate for income and development options that national leaders must regard mining as a strategic resource. Mining produces critical foreign exchange for many Latin American and Caribbean nations, employs and concentrates sectors with immense political potency, and holds the promise of critical revenues in these tax-poor countries.

Nowhere is the predicament quite as stark as in Bolivia, a nation at this writing regarded by its external creditors as near default on its foreign debt. David Fox's essay (Chapter 7) describes the texture of mining in that beleaguered nation and, even under national ownership, the mine's voracious distruction of national aspirations. Somewhat more manageable situations are described by Marshall Eakin (Chapter 2) for nineteenth-century Brazilian gold mining, by William Culver and Cornel Reinhart (Chapter 5) for a similar period in Chile's early capitalist copper industry, and by Cuauhtémoc Velasco for Mexican silver (Chapter 4). In each case, national development policy strongly attended to the bountiful visions of foreign capitalists and their investments, only to see them wither. In Chile's case, the decisions of entrepreneurs brought the industry to near collapse while profitable ore was yet available. In Mexico, the eventual winners were not the foreigners, but Mexican investors. Thus, while mining can be viewed as a productive process unto itself, its continuous role as a national strategic resource cannot be put aside.

The conjunction of politics and mine communities

The map of the Americas is dotted with the names of once well-known mining communities now but a shadow of their former eminence: Potosí, Real del Monte, Comstock, Chanarcillo, Keweenaw Peninsula, Tamaya. While some continue as residences for miners, and the production of mineral, the ores have declined, and with them, the social organisations for extracting them. When the mines in these regions opened they promised to lift individuals and even whole nations into the bosom of material progress, but in time the mines spawned communities of marginality, if not poverty. Was all this inevitable? Were the expectations for the power of mineral wealth exaggerated or misconceived? Were errors made in the application of the wealth the mines produced?

These questions animate the consternation not only of the miners, whose stake is personal and immediate, but also of the larger nation, whose fortunes, national aspirations and patriotic sensibilities turn on events in the mine centres. The mining community, is, thus, an arena in which all are players, none are spectators, and the outcomes are uncertain and real, rewarding or painful. In those countries and at those times, when national options rest on minerals, the combat is literally bloody. Tragedy, usually for the miner, marks the end of each inning.

David Becker's research (Chapter 13), examines the combat through half a century at Peru's Cerro de Pasco mine. John Mayo's work (Chapter 3) does the same for Chile in the nineteenth century. Attention is also paid to the local mining arena in the Greaves–Albo–Sandoval contribution (Chapter 10) for Bolivia, and accounts along the same lines can be found in several other chapters. Like other sites of contemporary social science research – factories, ghettos, peasant villages, corporations – their actions and histories are un-intelligible without understanding their role as arenas, without apprehending both the local and the external contestants, without considering the objects of dispute.

II Saliences of mining

The international aspect of mining lends itself to consideration of the questions of world systems, trade and dependency. In mining, events affecting production in one country are relevant to producers in all countries. It is a system full of consequences, intended and unintended, stemming from reforms, investments, strikes, discoveries and inventions. Little is predictable for long. This thick mixture of human efforts provides the grist for a number of areas of scholarly milling, some of which are suggested below.

Political questions

The study of mining implies linkage to a series of important political questions. Mining labour has a long standing reputation for radical politics, and for being a leading element in the unionisation of working people. Several of the chapters of this book examine the contemporary role of mine labour, with findings not entirely consistent with the reputation. Especially the large, highly capitalised mines of Peru and Chile have recently seen militant mine unions act carefully in strict attention to their own narrow interests.

Another political theme in these chapters is nationalisation. Ownership of mineral wealth, and its distribution are topics as old as mining itself. In Hoover's excellent notes to his translation of Agricola's *De Re Metallica*, the history of this theme is traced. Mineral wealth has always belonged to the most powerful social sector: the prince, the landowner, the merchant or, theoretically, the miner. In the United States the issue was settled soon after the Civil War by a series of capitalist mining policies put in place by Congress. As early as the acquisition of Mexican lands after the war with Mexico, mining commissioners recommended scrapping the Mexican code as being anti-capitalist. That same code, or a close variation, was the governing law in all of Latin America until much later, and became the source of much agitation. But as North American capital began entering and dominating mining after the turn of the century, nationalisation calls began to be made. Mexico after 1917, Bolivia in 1952, Peru after 1968, Chile after 1971 all had some form of nationalisation of major mining properties.

The anonymous industry

A related facet of the above reasons for increased attention to mining and miners is the generally low profile of the industry among scholars. Most scholars find themselves ignoring, if not ignorant of, mining. Not only does mining take place in remote places, but few people are engaged in it. Mining has become to most of society an invisible industry. Yet the exploitation of minerals is the foundation of modern industrial society. Additionally, at the very time when society may be overlooking the contributions of miners and of the mining industry, the owners and management of the industry are in the midst of a crisis of defensiveness. In recent years the public criticism of environmental consequences of mining operations the world over has made the industry shy. However well justified the criticism of the industry's environment record, intellectual isolation helps no one. There is need for mining engineers and metallurgists to dialogue with historians and social scientists. The low profile of mining needs not to be lower, but to become higher. At recent social science and Latin American studies conferences, panels on mining have included engineers and government mining officials. Theory or policy, the studies in this collection continue this dialogue.

The mining cycle

Mining, as compared to miners, is a concept that has not evolved sociologically. It has, however, responded to changes in technology. Technology in turn has continually evolved the actual process of mining. At its most general level mining is best understood as a process. At any one time a mine may be at one of several stages: discovery, development, production, suspension or closing. The exact nature of each stage varies with decades, geography and geology, and the nature of each mineral. The most dramatic trend over the past several centuries has been for each stage to become more involved in its own way with the allocation of capital, although the application of capital to mining has not been at the same rate in all countries, or evenly distributed among segments of the industry in individual countries. In fact, in Latin America the capitalisation of mining has led to a sort of social stratification of mines tied to the access to capital, while in North America small-scale mining has been largely absent since the 1880s. The role of labour in the mining process is dependent on the amount of capital invested in the mine project, as is productivity and often the resulting wages paid to labour. At the managerial level, control has shifted from the backer of prospectors, to the mine engineer, to the metallurgist and to the financier. Today, access to financing makes or breaks all but the smallest of mines.

III Structure and debt

For the convenience of manageable comparison, we have divided the chapters into two broad sets. In rough terms, the first set (Part I) deals with mining 'from above', exploring the ways in which the mine industry impacts both the lives of miners and the lives of countries. Unlike the approach to mining taken by the engineer or the resource economist, these chapters do not search for ways to combine labour, capital and technology more efficiently. Rather, the focus is on social relationships between mines as a place of employment and mines as a source of wealth. The time span of the first set of chapters ranges from the 1830s to the present, although most are focused on the last half of the nineteenth century, the years of intense world industrialisation of mining, and the period of transformation of world mining from a labour-intensive industry to a capital intensive one.

The second set (Part II) reflects more direct and contemporary observation of communities of miners. The commonality among them is their concern with how miners respond to the strictures and exactations of the mine and the industry. Observation and research among miners does not yield simply case-level sorts of information, oblivious to the distant action centres which delimit the options at the minehead. When ethnography is treated as text, the larger system can be mapped in the political and expressive behaviour of those groups that must act within it (*cf.* Hebdige 1979).[3] Seen in this light, the tired obeisances to the superiority of regional/system-level approaches to the ethnography of local communities lose cogency.

In Part III we offer an annotated bibliography of mining sources produced since 1970. It is our hope, shared with our colleague, Sue Turner, that this multidisciplinary and indexed source will serve the collaborative efforts of scholars across many disciplines and many countries.

Debts are a companion of the mining life, and also of a book about mining. Our debts, however, we acknowledge with gratitude and the hope that this volume will be received by them as a reasonable product of their collaborative good will. The unstinting support of Dr Roberto Etchepareborda, scholar and director, Department of Cultural Affairs of the Organization of American States, and of Gloria Loyola, chief of the Departamento's Technical Unit of Cultural Interdisciplinary Researches, were essential to the success of one of the two panel-workshops at Manchester from which this book originated.

We also acknowledge our respective universities, Trinity University and the State University of New York, Plattsburgh, for fostering an environment where research and learning is supported with graciousness, and to the several staff associates and our own family members who lent time and toil to this project: Wanetta Murtaugh, Doris McKinstry, Thomas Johnston, Sue Turner, Jo Ann Wadkins, and Margaret, Mark and Adrienne Greaves. Through the efforts of all, the book is done.

We also wish to thank the staff of Manchester University Press for their patience and support.

Notes

1 US historians have, since early in the twentieth century, written about mining, particularly regarding the bonanza strikes of the western frontier and the mining towns they generated. Among Latin American historians a similar pattern of studies of mining enterprises exists.

2 See the first section of Laite's paper, Chapter 6, for further context and citations.

3 Dick Hebdige, *Subculture; the Meaning of Style*, London, Methuen, 1979.

PART ONE

National and multinational underpinnings to mining

CHAPTER 2

The role of British capital in the development of Brazilian gold mining

MARSHALL C. EAKIN

Eakin's chapter places the British mining company, St John d'el Rey, within the context of the little understood nineteenth-century Brazilian mining. The St John d'el Rey is judged the most successful British mining operation in nineteenth-century Latin America. Eakin greatly expands our perspective on mining during the last century through his coverage of Brazilian mining policy, British capital, the British entrepreneur and the limits of Brazil's geology. He concludes that no matter how favourable the policy, abundant the capital, or skilful the entrepreneur, a lack of economic mineral deposits cannot be overcome. British capital in nineteenth-century Brazilian gold mining was an enormously powerful investor in an extremely weak industry.

Marshall C. Eakin is Assistant Professor of History, Vanderbilt University, Nashville, Tennessee 37235. Currently he is working on a study of the relations between business entrepreneurs and commercial development in Brazil, continuing his focus on the St John d'el Rey Mining Company.   —*The editors*

In the eighteenth century Brazil experienced the western world's first great gold rush, foreshadowing by more than a century the great rushes to California, Australia, Alaska and South Africa in the nineteenth century. Brazil became the world's leading producer of gold bullion before the gold mining industry fell into decline in the last quarter of the eighteenth century. Much has been written about Brazil's 'century of gold', but Brazilian gold mining fades from the history books as soon as the narrative enters the nineteenth century. Despite the neglect of historians, the Brazilian gold mining industry did not disappear with the advent of the nineteenth century. Historiography to the contrary, Brazil experienced a gold mining 'boom' and the revival of the industry in the nineteenth century. With Brazilian independence and the arrival of foreign capital in the early nineteenth century, the moribund gold mining industry came back to life and Brazil once again became one of the world's important producers of gold bullion. Although not as impressive nor as significant in world historical terms, gold mining in Brazil did once again rise to prominence in the nineteenth century (Phillips 1867:127; Curle 1905:1–3).[1]

This chapter focuses on the role of British capital in this gold mining boom.

The role of the British must be viewed in its larger context as part of the world-wide economic expansion of Great Britain in the nineteenth century, especially as part of the search for precious metals in Latin America. The aim of this study is to contribute to the growing literature on the history of British involvement in metals mining in nineteenth-century Latin America by reconstructing an important and neglected episode in that history. This study will shed new light on the role of British capital in metals mining not only in Brazil, but in the rest of Latin America as well. After briefly surveying the development and decline of gold mining in eighteenth century Brazil this chapter focuses on the size, nature and directions of British investment in the Brazilian gold mining industry. The British presence in nineteenth-century Brazilian gold mining will be broken down into three major periods of peak activity: the 1820s and 1830s, the 1860s, and the last quarter of the century. A detailed description of British companies active during these periods forms the core of the essay. Special emphasis is placed on the failures and major successes of the British efforts and an analysis of the reasons for failure and success forms the next section of the chapter. Finally, an overall vision of the effects of British capital on Brazilian gold mining closes this study.

I

Human society in the Brazilian interior has taken shape on top of some of the oldest geological formations known to man. The bedrock which underlies most of the region dates back some 2.8 billion years. During the last billion years the geological structure has been alternately uplifted, heavily eroded, buried beneath the sea, covered with layers of sediment, and crumpled and crushed by movements of the earth's crust. About 500 million years ago the intrusion of superheated magma from deep within the earth recrystallised minerals in the ancient bedrock, forming major deposits of iron, manganese, gold and other valuable minerals (Dorr 1969; Gair 1962).

During the hundreds of millions of years up to the present, the forces of nature have heavily eroded the uplifted portions of Minas Gerais producing a rugged, mountainous terrain, and gradually exposing the deposits of gold. Heavy winds and rains have cut deep valleys making access, except that by river, exceedingly difficult until recent years. The highlands produced by uplifting and erosion rise up quickly from the Atlantic coast, averaging between 300 and 1,000 metres at their highest point. The principal mountain range is the Serra do Espinhaco (Backbone Range) which runs roughly north to south on a line through Diamantina and Ouro Preto. This range separates the two river systems of the area. To the east the Rio Doce drains south and east into the Atlantic, and to the west the São Francisco and its tributaries (one of Brazil's three major river systems) drains to the north and east. Branching off from the Serra do Espinhaco are a number of smaller ranges which crisscross the gold-bearing area of central Minas Gerais.

Much like the gold rushes of the nineteenth century, the influx of thousands of prospectors to this mountainous region of central Brazil in the early eighteenth century gave rise to severe social and economic problems as well as political confusion. The foremost dilemma facing Portuguese authorities was how to regulate and tax effectively the production and circulation of gold. As in Spanish America, the Portuguese Crown demanded its royal fifth (quinto), but was hard pressed to find ways to collect it from the fiercely independent and widely dispersed miners. As the Spanish had two centuries earlier, the Portuguese immediately acted to construct a political and fiscal apparatus to control the mining region. Unlike the silver mines of central Mexico and Andean Peru, the gold mines of central Brazil were widely scattered and separated by rugged terrain, which made policing of the area practically impossible. Consequently, attempts by the Crown to establish central mints and to regulate production and circulation met much less success than in Spanish America. The shifting and transitory nature of placer mining demanded solutions radically different from those of the centralised, deep-shaft mines of the Spanish empire in the New World. In the first three decades of the eighteenth century the Crown elevated several of the mining settlements to the status of towns, and in 1720 the area of the mines was separated officially from the captaincy of Sao Paulo. By the 1730s royal correspondence regularly referred to the new captaincy as Minas Gerais (General Mines).

The towns of Sabara and Mariana sit roughly on the south-west and north-east corners of an imaginary rectangle which encompasses the principal gold-bearing deposits of Minas Gerais. The early prospectors followed the rivers of the region upstream, tracing the alluvial gold back to its origins in search of the mother lodes. Mining in this manner consisted of little more than placer washing techniques. The prospector would gather up the gravelly sands (cascalho) of the rivers in a wooden bowl (bateia), and then by gently rotating the bowl the lighter material would be tipped out leaving the heavier gold particles on the sides of the bateia. This backbreaking and tedious procedure dates back to ancient times in Europe and the Middle East, and may have been known to many of the slaves brought to Minas Gerais from the gold-bearing regions of Africa (Rolff 1976).

As the alluvial gold gave out and the colonists reached the hillsides from which the gold had been washed and weathered down the miners adapted their techniques. Employing groups of slaves, mining empresarios cut trenches and terraces in hillsides and then diverted streams and rivers to wash down through the cuts. The slaves kept the loose gravel and water moving across the cuts until the cascalho reached either crude stamps and sluices or workers using bateias. Waterwheels powered the wooden stamps (sometimes with iron heads) which crushed the cascalho into finer particles for the sluices or the washers with bateias. Animal hides (or fleece to which the gold particles adhere because of the vil) caught the heavy gold particles and systematic washing of these hides removed the accumulated gold. Essentially this type of hydraulic mining was a more highly evolved form of alluvial panning. Unlike the labour intensive, low technology panning process, hydraulic mining required more capital investment, more machinery (stamps, tools for excavating terraces and hillsides and aqueducts), and the management of collective labour. From this standpoint, hydraulic mining represented a large step forward in mining technology. A century later the same techniques would play a prominent role in the California gold rush (Mawe 1815: 78–9; Eschwege 1979: 167–80).

The third major type of mining to develop in Minas Gerais was also the least common – shaft mining. Deep shaft mining had been underway in Mexico and Peru for over two centuries. The Portuguese had little mining experience and apparently knew little about deep-shaft technology. The Spanish Hapsburgs controlled the silver-rich mining regions in central Europe and with the discovery of silver in the New World the Spanish Crown merely transferred its German experts and their expertise to the overseas colonies. Although some historians have attributed the lack of shaft mining on a large scale to the nature of the rock (friable) in Minas Gerais, the success of deep shaft mining in the nineteenth century compels us to reject this hypothesis. More likely, two factors hindered the development of shaft mining in eighteenth-century Brazil. First, shaft mining forms the last historical and evolutionary phase in gold rushes, and takes place after alluvial and easily recovered surface gold become exhausted spurring the search for subterranean sources. Subsequent centuries have shown that few well-developed lodes exist in Minas Gerais, making the possibilities for shaft mining in the region limited, especially for eighteenth-century miners with their scant knowledge of shaft mining techniques.

Second, the lodes which did exist could only be developed on an economically viable scale when the mining techniques of the Europeans (especially the Cornish) became readily available to Brazilians. Until these techniques became widely available in the nineteenth century the typical lode in Minas Gerais consisted of a shallow cut or open pit worked by slaves using crude tools. For unknown reasons Brazilian mining entrepreneurs showed little interest in shaft mining technology, and received little encouragement to employ new technology. Consequently, mining went into a long-term decline as soon as the miners had exhausted the major sources of alluvial and surface gold. After 1770 gold production rapidly fell off and the 'Golden Age of Brazil' fell victim to technological backwardness.

The erratic fiscalisation of gold and the probability of widespread contraband in eighteenth-century Brazil make estimates of gold production a guessing game. The most widely cited figures come from João Pandia Calogeras' early twentieth-century study, in which he carefully reviews all previous sources and calculations (Calogeras 1904–5). Calogeras estimated that Brazil produced one million kilograms of gold from 1700 to 1820. Apart three-quarters of the total production came from Minas Gerais, and approximately three-quarters of this bullion was produced prior to 1780. Traditionally, historians placed the beginning of the mining decline around 1750. More recently, however, experts have placed the onset of the decline two decades later. Whichever date one chooses, it is clear that by the beginning of the nineteenth century gold production in Brazil had declined drastically (Pinto 1979).

II

By the beginning of the nineteenth century the once productive Brazilian gold mining industry had settled into near-total stagnation. W. L. von Eschwege, a Prussian mining engineer in the employ of the Portuguese Crown, surveyed the mining industry in the second decade of the century and found it severely deficient in technology, capital investment and entrepreneurial talents. According to Eschwege's survey, the gold-mining industry in Minas Gerais contained many small and medium-sized works operating on a shoestring. He catalogued 555 mining claims (lavras), the vast majority employing ten slaves or less. A mere seventy employed more than twenty slaves, with but two employing over 100 slaves. Less than 200 free labourers worked on these lavras. Eschwege attempted to stimulate interest in the more important mines and found the Brazilians unwilling or unable to invest in mining. He scorned the Brazilian mine owners for refusing to upgrade their primitive technology. Eschwege found the use of water power extremely limited, and mining operations wasteful of labour power. No steam engines were in use. Gold production, which had averaged well over seven million grams annually in the mid-eighteenth century in 1814, barely reached 400,000 grams from mines with an equal amount coming from placering. In the 1850s a single mine in Minas Gerais would be producing at this level (Eschwege 1979: 49).

The transfer of the Portuguese royal family and court to Rio de Janeiro in 1808 began the transformation of the gold-mining industry which would result in the nineteenth century boom. The Regent and later King João VI opened the country to foreign commerce and merchants in early 1808, and by the following decade more than one hundred British merchants resided in the major towns of the coast and the mining zone. One of the most famous of these merchants, John Mawe, traversed the mining region and produced a perspicacious travel account publicising its potential to British investors. Restrictions remained, however, on foreign investment in mining (Mawe 1815). Following the declaration of Brazilian independence in 1922 Dom Pedro I struck down the last of the old colonial obstacles to foreign investment in Brazilian gold mining.

In the colonial period all subsoil rights legally belonged to the Crown and the miners were granted the use and benefits of the subsoil by the agents of the king. In return for this concession on mineral rights miners were required to pay the royal quinto which, despite the name, varied widely from the theoretical one-fifth of production. This policy of granting mining concessions was continued under the empire with the major difference being that foreigners as well as nationals were allowed to receive concessions. The constitution of 1824 made the entry of foreign capital into Brazilian mining a legal possibility for the first time in Brazilian history. Foreigners receiving mining concessions did have to meet certain special requirements. First, they had to pay 5 per cent higher taxes on production than Brazilians. Second, they had to offer up to one-third of mining company stocks to Brazilian investors. Finally, the mining companies were required to pay a deposit of 150,000 milreis (about $75,000) to the treasury as a guarantee against payment of taxes and duties. These qualifications were minor and did not hinder the entry of foreign capitalists in the least. The constitutiton of the empire opened the gate to foreign capital and the British stood in the best possible position to pass through the gate (Pasta Histórica).

Great Britain by the third decade of the nineteenth century had begun to ride a wave of economic expansion and prosperity growing out of a half-century of industrial revolution and the peace of post-Napoleonic Europe. The former had brought about an enormous rise in capital formation and industrial production, requiring the British to crisscross the globe in search of markets for their goods and opportunities for their capital. The latter made access to those markets less problematic. The revolutions in Latin America in the 1820s provided the perfect answer to British needs, and in the century following the wars of independence Great Britain inundated the Latin American republics with investments and manufactured goods. British investors pumped some twenty-five to thirty million pounds sterling into Latin America in the 1820s with Brazilian projects taking the single largest share of this capital. The financial and speculative 'bubble' burst in the late 1820s and investment slowed considerably for the next quarter-century. Investment accelerated after mid-century and by 1880 British investments in Latin America probably totalled more than £180 million sterling with Brazil receiving one-fifth of that investment (Rippy 1952).

The major portion of British capital investment funded the construction of railways, ports, communication systems, public utilities and other infrastructural improvements. In 1880, for example, one-fifth of the total capital went to railways, about 6 per cent to public utilities, and about 2 per cent each to shipping, banking and mining ventures. The total British investment in Latin American mining hovered around three and one-half million pounds sterling for most of the nineteenth century, ascending to more than twenty million in 1890 and staying in that vicinity into the 1930s. The major focal points of British mining investment in Latin America were Brazil, Colombia, Mexico and Chile, with Peru joining the group in the later part of the century. Gold mining companies were concentrated in Brazil, Mexico and Colombia. Almost without exception the mining companies funded by British investors in nineteenth-century Brazil were gold mining enterprises. Brazil accounted for some 6 per cent of British capital invested in Latin American mining ventures in the 1820s and probably half that in the 1890s (Rippy 1952: 41; 1954: 43; 1959: 24 and 50).

British investment in Brazilian gold mining during the nineteenth century flowed in three major waves. The first wave appeared in the 1820s and lasted into the 1830s. Great Britain had adopted the gold standard in 1816 and other European nations followed her lead in the succeeding decades. The British eagerly sought to stimulate the production of Latin America's precious metals which had declined in the late colonial period and practically ceased during the revolutionary upheavals. This rush for gold and silver began with the floating of joint-stock ventures, the first great venture being the Real del Monte in Mexico. By 1824–5 speculation ran rampant in London financial markets as everyone from clerks to the highest politicians scrambled to get into the American mining ventures. The Lord Chancellor himself speculated in these ventures and a young broker's clerk named Benjamin Disraeli published pamphlets on the American mining companies (Randall 1972; Jenks 1927; Disraeli 1825). This wave of investment ebbed in Latin America after the financial crisis of 1825 in London, although investment in Brazilian gold mining companies continued into the early 1830s.

A second and somewhat more intense wave swept into Minas Gerais in the 1860s, primarily on the success of previous ventures. The third, and least intense, wave of investment stretched across the last quarter of the nineteenth century. The mining ventures which arose out of these waves, by and large, did not last long. Of those companies established in the first wave only two survived mid-century. None of those companies founded in the 1860s survived the century, although two lasted into the 1890s. Those founded in the third wave (with one exception) disappeared by 1905. Thus only two of the companies founded in the nineteenth century survived the first decade of the twentieth century. It should be noted that Brazil was not exceptional in the failure of mining ventures. Mining has always been a high risk enterprise and Latin America has been the graveyard of many a British mining venture. J. Fred Rippy estimated that Englishmen invested in approximately 400 mining companies in Latin America after 1820. Less than a score became profitable! The nineteenth-century Brazilian gold mining boom gave birth to several of those profitable companies, including the most successful of all the British gold mining ventures in Latin America (Rippy 1952: 32; 1948: 72).

III

The first British mining venture in Brazil also proved to be one of the most profitable. Edward Oxenford, an English merchant, had installed himself in Ouro Preto in the first wave of British merchants arriving in Brazil. Oxenford served as the agent to a number of British mining companies during the next two decades. In the 1820s he explored an old claim known as Gongo Soco located between Caete and Santa Barbara. In 1824 Oxenford obtained the first mining concession granted to a foreigner and purchased the Gongo Soco from the Baron of Catas Altas for the newly-formed Imperial Brazilian Mining

Association. This British company had been established in London with an initial nominal capital stock of £350,000. The Association agreed to pay government taxes of 25 per cent on production, an amount reduced to 20 per cent in 1837, 10 per cent in 1850, and 5 per cent in 1853. In 1826 the British took control of the Gongo Soco property as well as a number of smaller and less important mines in the surrounding area.

The Imperial Brazilian set the general pattern which most of the subsequent British mining companies would follow for the next century. After issuing a call for investors the company sent a group of Cornish miners to Gongo Soco with a commissioner to direct operations. The core of British miners and managers and their families established a small but vigorous British settlement complete with Anglican parson and chapel. By the 1840s the company employed some 200 Europeans who led a workforce of more than 500 slaves and 200 free Brazilians. The company introduced the deep-shaft mining techniques of Cornwall, and the latest milling procedures of the era. The cost of fuel and the difficulties of transport prohibited the use of steam engines, but the abundance of water more than made up for the excessive costs of steam power. Water was utilised to power stamping mills, arrastras, and amalgamation equipment. In European mining this technology was not novel; its application on a rational, large-scale and systematic basis in Brazilian gold mining was (Ferrand 1913: 104–19; Eschwege 1979: 51–82).

The key to the success of the Imperial Brazilian was not only the influx of British capital and technology, but also the existence of a rich lode. The Gongo Soco had been one of the legendary mines of Brazil. Under the British it produced consistently large amounts of ore rich in gold. Nevertheless, the mine works flooded and collapsed in complete ruin in 1856 ending three decades of extremely successful operations. According to the company account books, the mine produced almost thirteen million grams of gold worth £1,697,295. Approximately one-fifth of this amount went to the Brazilian government in the form of taxes and duties. Another one-fifth went to investors in the form of dividends, and expenses gobbled up the remaining three-fifths (St John d'el Rey Archives: Gongo Soco).

Although the financial collapse in 1825 in London slowed considerably the flow of British investment to Latin America in the following decades, it seems to have done little to deter investment in Brazilian gold mining. The success of the Gongo Soco spurred a handful of gold-mining ventures in Minas Gerais during the next decade. Two of these ventures quickly appeared and then vanished from the scene leaving little trace. The General Mining Association and the Serra da Candonga Gold Mining Company, initiated in 1828 and 1834, respectively, never became operational. Three more mining companies were formed in the early 1830s, two achieving limited success and then withering away in the 1840s. One, the St John d'el Rey Mining Company, became the most successful British gold mining company in nineteenth and twentieth century Latin America (Leonardos 1970; Ferrand 1914; Jacob 1911).[2]

The Brazilian Company Limited, founded in London in 1833, worked the Cata Branca mine on the western flank of the famous Pica de Itabira. An agent of the British company purchased the mine from the Count of Linhares and, following the pattern of the Imperial Brazilian, the Brazilian Company imported a crew of British managers and Cornish miners to guide the work of a largely slave labour force. For a decade the mine produced substantial quantities of gold. Before the collapse of the works in 1844 the company extracted over one million grams of gold bullion. The initial capital investment had been a mere £60,000. The National Brazilian Mining Association, also founded in London in 1833, worked the Cocais mine to the northeast of the Gongo Soco. Beginning with an initial capital stock of 200,000 pounds, the company did not equal the productivity of the Imperial Brazilian nor the Brazilian Company. This moderately successful operation produced just over 200,000 grams of gold before it too collapsed in ruins in 1846. Both these mines were worked on a much smaller scale than Gongo Soco. A British traveller reported approximately three dozen English miners and some 300 slaves at Cocais during its heyday (Gardner 1975: 219).

With one major exception all the mining companies founded in the 1820s and 1830s failed to survive the 1850s. That one exception, the St John d'el Rey Mining Company, Limited, not only survived, but went on to become the most famous and most successful British gold mining operation in Latin America. Founded by a group of London merchants in the late 1820s, the company became a functioning reality in 1830 with the purchase of some mines near Sao Joao D'el Rei, hence, the company's anglicised name. When these mines quickly turned barren the British superintendent purchased the Morro Velho estate some twenty kilometres south-east of the present-day Belo Horizonte. Thus began a long and productive enterprise. The Morro Velho had been worked since the early eighteenth century by Brazilians with sporadic success due to capital shortages and lack of vision on the part of owners. The British introduced the latest deep-shaft mining techniques and technology and sent out a series of unusually capable Cornish mining captains and British superintendents. By the 1860s the British community at Morro Velho housed more than 200 Europeans, a chapel, school, cemetery, alongside a small Brazilian town of 5,000. By the 1840s when the Cata Branca and Cocais mines were crumbling in ruins, the Morro Velho had just begun to pay regular and princely dividends (Eakin 1981).

Beginning with an initial capital of £50,000 the company added to that stock periodically reaching just over £250,000 in the 1870s. By the beginning of this century that stock had risen to over £800,000, where it would remain into the 1950s. The company paid regular dividends until the 1870s with a few exceptions. Cave-ins in the late 1860s and in 1886 brought hard times and no dividends. Unlike the other mining companies the St John d'el Rey managed to overcome these crises and re-open the mine, ultimately setting up an even more prosperous and successful operation. From the late 1890s until the 1950s the company paid regular dividends of 5 per cent twice a year without fail. Over a century and a quarter the St John d'el Rey racked up profits totaling over £8,000,000. Approximately one-fifth of this total accrued prior to the collapse of the mine in 1886. The company accumulated the remaining four-fifths between 1892 and 1957. On an annual average this makes the twentieth-century mine almost three times as profitable as the nineteenth-century mine. Taking into account good years and bad, and beginning with the formation of the company in 1830, the St John d'el Rey annually averaged better than 60,000 pounds in profit (Eakin 1981: 130–2).

The company became a financial and economic power in central Minas Gerais as operations expanded. By the late nineteenth century the St John d'el Rey had become the single largest taxpayer in the state, the single largest employer and the most modern industrial enterprise. In the 1930s the company employed more than 8,000 persons, dominating the surrounding community of Nova Lima. The company owned thousands of acres of watershed, woodlands, mineral lands (diamonds, iron, manganese), and real estate, ruling over a veritable empire in the Brazilian interior. Shortly after the turn of the century the company even loaned £50,000 to the financially strapped state government. The company budget at this time regularly ran higher than that of the state of Minas Gerais (Eakin 1981: Ch. 2).

The success of the St John d'el Rey rests on the unusual combination of circumstances which converged at Morro Velho. First, the company was blessed with the finest mining claim in Brazil. The Morro Velho lode ultimately made everything else possible. In the 1840s the mine produced between 100,000 and 400,000 grams of gold per year, rising to over one million grams in the 1850s and two million in the 1860s. With the re-opening of the mine in the 1890s the company regularly produced more than three million grams *per annum*, surpassing four million in the forties. Clearly, the favourable geology of the Morro Velho lode gave the St John d'el Rey an advantage and longevity unshared by any other British mining venture. Second, the capital investment from British stockholders made it possible to develop the lode with the most advanced technology of the times. The gold ore would have lain idle as it had for decades previous had it not been for the capital which made possible the third factor – technology. After 1890 the British superintendent exploited the lode with dynamite, compressed air drills, electric and hydraulic hoisting and hauling, and indigenous excavating techniques. The Morro Velho even became the first deep mine to utilise air-cooling through a refrigeration plant (Eakin 1981: Ch. 3). Above ground the mill was powered by a series of company-built hydroelectric plants and the newly-perfected cyanide process raised productivity. The Morro Velho at the turn of the century was not only Brazil's most advanced mine, it was one of the most advanced, and certainly the deepest, in the world (Rickard 1920: 477–8).

Capital, technology and geology, moreover, converged primarily due to the final and most important factor – entrepreneurial talent and vision.

The administrators of the mine in both London and Morro Velho pulled the company through crises and built a strong operation during normal years. The mine was worked carefully and rationally by a series of superintendents and managers, and when the mine did collapse (as had other British mines in Brazil) strong, visionary administrators chose to rebuild instead of withdrawing. In 1886 when all appeared lost with the complete collapse of the mine for the second time in twenty years, superintendent George Chalmers laid out plans to build a completely new mine and mill. In the four succeeding decades Chalmers constructed one of the most efficient and modern gold mining enterprises in the world. Clearly, this type of strong and capable leadership provided the St John d'el Rey with an advantage over other British mining companies.

IV

By the 1830s the first wave of British investment in Brazilian gold mining had passed. The initial mines, with the notable exception of the Morro Velho, had all disappeared from the scene. As the 1860s approached the St John d'el Rey stood as the sole surviving venture of the boom of the 1820s and 1830s. The success of the St John d'el Rey spurred another half-dozen gold-mining ventures in the 1860s. Only one of these ventures achieved more than a moderate success. All of them were founded between 1861 and 1868. The briefest lasted but six years and the longest nearly three decades. The most successful single mine changed hands to re-appear under the control of another company in the 1880s. The initial nominal capital stock of these companies approximated £600,000, about that of the British companies formed in the 1820s and 1830s. As had the earlier mining companies, these Englishmen purchased old mining claims, poured capital, expertise and new technology into them and, for a short time at least, managed to pull out some profit.

Two of the new companies attempted to exploit the fame and renown of the St John d'el Rey by taking names similar to this successful mining concern. Founded in 1861, the East Del Rey Mining Company Limited worked the Morro do São Vicente and Morro das Almas mines close to Itabirito and Santa Barbara. With an initial stock offering of £90,000 the company encountered moderate success utilising the expertise of several British mining men who had years of experience in Brazil. The manager of the Cocais mine acquired these mines and secretly received aid from the superintendent of the St John d'el Rey, much to the later displeasure of his board of directors. The mine did well initially, but closed in 1875 due to low productivity (Eakin 1981: 74–5; Ferrand 1913: 133–4).

The Don Pedro North Del Rey Mining Company Limited also attempted to build on the name of the Morro Velho owners and the mining expertise built up by the British with experience in Minas Gerais. Founded in 1862 with an initial capital stock of £150,000, the company worked the Morro de Santa Ana mine near Ouro Preto. The company employed the former mining captain of the Morro Velho mine who had more than two decades experience at Morro Velho and Cata Branca. The mine produced nearly two and one-half million grams of gold in the 1860s before the productivity dropped off and the company began to encounter financial difficulties. Operations were moved to the nearby Maquine mine where another five million grams of gold were extracted by the late 1890s. The company paid out dividends with some regularity, but eventually in the 1880s the mine would be bought out by another company with greater financial resources (Ferrand 1913: 134−6).

A third company which thrived briefly around a good mine was the Santa Barbara Gold Mining Company Limited which controlled the Pari mine just west of Santa Barbara. A fine lode whose history also dated back to the eighteenth century, Pari had an initial stock of £60,000 of which nearly a third went towards the installation of modern mining works. Work continued at this mine until the late 1890s and the company produced well over two and one-half million grams of gold in its three decades of operation. In the 1890s the property passed into the hands of the St John d'el Rey. Two companies appeared and disappeared in the 1860s with little fanfare. The Roca Grande Brazilian Gold Mining Company near Caete, and the Anglo-Brazilian Gold Syndicate Limited near Itabira, were both ventures with short lives. The former, founded in 1864, and the latter in 1868, apparently achieved little more than exploratory work at the mines, and neither survived the early 1870s (Ferrand 1913: 136−45).

The final venture of this unlucky half-dozen was the Anglo-Brazilian Gold Mining Company Limited formed in 1863. Although the Anglo-Brazilian finished in bankruptcy exactly two decades after its birth, the company is noteworthy for having exploited the second most important gold mine in Brazilian history − the Passagem near Mariana. This mine, like many of the others, dates back to the gold rush period. It had been re-opened in the second decade of the nineteenth century by Eschwege himself who could not raise enough capital to make the mine profitable. The mine passed through the hands of a number of investors eventually coming into the possession of this British company in the 1860s. Beginning with an initial capital of £100,000 the mine produced gold valued at £90,000, but had outlays totalling more than £115,000 during its twenty years of operation. In the 1880s the company sold its rights to the mine to another British mining company which arose in the third and final wave of British investment in nineteenth-century Brazilian gold mining (Ferrand 1913: 136−45).

V

As in the first and second waves of British investment in Brazilian gold mines, the third wave brought with it the creation of another half-dozen companies. As before, very little came of these ventures with one exception. Most of these ventures came into existence after 1880, although two companies did take shape in the 1870s serving as a bridge between the second and third waves. Neither enterprise produced anything substantial. The first, the Brazilian Consols Gold Mining Company Limited, acquired the Taquara Queimada near Ouro Preto in 1873 and raised an initial capital of £100,000. The operation lasted less than two years, closing down in 1875 with the substantial production of nearly five million grams of gold, but insurmountable financial problems. The Pitangui Gold Mines Limited purchased the Pitangui mine from the liquidating Anglo-Brazilian Gold Syndicate in 1875 and worked the lode until the works flooded in 1887. The venture also suffered from lack of capital and produced an insignificant 285 kilograms of gold (Ferrand 1913: 145–6).

Two more ill-fated ventures operated briefly in the 1880s. The Brazilian Gold Mines Company Limited organised in London in 1880 with a capital stock of £80,000 took control of the Descoberta mine near Caete, and ran the operation for seven years. The mine produced a mere fifteen kilograms of gold during those years and the company liquidated assets in 1887. Another company, the São José d'el Rey Mining Company Limited, bought the Cacula mine near Lagoa Dourada, but the operation apparently never got off the ground. Two mining ventures during this last wave of nineteenth-century British investment achieved short-term success. The São Bento Gold Estates Limited ran the São Bento mine near Santa Barbara. Founded in 1898 with an initial capital of £250,000 the company met with limited initial success after completely modernising the mine and mill. Nevertheless, the mine did not produce enough to justify the capital investment. The mine closed down in 1905 after producing almost one million grams of gold (Ferrand 1913: 146–7).

The most successful of the mining ventures of this last wave of investment was also the second most important mining operation in Brazilian history. Formed in 1880, the Ouro Preto Gold Mines of Brazil Limited owned the Passagem, Espírito Santo and several smaller mines. The Passagem had earlier been exploited by other companies and had been purchased as the centrepiece of the Ouro Preto Gold Mines operations. The smaller and less productive Espírito Santo mine was located near Raposos, a few kilometres from the Morro Velho mine of the St John d'el Rey. The Passagem was the second deepest mine in Brazil and the second largest mining complex after that of the St John d'el Rey. The Passagem mine was blessed with good ore and a fine mining complex constructed with a capital investment of £140,000. From 1905 until the liquidation of the company in 1927, the mine was directed by Arthur Bensusan, one of the most capable British mining men ever to appear in Brazil. Under Bensusan's guidance the company managed to pay fairly regular dividends and to turn a profit, but never a very large one. Between 1884 and 1914, for example, the company averaged a mere £2,193 *per annum* in dividends on a capital stock of over £140,000. A Brazilian banking family bought the operation in 1927 and the mine has been operating sporadically over the past half-century (St John d'el Rey Archives: historical notes).

VI

As the preceding account clearly demonstrates, British gold mining companies fared poorly in Brazil during the nineteenth century. Of eighteen companies only ten achieved even minimal production levels. These ten companies produced just over 102 million grams of gold in the last three-quarters of the nineteenth century. Fully 73 million grams came from the Morro Velho mine, and another 13 million from the Gongo Soco. The Passagem mine produced just over 8 million grams, and continued to produce well into the twentieth century along with the Morro Velho. The Don Pedro North Del Rey (Morro de Santa Ana and Maquine mines) and the Santa Barbara (Pari) companies both produced close to two and one-half million grams of gold and the National Brazilian (Cocais) just over one million. Four more companies (East Del Rey, Anglo-Brazilian Gold Mining, Pitangui, and São Bento) produced between 200,000 and 900,000 grams during their existence accounting for the remaining two million grams (Jacob 1911: 160).

The turn of the century did not mark the end of British investment in Brazilian gold mining. A smattering of companies formed in the first three decades of the twentieth century, but none did much more than exploratory work. The high-water mark of British investment in Brazilian gold mining, and in Latin America as well, had passed by the early twentieth century as the United States gradually replaced Great Britain as the major supplier of foreign capital to Latin America. Through the nineteenth century Brazilian gold mining companies had been almost the exclusive reserve of British investors. A handful of Brazilians had attempted to develop mines with very little success, and a short-lived French venture represented the only other major foreign investment in the gold mining industry (Jacob 1911).[3]

The overall record of British capital in Brazilian gold mining fits into the general picture presented by British gold mining ventures in the rest of Latin America in the nineteenth century. At the beginning of the 1880s no more than three British mining companies organised prior to 1850 still operated in all of Latin America – one in Mexico, one in Chile and the St John d'el Rey. The Mexican and Chilean companies did not mine gold. British gold mining investment in Mexico and Colombia seems to have followed the same general pattern of that in Brazil. Probably the single most important gold mining operation outside of the St John d'el Rey was the Frontino gold mines in Colombia, founded in the late nineteenth century and still operating (Rippy 1954: 96).

The only truly successful British mining company in Brazilian history was the St John d'el Rey Mining Company Limited. Lasting for a century and a quarter (and still operating under South African control today) the company was a success by any standard. The St John had a number of advantages over its fellow British companies in Brazil. First, it was blessed with the finest deep lode in the country. Second, the company had the good fortune to have farsighted and extremely capable management in London and Morro Velho. Finally, in times of financial need the company managed to raise the capital it needed to forge ahead and meet the crisis of the moment.[4]

When surveying the overall role of British mining companies in nineteenth-century Brazil one can quickly see that in global terms these companies had little impact on the national economy. Gold production lagged far behind other industries in the value of output, and in terms of total export earnings gold normally accounted for a few per cent of the value of all exports. Almost all gold produced by the British was exported to London in the nineteenth century. The vast majority of production came from the Morro Velho mine which at times accounted for more than 90 per cent of Brazil's official gold production. (Alluvial gold production was probably quite high but was not easily fiscalised, and contraband in this portion of the industry is a Brazilian tradition.) The amount of capital invested in gold mining probably never reached more than a few million pound sterling, a small portion of total British investment in Brazil. The taxes generated by these companies also tended to be low as production was minimal in most cases. The St John d'el Rey had the greatest impact, and that mainly on a regional level. The company truly wielded immense economic and social power in central Minas Gerais, particularly prior to the construction of the capital city of Belo Horizonte in the 1890s.

These British mining companies did stimulate the growth of the mining industry in Brazil, however imperfectly. The companies brought into the country the latest mining techniques and technology, and the British super-intendents attempted to stimulate as much interest in mining as possible. The technology and techniques do not appear to have had much impact in the long run due to the lack of deep mines in Brazil. Little technology transfer to other industries seems to have occurred. As the companies failed, the mining machinery rusted and decayed and the British technicians returned to their homeland. The lack of Brazilian interest in developing gold mining properties, the lack of local capital, and the scarcity of Brazilian mining engineers made the adaptation of foreign techniques and technology highly improbable.

Four main factors seem to account for the numerous failures and scarce successes in nineteenth-century Brazilian gold mining. First, the government of Brazil played an important and changing role in the mining industry. At the beginning of the century the government opened the country to foreign investment making the development of the gold mining industry possible, and ensuring its domination by the British. The government provided tax incentives by gradually lowering duties and allowing the importation of mining machinery with minimal tariffs. On the other hand, the government did very little to encourage the development of nationally-owned mining companies. True, the government did create the Escola de Minas in the 1870s, but the majority of the graduates faced a small job market, and the school suffered severely from insufficient funding. In addition, during the last quarter of the century the tendency toward higher taxes, duties and tariffs squeezed the gold mining industry which was already operating on thin ice. In short, by the late nineteenth century the Brazilian government had begun to restrict foreign capital in gold mining, while doing very little to stimulate national efforts in the mining industry.

The second factor in the development of gold mining was capital investment. The influx of British capital had made the nineteenth-century mining boom a reality. The moribund gold mining industry suffered enormously from lack of capital in the late eighteenth and early nineteenth centuries. This British capital created the modern gold mining industry in Brazil. Along with this capital came a third factor of signal importance – entrepreneurial and managerial talent. The knowledge and skills of the British miners and managers placed Brazilian gold mining on a footing equal to that of any area of the world in the nineteenth century. One must note, however, that this talent and skill does not seem to have been transferred to the Brazilian in any significant manner. It is true that the British preferred to hire foreign technicians, but it is equally true that they wished to see the creation of trained Brazilian engineers and technicians. The more farsighted of the British recognised that their industry and its survival depended on the training of local people who could direct more energy and investment toward gold mining.

The final and crucial factor in the development of the gold mining industry in nineteenth-century Brazil was geology. No amount of government aid, foreign capital or talent can make up for the lack of gold in the ground. Brazil had been the world's greatest gold producer in the mid-eighteenth century, much as California was to be in the mid-nineteenth century. As in California the gold rush consisted mainly of alluvial and surface gold production followed by the search for deeper lying deposits. The critical question in nineteenth century Brazil and California was whether deep deposits of gold existed which would lead to the revival of the industry. As in California, it appears that Brazil is not blessed with substantial numbers of significant deep veins. What the British experience in nineteenth-century Brazil proved more than anything else was that there were few deep deposits in the country which could be economically exploited. Ultimately, it was the absence of truly economical deposits which hampered the development of Brazilian gold mining. The Morro Velho was the exception that proved the rule. Economically viable deep shafts just did not exist in nineteenth-century Brazil. To be sure, gold was there, but not in high enough grade to achieve profitability given the technology and conditions of the nineteenth-century mining industry.

The British capital influx into nineteenth-century Brazil forms an interesting chapter in the larger story of British investment in precious metals mining in nineteenth-century Latin America. With the single and notable exception of the St John d'el Rey, however, the story of British capital in Brazilian gold mining is a litany of failures and small successes. The epic story of Brazilian gold mining had run its course by the late eighteenth century, and the British effort to revive the gold mining industry ultimately proved a failure. In the end, British capital could not surmount the geological challenge. The history of British capital in nineteenth-century Brazilian gold mining is that of an enormously powerful investor in an extremely weak industry. Nineteenth-century Brazil proved to be another California, with British capitalists appearing on the stage after the major scenes had been played, and after the climax of the drama had passed.

Notes

1 Prior to the discovery of gold in California at mid-century Brazil was probably producing about 15 per cent of the world's gold. Production in the eighteenth century probably ran as high as three to four times that amount.

2 See the complete listing of British companies in Appendix A.

3 The Société des Mines d'or de Faria worked the Faria mine near Nova Lima from 1887 until liquidation in 1906. The mine produced approximately 150,000 grams of gold (Jacob: 160).

4 The British presence in Brazilian gold mining ended in 1957 when American investors purchased the stock of the St John d'el Rey. From 1960 to 1980 the Morro Velho was in the hands of Brazilian investors. At the present the mine is controlled by the Anglo-American Corporation of South Africa.

Appendix A

Year founded	Year terminated	Company name (principal mine)	Capital (pounds)	Production (grams)
1824	1856	Imperial Brazilian	350,000	12,887,000
1828	?	General Mining Association (São Jose D'el Rei)	?	?
1830	1957	St John d'el Rey Mining Company, Limited (Morro Velho)	252,000	72,840,000
1833	1844	Brazilian Company Limited (Cata Branca)	60,000	1,181,290
1833	1846	National Brazilian Mining Association (Cocais)	200,000	207,900
1834	?	Serra da Candonga Gold Mining Company Limited (Serra da Candonga)	?	
1861	1876	East Del Rey Mining Company Limited (Morro das Almas)	90,000	
1862	1896	Don Pedro North Del Rey Gold Mining Company, Limited (Morro de Santa Ana and Maquine)	150,000	2,427,000
1862	1898	Santa Barbara Gold Mining Company Limited (Pari)	60,000	2,682,453
1863	1882	Anglo-Brazilian Gold Mining Company, Limited (Passagem)	100,000	753,500

1864	?	Roca Grande Brazilian Gold Mining Company (Roca Grande)	100,000	
1868	1874	Anglo-Brazilian Gold Syndicate, Limited (Itabira)	100,000	
1873	1875	Brazilian Consols Gold Mining Company (Taquara Queimada)	100,000	
1876	1887	Pitangui Gold Mines Limited (Pitangui)	8,000	285,000
1880	1887	Brazilian Gold Mines Company, Limited (Descoberta)	80,000	
1880	1927	Ouro Prêto Gold Mines Company, Limited (Passagem)	140,000	8,210,119
1886	?	São Jose D'El Rey Gold Mining Company, Limited (Cacula)	?	
1898	1905	São Bento Gold Estates, Limited (São Bento)	250,000	922,739
Totals			2,040,000	102,397,002

Sources: Ferrand 1913; Jacob 1911; Leonardos 1970; St John d'el Rey Mining Company Archives.

References   * *See annotated bibliography for full details of all asterisked titles*

Boxer, C. R.
1969 *The Golden Age of Brazil, 1695–1970: Growing Pains of a Colonial Society.* Berkeley: University of California Press.

Calogeras, João Pandía
1904–5 *As Minas do Brasil e sua Legislação.* São Paulo: Companhia Editora Nacional.

Curle, J. H.
1905 *The Gold Mines of the World.* London: George Routledge.

Disraeli, Benjamin
1825 *An Inquiry Into the Plans, Progress and Policy of the American Mining Companies.* London: John Murray.

Dorr, John Van N.
1969 *Phisiographic, Stratigraphic and Structural Development of the Quadrilatero Ferrifero Minas Gerais, Brazil.* Washington, DC: United States Government Printing Office.

Eakin, Marshall C.
1981*

Eschwege, W. L. von
1979 *Pluto Brasiliensis.* Domicio de Figueiredo Murta, transl. Belo Horizonte: Livraria Itatiaia Editoıa Limitada.

Ferrand, Paul
1913 *L'Or a Minas Gerães, Brésil.* Belo Horizonte.

Gair, Jacob E.
1962 *Geology and Ore Deposits of the Nova Lima and Río Acima Quadrangles, Minas Gerais, Brazil.* Washington, DC: United States Government Printing Office.

Gardner, Jacob E.
1975 *Viagem ão Interior do Brasil*, Milton Amado, transl. Belo Horizonte: Livraria Itatiaia Editora Limitada.

Jacob, Rodolpho
1911 *Minas Geraes No XX seculo*. Rio de Janeiro: Gomes, Irmão and Co.

Jenks, Leland Hamilton
1927 *The Migration of British Capital to 1875*. New York: Knopf.

Leonardos, Othon Henry
1970 *Geociencias do Brasil: A Contribuição Británica*. Rio de Janeiro.

Mawe, John
1815 *Travels in the Interior of Brazil*. London: Longman, Hurst, Rees, Orme and Brown.

Phillips, J. Arthur
1867 *The Mining and Metallurgy of Gold and Silver*. London: E. and F. N. Spon.

Pinto, Virgílio Noya
1979 *O Ouro Brasileiro e o Comercio Anglo-Portuguese*. São Paulo: Companhia Editora Nacional.

Randall, Robert W.
1972*

Rickard, T. A.
1920 *The Deepest Mine*. Mining and Scientific Press.

Rippy, J. Fred
1952 *'Early British investments in the Latin American republics'*. *Inter-American Economic Affairs*. 6(1): 40.
1954 'The most profitable British mining investments in the Hispanic world. *Inter-American Economic Affairs* 8(2): 43.
1959 *British Investments in Latin America, 1822–1949: A Case Study in the Operations of Private Enterprise in Retarded Regions*. Minneapolis: University of Minnesota Press.
1948 'The British investment 'boom' of the 1880s in Latin American mines'. *Inter-American Economic Affairs* 1(4): 72.

Rolff, Paulo Aníbal Marques de Almeida
1976 'Sintese Historica da Bateia'. *Mineração e Metalurgia* 40(378): 48–58.

St John d'el Rey Archives

n.d. Gongo Soco Accounts. Nova Lima.

n.d. St John d'el Rey Mining Company Limited: Historical Notes. Nova Lima.

CHAPTER 3

Commerce, credit and control in Chilean copper mining before 1880

JOHN MAYO

Mayo offers in this chapter an interpretation of the decline of the nineteenth-century Chilean copper industry. He emphasises the lack of a constituency of capitalists to fight for the survival of the industry; a lack of foreigners or nationals committed to develop the copper industry in the face of competitive challenges. When more attractive investments in nitrates became possible, Mayo finds the individual profit motive dictating business decisions. He believes this was harmful to Chilean national interests, and certainly did little to contribute to Chilean economic or social development. The chapter also contains fascinating details on the operation of British merchant houses.

John Mayo is a member of the History Department at the University of the West Indies, PO Box 64, Bridgetown, Barbados, West Indies. He is currently researching nineteenth-century silver trade in Mexico. A related recent publication is: 'Britain and Chile, 1851–1866: anatomy of a relationship', *Journal of Interamerican Studies and World Affairs*, 23: 1 (February 1981).   — *The editors*

I

The progress, or lack of it, of Latin America during the nineteenth century, and the role of foreigners in the process, has been a subject attracting increasing interest in recent years. This was the century during which the ideology of European liberalism achieved widespread acceptance, at least in political spheres and at rhetorical levels, in most of the republics. Yet this acceptance in practice changed little on the ground. Whatever the form, the substance remained intractable, and change cosmetic, as reformers all too often discovered when they stood back to observe their work.

The reasons for this were many, interconnected, and varied. The absence of suitable political institutions, the structure of society, poverty, and ignorance all contributed to the failure of efforts to follow in Latin America nineteenth-century Europe's political, social and economic development, though the extent of material success or failure varied in different degrees at different times from country to country. However, a basic necessity for progress, and a measurement of it, was economic advance: the expansion of production and the marketing of produce, in order to raise the funds needed to purchase the apparatus of 'progress', from railways and telegraphs to potable water, parliamentary democracy and universal education.

Some countries were notably more successful than others in achieving this necessary increase in production, and there proved to be similar variety in the success achieved in exploiting these gains in advancing toward the social and political goals of nineteenth-century liberalism. Economic expansion was not synonymous with advance in other areas. Generally, the gains were based upon the production of staples for the world market, which for much of the nineteenth century meant Britain, either as consumer or as middleman. In return, Britain exported goods and services, and the result of the two-way process was the integration, to varying degrees, of the Latin American economies into the North Atlantic economy, and their increasing dependency upon foreign trade for the financing of government (from import and export taxes), and upon foreign sources for the technology, services and manufactured goods needed to sustain their economic development.

In Latin America, this process nowhere resulted in the formal loss of sovereignty that occurred in most of Africa and much of Asia as similar relationships unfolded. Each individual case has its own reasons, but generally it may be accepted that the absence of great power, i.e. European, rivalry or vital interests in the hemisphere played its part; so Argentina and Uruguay were largely left to sort themselves out in the first half of the nineteenth century. Another factor was the presence of what Ronald Robinson has called 'collaborating or mediating elites' (Robinson 1972). These groups, in government, politics, society and the economy, believed in the advantages of constructive intercourse with foreign powers, or realised the danger of full-scale conflict, and except in moments of complete breakdown in order, maintained tolerable conditions for foreigners and their property even in periods of governmental dispute, as during Rosas' disputes with the British and French governments (Lynch 1981: Ch. VII). But though formal political independence was everywhere retained, similar achievement was notably absent in other areas. So the accomplishment of orderly constitutional changes of government did not mean the creation of an educated electorate or even of a school system to produce one, nor did economic expansion, even modernisation, mean the wider distribution of wealth. In other words, success in one sphere implied nothing about other areas; indeed, such success might actually impede change elsewhere. Generally, it can be said that spin-offs were limited, as were trickle-down benefits.

Given that some form of economic expansion was a fundamental underpinning of nineteenth-century 'progress', however little discussed at the time, this chapter examines the organisation of the copper industry − an industry notably competent at increasing production in a country which by the last quarter of the nineteenth century was attracting to itself the approving attention of Europe for what appeared to be a singularly successful progress on many fronts. If Englishmen smiled when they heard Chileans describe themselves as 'the English of South America', still, they were not insulted by the comparison (Parl 1876). Undoubted economic success did not lead to similar advances in other areas, despite sustained, profitable production for the world market, though the copper industry remained predominantly in Chilean hands throughout, and agricultural production entirely.

Examination of the organisation of copper mining and of the marketing of its production offers some explanation for this result. Basically, it can be demonstrated that the essentially limited nature of the foreign role in the industry resulted in the failure to establish any constituency of Britons or Chileans committed to the continued development of Chilean copper and with the means to support this development; and that foreigners were able to move on to new fields of endeavour without loss or regret (as could Chileans).

II

During the nineteenth century before 1880, Chile moved from fourth to first place among the world's copper producers, displacing Britain from the position of largest producer in the decade of the 1850s. As the republic's own demand for the metal was negligible, it follows that external demand provided the stimulus that made this expansion possible. This demand arose in Europe, and particularly in Britain, the leader in the industrial revolution. Before this, sustained trade on any considerable scale had been limited, though trade with Calcutta for a few years after independence was of importance for the industry. Not only did this exchange raise prices, but it led to the injection of new men, some expertise, and considerable capital from Britain into copper mining. The capitalists lost their money, but many of the men stayed on and deployed their skills, certainly to the benefit of the industry, and in a few cases to their own enrichment (Keeble 1970: 20–3; Vicuna Mackenna 1883: Ch. VI). Among Britons who arrived in these years, and who were to become closely and for long periods involved in producing copper were the Calcutta merchants John Sewell and Thomas Patrickson, and the mining engineer Charles Lambert.

Mining was not the only area that attracted foreign interest, nor was it the principal one. Indeed, throughout the period here considered, Chileans dominated copper mining, owning most of the mines, including the largest, providing most of the labour force (though foreigners made considerable contributions in the skilled job), and even providing much of the fixed capital. Rather, foreigners made their most important inputs into the area of commerce: financing, importing and exporting, organising shipping and insurance, and generally providing services related to the import-export sector not readily available from indigenous sources.

It was not an accident that foreigners moved into trade. Rector has compiled from the Santiago and Valparaiso customs statistics lists of the forty most active merchants or firms for the decades 1808–18, 1818–28, and 1828–38. The results are summarised in Table 3.1. Clearly demonstrated is the dominant position of non-Chilean interests.

Table 3.1

		Per cent of total	
Origin of firm	1808–18	1818–28	1828–38
Chile	17.5	27.5	17.5
Spain	70	7.5	2.5
Great Britain		40	32.5
USA		5	7.5
Argentina		17.5	5
France			12.5
Other and unknown	12.5	2.5	25

Source: John L. Rector 'Merchants, trade and commercial policy in Chile 1810–1840' (unpub. diss. Indiana University 1976) apps. 1, 2, 3.

Peninsulares controlled trade before independence, but their withdrawal (or expulsion) left a void which Chileans proved unable or unwilling to fill. This is not surprising, considering the small role actually played by Chileans in the trade, and the resulting lack of contacts with the rest of the world. Consequently Chileans refused to rush into a field which had been of relatively little importance to them before independence. They still had, after all, their traditional agricultural and mining pursuits as well as politics and the professions, and it was natural that they should turn to others to organise the marketing of their produce. If the destinations of exports and sources of imports were different from colonial times, the means of trading remained the same, and the system continued to operate as before. What had changed were the middlemen, and what was to change, greatly, was the scale of operations.

III

Copper mining in Chile before 1880 was not a particularly capital-intensive operation. The richness and abundance of the ores made for relatively simple and cheap extraction processes, and help explain the fact that 'the vast majority of Chile's mining entrepreneurs were incapable of adopting more than the simplest of the new techniques, such as oil lamps in replacement for tallow candles' (Pederson 1966: 193). Pederson estimates that Chilean ores averaged between 10 and 15 per cent during the nineteenth century, ranging from 3 or 4 per cent to over 50; during the 1860s Cerro Blanco mined oxidised ores that were 35–40 per cent, and in the 1860s still ranged between 12–14 per cent

(Pederson 1966: 186–7). The British-owned Panulcillo Copper Co., which was working 4.6 per cent ores in 1872, was exceptional; the firm had introduced modern coke-fired blast furnaces the year before, which made this feat possible (*Valparaiso and West Coast Mail* 1871, 1872). However, most miners neither felt the need nor had the resources for such capital investment. When the *Chilean Times* recalled with amusement in 1880 conditions in the 1850s, the newspaper unwittingly drew attention to the limited nature of technical progress:

steam power had not even been dreamed of then, and even whims, or horse-power drawing machines, were looked upon as costly and probably unsuccessful innovations, recently introduced by Cornish Mining Captains: what might answer very well in *Inglaterra* they thought might not answer in Chile at all. (*Chilean Times* 1880)

But if the provision of fixed assets made small demands upon the republic's capital resources, working the mines was more demanding. Under the mining *ordenanza*, essentially still the colonial ordinance, anyone over twenty-five could claim a mine. (This included foreigners. Lt. Gillis USN, a visitor in the early 1850s, observed that 'custom has abrogated law, and foreigners now enjoy equal rights with natives', a view confirmed officially by the Chilean government in correspondence with the British chargé in 1856. See Gillis 1855: 267; Archivo Nacional de Chile 1856.) Once claimed, the miner had to work his mine in order to keep it. The result was an immediate need for funds to pay workers and credit to buy supplies. The role of the *habilitador* was to fill this need. The basis of the habilitador's position was financial, but often he also provided, indeed insisted upon providing, goods and other services as well.

Just as foreigners were allowed to own and work mines, they also entered into the business of financing miners. Before 1880, as mentioned, domestic interests retained their grip on the ownership of the industry (Centner 1942: 78). Individual foreigners were important producers in particular localities, and alongside some of the large Chilean producers, but no more. It is impossible to establish exactly the foreign stake as against the Chilean one, but a few figures may be indicative. In the early 1850s, Charles Lambert employed 189 men in his Brillador mine, and some 300 in his smelter, but J. T. de Urmeneta employed 430 in just three of his mines. In 1864, Lambert produced 1,575 tons of copper from 16,840 tons of ore worth $630,000; Chileans 13,000 tons of copper from 67,528 tons of ore worth $5,220,000. British consuls reported that Lambert's interests were greater than those of all other foreigners combined in Coquimbo province, as were those of the shareholders in the Copiapo Railroad and the Copiapo Mining Co. in Atacama province (Intendencia de Coquimbo 1854; *El Mercurio* 1863; Parl 1865: 8, 1873: 150). While this was not negligible, it was obviously far short of dominance, and foreigners deployed their capital much more in financing trade than in owning mines; in this area, their dominance appears to have been complete, through simple advances, habilitaciones and the provision of services.

Habilitacion agreements, which could lead to ownership, were not complex arrangements. Typical was the contract between Roberto F. Budge, habilitador, and Alberto F. Blest, of the mineral Las Animas, Copiapo, notarised in Valparaiso on 24 May 1855. Under it, Blest (who had already received $7,612.21 from Budge) undertook not to sell or to make contracts to sell metal without the habilitador's consent, and also agreed to deliver on the beach 1,000 *quintals* of copper of a ley of not less than 20 per cent (a quintal was either 46 kilograms or 101.4 pounds). In return, Blest was to receive provisions and an advance of $500 a month. Budge was to be charged one per cent a month on the cash advance, the same on outlays for supplies, and receive a 2.5 per cent commission on the value of metals, and 5 per cent on supplies. Blest could sell the mines, but had to settle his debt first, and pay a 2 per cent commission on the sale price. Security was provided by mortgaging the mines (*Notorial de Valparaiso* 1855).

Such small arrangements were very frequent, and many of the large producers operated in the same way though they secured better terms from their 'friends', the commission houses or individuals that advanced the funds. The richness of the ores rendered the investment unnecessary, while the usual organisation of the industry and the attitude of banks led to the acceptance of advances, or habilitacion-type deals as the norm.

Most producers were either individuals or private partnerships, often based on family connections with consequently limited resources. So the firm of Urmeneta & Errazuriz consisted of J.T. de Urmeneta and his son-in-law, Maximiano Errazuriz. Joint stock companies were rare in copper mining, the two most important being the British-owned Panulcillo Copper Co. and the Copiapo Mining Co. These were able to draw upon the London money market, as the Panulcillo company did with debenture issues in 1874 and 1882 (*Notorial de Ovalle* 1874, 1882). Chile could offer no similar facility, and bankers were loath to fill the gap, not entirely without reason. Their natural desire for security made them cautious, and the uncertainties of mining and the international market had small appeal for them and their shareholders. Stephen Williamson, head of the commission house Williamson, Balfour & Co. which owned shares in the Banco Nacional, wrote to his firm's representative on its board: 'let people with means seek to develop the resources of the country themselves and not cumber you with working their spare capital' (*Williamson Letterbooks* IV 1877). In these circumstances, it was natural that copper producers should look elsewhere for what was effectively bridging finance, and in providing this, the foreign-owned commission houses in the import-export trade were in a strong position.

These firms were able to provide services on a scale and of a breadth purely Chilean houses could not, and indeed greater than a single trader such as Robert Budge could. Among them, Britons were particularly well placed, acting as they did as agents for the country which was both Chile's largest supplier and largest customer, as well as the possessor of the world's greatest merchant marine and the centre of the international financial system. Such firms could organise all stages of the disposal of producers' copper, from arranging insurance and supplying cargo space to delivery to, and payment from the customer. For these services, they charged a variety of commissions and interest on any advances made. Copper was security for both producer and commission house, guaranteeing the former return for his labour and the latter repayment of advances plus interest and commissions – if the business were properly organised, for it was not without risk of losses.

Such losses were of two kinds. One was the isolated deficit that might arise in the case of a single cargo, which the producer could shrug off, and which need not affect the habilitador at all, so long as he received his commission and his advances were covered. The other kind occurred when losses occurred over a long period, and the habilitador found himself locked into a consistently unprofitable deal that might continue for years. The requirement of the mining ordenanza that certain minimum activity be sustained, backed up by past success and optimism about the future proved a combination of motives that sometimes led to imprudent advances, particularly as the amounts involved were usually not very considerable in absolute terms. Neither producers nor the supporting commission houses were, by themselves, largely capitalised, despite the overall importance of the copper industry to Chile and in the world economy. Here, of course, the steady demand and the easy mining and treatment of the ores were important factors, reducing the need for expensive industrial processes and large cash reserves.

This may be demonstrated by a consideration of the operations of the merchant house now known as Gibbs & Co. This house began operations in Chile in December 1825, under the name of Gibbs, Crawley & Co., subsequently becoming William Gibbs & Co. before assuming its present name in 1880. Initially a commission house in the import-export trade, it entered the copper industry first on a commission basis, making advances to producers in return for their custom. Amongst its early constituents was the firm Sewell & Patrickson. Sewell & Patrickson itself began as a merchant business, and subsequently became the first foreign habilitador in Chile (Vicuna Mackenna 1883: 320). Doubtless ties of nationality helped the two firms gravitate toward each other, as is indicated by Sewell & Patrickson owing Gibbs some $27,000 at the end of January 1828 (Gibbs 1828). The position was formalised when Gibbs in turn became habilitador of Sewell & Patrickson in 1837 (*Notarial de Valparaiso* 1858: 119, 940). This meant that Gibbs was now effectively a producer, as well as a commission agent, and in the financial year 1 May 1839–30 April 1840, the firm mentioned for the first time purchasing copper, this for its parent house Antony Gibbs & Sons of London, which was charged a commission of 5 per cent (Gibbs 1840). So Gibbs entered all aspects of the copper industry well before 1850: mining, through the habilitacion agreement with Sewell & Patrickson, purchasing, making advances to secure business and charging commission for a wide variety of services rendered.

The merchant houses that financed the copper trade looked to make their profit from commissions. Advances of cash tied up capital, and were justified to the extent that they promoted business. The merchants closely scrutinised any more permanent relationship, particularly if it involved the formal responsibilities of habilitacion, though circumstances or the desirability of the business might lead to bending of principles, and the acceptance of an habilitacion agreement. In any case, the maintenance of good relations with a 'constituent' and the retention of his business frequently required the granting of special privileges.

Table 3.2 presents the copper sales and commissions of William Gibbs &Co.,for the years 1860−64 inclusive. The house charged each of these shippers 2½ per cent commission, plus one per cent guarantee (the rate for precious metals was one-half per cent). The size of shipments by Chileans clearly dwarfs those made by Britons, and the table clearly shows that commissions offered attractive business to the firms.

Table 3.2   Comparative copper sales and commission made by William Gibbs & Co. 1860−64 (pesos)

Shipper	Sales	Commission
Ossa & Escobar, Copiapo	8,724,530	218,113.25
Ossa & Escobar, Valparaiso	501,193	12,529.82
J. A. Moreno	1,137,084	28,427.10
Wm Gibbs & Co.	465,334	11,633.35
Abbott & Sewell		3,049.65
Maquina del Cerro	158,125	3,953.12
E. Abbott	18,988	474.70
Sewell & Patrickson, Copiapo	3,107,317	77,082.92
Sewell & Patrickson, Huasco	938,859	23,471.47
Blas Ossa & Varas	90,442	2,261.05
Mandiola & Co.	30,629	765.72
R. Ovalle & Co.	165,347	4,133.67
Salas Hermanos	23,074	576.85
Urmeneta, Errazuriz & Co. (to 30 April 1860)	619,879	15,496.97
A. Edwards	1,119,933	27,998.32
Edwards & Co., Copiapo	2,496,716	46,813.42
Urmeneta & Errazuriz	3,881,046	72,769.62
Ramon Ovalle & Co. (begin 30 April 1862)	531,104	9,958.20
P. Soruco & Co.	188,507	3,534.51
Robert Walker	468,340	8,781.38
S. M. Artola & Hijos	971,338	18,212.59
J. T. Ovalle	113,949	2,136.54
E. Hardy	721,895	18,047.57
J. M. González Velez	185,431	4,635.77
Ruden & Co.	9,990	249.75
J. Brown	362	9.05
J. Byers	5,880	147.00

Sources: Gibbs MS 11033/5 comparative commissions 30/4/1864.

Of the shippers in Table 3.2, only Hardy, Abbott, Abbott & Sewell, Sewell & Patrickson and the related Maquina del Cerro had no choice, by reason of debt or habilitacion terms, in selecting Gibbs as their agent. Just as no single miner ever succeeded in dominating production, so no commission house ever controlled trade, though Gibbs was perhaps the largest in the period considered. Not only was there competition amongst the British houses, but nationals from other nations actively pursued the business, through merchant houses organised similarly to the British ones, as partnerships with unlimited liability. Such were the North American firms Alsop & Co. and Hemenway & Co., and the German Vorwerk & Co., though Britain's position as the principal market obviously gave British house an advantage (Britons were often partners in the American houses). In these circumstances, Chilean producers could and did look for better deals. For example, in 1860, Ramon Ovalle abandoned Gibbs, apparently because he wanted larger advances which the house refused (Gibbs 1860).

Relations between producer and commission house were usually but not always good. The latter necessarily had considerable discretion in both the timing and the prices of the sales they made, and occasionally failed to satisfy their clients. Further, sometimes in acting in the interests of one client, they acted against others. In 1863, Augustin Edwards, Gibbs' largest constituent, received a very good price for £800,000 worth of copper, which he sold off through his own agent because Antony Gibbs held their own stocks off the market (Gibbs 1863a). Edwards' action, in fact, upset an arrangement Gibbs had worked out with Swansea smelters which they had expected would enable them to sell off their own stock at slowly increasing prices (Gibbs 1862). But such quarrels were rare, which may be taken as evidence that both sides found the situation generally satisfactory. Further, there were two ways of avoiding the pitfalls of consignment through a commission house, as well as the option of changing firms.

The method most frequently resorted to was to sell on the coast, so eliminating the consignment and commission element (though commissions might be paid to a merchant on the coast). This was the obvious course for small producers needing cash quickly, and the customers were there. Sellers of ores could sell either to smelters or for export in unrefined or semi-refined form. Frequently the purchasers were the by-passed commission houses, entering into what were called, with some reason, 'speculations' or 'adventures'.

The other way of avoiding foreign middlemen was to enter the trade oneself. Few Chileans did this. Lambert had his own smelter in Swansea, but he regarded himself as British and ultimately went home to manage his interests. Joshua Waddington and Augustin Edwards certainly had facilities at Swansea, but the former was in any case the Chilean end of a British commission house, while Edwards seems to have had his British interests more as insurance than in any attempt to create an integrated bi-national business. However, apparently alone among Chilean capitalists, Edwards did move to create a commission

house, allying his money with British expertise. In 1858 he formed a partner-
ship with Edward Logan (whom he may have lured from Gunston, Logan &
Co., a British firm). The firm was to be called Edwards Logan & Co., and
Logan would direct the Liverpool end of the business, putting in whatever
he got out of Gunston, Logan & Co. and receiving 60 per cent of any gains/
losses. Edwards, for his part, was to be a sleeping partner and to receive 40
per cent of any gains/losses, and to give the house management of his private
business in England and $100,000 capital (*Notarial de Valparaiso* 1858: 118).
Nothing much came of this venture, Gibbs eventually securing Edwards'
business (Gibbs 1863b). His next attempt was on a far larger scale.

In December 1872, Edwards and four Britons formed Sawers, Woodgate
and Co. in Valparaiso and James Sawers & Co. in Liverpool. Sawers had long
experience on the coast, and put up $200,000; he was to receive 32½ per cent
of the gains/losses. Edwin Woodgate in Liverpool was to put up $50,000 and
received 22½ per cent. Edwards was, again, to be a sleeping partner, and
receive 25 per cent but his liability was specifically limited to the sum he put
up: $500,000. The other two put up nothing, and received 10 per cent each.
Unusually, the articles of this firm made provision for speculative ventures
in imports or exports outside normal business: if Sawers did not approve in
writing, such ventures became the sole responsibility of the partner entering
them (*Notarial de Valparaiso* 1872: 173).

Again, Edwards clearly limited both his commitment, considerable though
it was for the company though not in terms of his own wealth and responsibility. Perhaps he regarded it as a useful means to keep his other connections
on their toes. Certainly his heirs demonstrated no great commitment to it, and
it did not long survive his death in January 1878.

IV

But if Chileans demonstrated the effectiveness of the laws of comparative
advantage and/or the efficacy of the system by their unwillingness to enter
it, this still left the commission houses the task of securing business. Credit,
in various forms, was the basis of the operations. Advances on production
and competitive commissions were the simplest means, but size and time
complicated the process. A large constituent could receive large sums and both
sides benefit, or the process could turn into an habilitación contract.

Gibbs' business transactions demonstrate all these processes. One of
their oldest connections was with the family of Urmeneta, which first appeared
on their books as a debtor of $4,404 in July 1828 (Gibbs 1828). Beginning
as a merchant account, it became an exceptionally large mining business
after J.T. de Urmeneta successfully exploited the copper of Tamaya from
the 1850s onward. In 1858, the Sociedad Chilena de Fundiciones which
mined Urmeneta's ores owed $442,063.54, Urmeneta, Errazuriz & Co. owed
$277,431,46, Maximiano Errazuriz (Urmeneta's son-in-law) had a personal

loan of $53,750, and Urmeneta himself owed $600,000 (Gibbs 1858). In the personalised world of Chilean mining, clearly the commission houses had to accept 'family' as business responsibilities when they attempted to tie constituents to them. Despite the size of the Urmeneta & Errazuriz business, it still sometimes caused Gibbs worry, which is indicative of the tight limits within which the relatively lightly capitalised commission houses operated. In 1876 Antony Gibbs & Sons set out the position of the Chileans' account. Because of the deficit, London wanted 500 tons of copper shipped to cover it, without the seller setting a price limit. (This provision reflected a change brought about

Table 3.3   Urmeneta & Errazuriz's account with Gibbs, 31 March 1876

Debit	£	
On advances on consignment	322,457/0/9	
Accrued interest	6,022/2/7	
	328,479/3/4 / 328,479/3/4	
Credit		
3,035 bars at present prices worth	208,053/10/6	
ingots	86,552/9/7	
	294,496/0/1 / 294,496/0/1	
	Deficit 33,983/3/3	

Source: Gibbs 1876a, Antony Gibbs & Sons to W. Gibbs & Co., 31 March–1 April 1876.

by the arrival of the cable in Chile. Urmeneta was now able to telegraph limits, rather than, as before, having Gibbs sell on arrival for a covering price if possible, but if not to sell at market rates, to prevent the accumulation of stocks.) Urmeneta & Errazuriz had also had a floating credit of £50,000, and a loan of £100,000 and Gibbs expected bills drawn upon it to be covered in 'moderate time' – a month – but found that the account was in fact being used as a permanent loan without security (Gibbs 1876a).

There were limits to what a commission house would accept. In 1876 Gibbs ended their long and mutually profitable relationship with Augustin Edwards, as they felt they were not getting enough for what they did. At the end of June 1876 they found themselves holding his drafts on various correspondents for a total of £541,772, and while accepting that Edwards was safe, the house found the size of the bills 'inconvenient' (Gibbs 1876a). As George Gibbs put it, 'we are not justified in accepting for such large amounts without any more tangible security than the liability to Messrs. Sawers in England, Mr. Bordes in France – of whom we have no intimate knowledge – and Mr. Edwards in Chile who it seems from Comber's letter is by no means strong in health'

(Gibbs 1876b). Edwards refused to deposit funds with Gibbs which would have solved the problem, unless he could get 5 per cent or more on any balance in his favour (Gibbs 1878). So the account was closed.

Habilitación agreements generally resulted from miscalculation, when a miner failed to cover the advances received over a considerable period. Normally miners were left to work their mines themselves, their credit sources being protected by their right to sell the copper produced. What these credit sources wished to avoid was involvement in the actual costs of mining, above that covered by advances, which would be recovered in the normal course of events, for their own capital was based on the requirement of trade, not production. As Antony Gibbs reminded their Valparaiso house in 1859, capital

should be actively employed in one side or the other, and not upon dead works which cause it to lie idle. It must therefore be sparingly employed, even at a good interest on what may be considered permanent investments ... and then only if it be necessary for getting back funds invested in this way.

This habilitación agreement with Sewell and Patrickson began for precisely this reason, and one cause of London's warning to the Chile house was that Sewell and Patrickson's advance had increased by $331,674, which London feared might lead them 'to enter into dangerous habilitaciones with miners', and they instructed W. Gibbs and Co. as a general rule to abstain from such agreements (Gibbs 1859a). There was reason for the warning, for in April 1859, the total capital of W. Gibbs and Co. was $2,750,796, of which habilitación consumed $548,182 (Gibbs 1859b). In 1863, W. Gibbs and Co. valued Sewell and Patrickson's debt at 60 per cent of face value (the reduced figure was still $265,806) and faced bitter criticism from London:

Now it does seem as far as possible from the business of a commission merchant to advance the full cost of the produce consigned; much more unmercantile, then, must it be to advance to people from whom there is but faint hope of recovering the deficits, a considerable sum beyond the actual cost.

It was safer to speculate themselves (Gibbs 1863c). Sewell and Patrickson was liquidated in 1868, when Gibbs took over its properties.

Experiences such as Gibbs' with Sewell and Patrickson help explain the unwillingness of knowledgeable British firms to extend their interests into production. Further, they did very well out of their commission business, Gibbs not least, as Table 4 shows. Chile's total exports and mining exports are included as an indication of the scale of Gibbs' operations. In considering these figures, two factors should be noted. Firstly, Valparaiso was an *entrepôt* for the whole west coast and indeed the Pacific, a fact of which Gibbs took advantage securing business from Bolivia and elsewhere. Consequently, shipments 'influenced' by them did not necessarily originate in Chile, though put to the credit of William Gibbs there. Secondly, nitrate enters the statistics at different times: Gibbs' in the 1870s, Chile's in the 1880s. What is demonstrated is that there was a large trade available, with opportunities for many. Given

Table 3.4   Gibbs' consignments and coast sales, and Chile's total and mining exports 1864–82

Year	Consignment	Coast sale	Total export	Mining export
1864	5,449,137	311,345	27,242,853	19,722,169
1865	6,069,566	516,038	25,712,623	16,257,895
1866	7,963,772	501,334	26,680,510	14,758,425
1867	4,722,280	966,926	30,686,930	18,539,763
1868	5,577,339	1,234,035	29,518,817	16,346,945
1869	8,025,475	1,045,768	27,725,778	18,067,018
1870	5,791,452	1,385,389	26,975,819	16,337,620
1871	8,097,532	1,448,698	31,981,693	16,444,715
1872	10,020,010	1,803,000	37,122,460	18,285,140
1873	15,175,950	1,623,844	38,810,271	18,140,984
1874	8,941,614	545,799	36,540,659	16,564,206
1875	5,407,556	500,670	35,927,592	18,523,876
1876	5,210,705	741,181	37,771,139	21,940,167
1877	4,399,305	328,699	29,715,372	16,759,070
May–Dec.				
1877	3,514,926			
1878	4,850,078		31,695,859	17,525,866
1879	7,923,136		42,657,839	26,248,731
1880	8,465,787		51,648,549	37,812,150
1881	6,720,232	670,812	60,525,859	47,145,757
1882	4,426,300	435,187	71,209,604	57,055,681

Sources: Gibbs MS 11033/A, Consignments 31 Dec. 1882; *Estadística Comercial* 1864–82; Resumen de la hacienda pública de Chile 1883–1914.

this, it is unsurprising that the commission houses, which represented the largest concentrations of foreign capital in Chile, were content to stick to the business they knew, as were their Chilean collaborators.

V

This organisation of copper mining had important implications for the country, some positive, some negative. First, it should be noted that it was a relatively cheap way for Chile to get copper onto the world market, given the republic's own relative deficiencies in skills, contacts and shipping. Commissions varied with the scale of operations and reputation of the miner, but were not high. Additionally, large-scale miners could secure loans from their commission houses at less than the going commercial rate in Chile. All in all, it seems leakages from the economy must have been small during this first heyday of Chilean copper, as foreign ownership was minor, meaning profit repatriation was similarly unimportant. Nor can interest charges remitted overseas have been great, as the system was largely financed by the rolling over of relatively small capital advances, secured by ores, so in a sense being almost self-financing, just as its technology remained essentially static.

It is this element of self-finance that raises questions concerning the ultimate suitability of the system as it evolved before 1880 for the modernisation of Chile. Would the development of the republic have taken a different course had foreigners not been so available and so heavily involved in the financing and marketing of Chile's copper? Could Chileans have done more?

It must be admitted that foreign participation was inevitable in the years after independence. Chile simply did not have the equipment and services to organise and operate copper mining on the scale world demand warranted. However, as the industry expanded, so did the range of services Chile could provide. In the 1850s, Chilean-owned railways, banks and insurance companies all began operations. The government raised a large loan in London in 1857, a visible vote of confidence in the country's progress and stability. Yet copper miners remained firmly wedded to largely traditional methods, both in technology and in finance and marketing, when there were clearly among them individually or collectively, many who could match at least the financial resources of the commission houses. That they chose not to challenge them is probably more an indication of their judgement of where the profits were than of a snobbish dislike of trade, but it meant that a principal source of working capital for the industry remained in foreign, non-mining, mercantile hands; hands which also were profit-oriented and by no means bound to stay in the industry if times turned bad. Merchants dropped bad lines, rather than throw good money after bad.

In the same way, foreign merchants had no incentive to fill gaps in Chile's economy, if they could find resources elsewhere. A case in point is shipping. Chile's movement toward *laissez-faire* policies reduced the protection given the mercantile marine during the 1850s and ended it in the 1860s. Foreign competition and the flight to foreign flags during the 1866 war with Spain and the war of the Pacific (1879–83) led to a quick decline (Veliz 1961: Chs. 3 and 4). The ready availability of foreign ships meant that there was no need to re-establish the national merchant fleet, at least in economic terms, and the absence of Chileans from the commission business meant that there was no group interested in pushing trade toward national carriers.

The limited nature of foreign interest in the copper industry had, then, disadvantages. Though the expansion of copper mining led to the modernisation of aspects of the import-export industry, paradoxically it least influenced the industry itself. This remained technologically backward and failed to attract the capital needed to introduce new processes. Indeed, it might be argued that the small scale of foreign investment was a significant disadvantage, for it was in foreign-owned mines that new methods were introduced into the industry, as for example by Charles Lambert and the Panulcillo Co. During the nearly three decades of copper's eclipse in Chile, after 1880, 'two conspicuously successful English companies were demonstrating that sustained investment in Chilean copper mines, with vertical integration of the industry to include smelting, could be profitable over a long period' (Mamalakis and Reynolds 1965: 212). The companies were the Copiapo Mining Co. and the Panulcillo Co., now called the Central Chile Company. Chilean miners, when successful, found other more profitable and quicker uses for their capital outside the industry.

What happened in the decades before 1880 was the massive expansion of a traditional industry along largely traditional lines. An infrastructure of facilities and services came into being, to serve the import-export trade, which was done inexpensively and efficiently. In this process, foreign interests assumed a dominant role, initially through need, later by default, and always with a commitment to providing services, not to production. In the absence of investment in production, and because ready supplies were available from other sources, both foreigners and natives switched their skills and capital to other areas of activity. In the *laissez-faire* climate of the times, and given the sudden acquisition of the nitrate monopoly, this was inevitable. However, it seems clear that the organisation of copper mining, with the ready availability of credit and other services, but provided by uncommitted outsiders, contributed to the failure of the industry to modernise, and so paved the way for its denationalisation in the twentieth century. Indeed, it might be argued that the absence of a large number of joint-stock mining companies, foreign or domestically controlled, contributed to the decline, for there were neither shareholders to criticise management, nor to put up funds needed to develop or purchase new technology. In this connection, it is interesting to note that when the revival of copper came about, it was financed by the owners, who happened to be North American – 'a miracle introduced by foreigners' (Mamalakis 1976: 41).

What happened, then, was that Chile made a great contribution in supplying the demand for one of the world's staple metals during the period when the republic's copper resources and technology were adequate to meet demand with minimal changes in a traditional industry. Despite the importance of Chile as a producer, the republic never possessed anything approaching a monopoly in the metal, which reduced the attraction to foreigners of becoming producers themselves, especially given the success and ubiquity of indigenous miners. Promising sites never lacked claimants. Consequently, though foreign factors played a crucial role in the industry, and assisted in the establishment of an adequate infrastructure to serve the import-export sector (and incidentally to the benefit of the domestic economy), they limited themselves largely to this sector in which they aimed to keep their commitments as mobile and as unfettered as they liked to keep their capital liquid.

The decline of copper met the rise of nitrate, requiring the same services, and with the bonus of a monopoly, so the commission houses could give up copper without regrets, as apparently did Chilean capitalists. Some firms became producers. In this, the profit motive governed action, leading sober merchants and aspiring capitalists to get out of copper. This was a responsible business decision, debatable in terms of Chile's national interest, but one which the republic could not influence. Chile's copper industry before 1880 created remarkably few hostages to fortune among its organisers: miners and merchants alike proved able to move their capital out, as mines became exhausted or better opportunities emerged, with what now appears considerable flexibility and ease, whatever the anxieties felt at the time. This facility, clearly a boon to business, contributed to the limited effects of the copper boom upon the republic's wider economy and upon society. Ironically, an industry that avoided the pitfalls of foreign control and the ill-effects of 'enclave' development contributed little more to the country at large than did the economy's next 'saviour': the foreign dominated exploitation of nitrate.

So for all the undoubted success of the miners and merchants in producing wealth from copper, they failed to create an industry whose self-sustaining technology could overcome the challenge of declining ore grades and differing ore bodies. They did not even try: instead they cut their losses and moved into new sources of money-making. For many there were no losses anyway. In these circumstances it is then no wonder that the wealth earned from copper had small effects in the economy and society at large. This was not solely the fault of the miners and merchants who produced the wealth, but their business attitudes and methods were no help (nor it may be argued, were the wages they paid to their workers: the many supported the few), while their introduction of modern technology and techniques into the infrastructure of the import-export sector also had only limited benefits. It remains true that economic success is no passport to social or political development, and in Chile, the success involved little of the society-shattering processes of European industrial capitalism that turned the industrial revolution into a great engine of social change, the mainspring of what nineteenth-century European liberals saw as 'progress'.

References   * *See annotated bibliography for full details of all asterisked titles*

Archivo Nacional de Chile (Ministerio de relaciones exteriores 88)
1856

Centner, C. W.
1942 'Great Britain and Chilean mining 1830–1914'. *Economic History Review* 12 (1 and 2).

Chile Times
1880

Encina, F. A.
1912 *Nuestra Inferioridad Económica*. Santiago.

Gibbs archives
1828 Guildhall Library, MS 11033/1. London.
1840 Guildhall Library, MS 11033/2. London.
1858 Guildhall Library, MS 11033/4. London.

1859a Guildhall Library, MS 11471/1. London.
1859b Guildhall Library, MS 11033/4. London.
1860 Guildhall Library, MS 11471/1. London.
1862 Guildhall Library, MS 11036/3. London.
1863a Guildhall Library, MS 11037/1. London.
1863b Guildhall Library, MS 11036/3. London.
1863c Guildhall Library, MS 11471/1. London.
1876a Guildhall Library, MS 11471/2. London.
1876b Guildhall Library, MS 11471. London.
1878 Guildhall Library, MS 11470/1. London.

Gillis, J.M.
1855 *The United States Naval Astronomical Expedition in the Southern Hemisphere During the Years 1849, 1850, 1851, 1852*. Washington.

Intendencia de Coquimbo, Archivo Nacional de Chile, Santiago, 232
1854

Keeble, T.W.
1970 *Commercial Relations Between British Overseas Territories and South America 1806–1914*. London.

Lynch, J.
1981 *Argentine Dictator: Juan Manuel de Rosas 1829–1852*. Oxford.

Mamalakis, M.
1976*

Mamalakis, M., and Reynolds, C.W.
1965 *Essays on the Chilean Economy*. Homewood.

El Mercurio
1863 Santiago.

Notorial de Ovalle
1874 54
1882 91

Notorial de Valparaiso
1855 107
1858 119(140)
1858 118
1872 173

Parl Papers
1876 *Report by Mr. Rumbold on the Progress and General Condition of Chile.*
1865
1872

Pederson, Leland R.
1966 *The Mining Industry of the Norte Chico, Chile*. Evanston.

Robinson, Ronald
1972 'Non-European foundations of European imperialism: sketch for a theory of collaboration'. In *Studies in the Theory of Imperialism*. R. Owen and B. Sutcliffe, eds. London.

Valparaiso and West Coast Mail
1871
1872

Veliz, C.
1961   *Historia de la Marina Mercante de Chile*. Santiago.
Vicuna Mackenna, B.
1883   *El Libro del Cobre i del Carbon de Piedra en Chile*. Santiago.
Williamson, Stephen
1877   *Williamson Letterbooks*. London.

CHAPTER 4

Labour relations in mining: Real del Monte and Pachuca, 1824–74

CUAUHTEMOC VELASCO AVILA
translated by Clara García Ayluardo

Velasco's work is a part of a collaborative research project based on the extensive archive at the Real del Monte y Pachuca Mining Company — a complete archive dating back to the late eighteenth century, located in Pachuca, Mexico. The present work focuses on the refining mill, examining labour relations in terms of labour organisation and division of labour, and by comparing salaried labour with other systems. During the time period of the study he does not find a trend towards salaried and away from forced labour, rather there existed a heterogeneous labour system tied to different types of work.

Cuauhtémoc Velasco Avila is carrying out his research at the Dirección de Estudios Históricos, Instituto Nacional de Antropología e Historia, Apartado Postal 5–119, Mexico, DF, Mexico CP 06500. He has recently published on the same topic 'Los trabajadores mineros en la Nueva España, 1750–1810', in *La Clase Obrera en la Historia de Mexico*, Vol. I, Mexico City: Siglo XXI, 1980. Currently he is studying work and workers in Mexican silver mining.
— *The editors*

I Introduction

Mexico has a long silver mining tradition. From the very start of the colonial period, New Spain sent huge quantities of this precious metal to Europe, and through Europe, to the rest of the world. Colonial silver mining became a fundamental productive activity and was used to justify, from Spain's point of view, New Spain's condition of subjugation.

Indians, mestizos and blacks worked in the extraction and refining of silver. Through their arduous daily routines, they collectively began to experience new patterns of labour organisation in the mines according to the particular conditions of the mineral deposits and of their particular social environment. A complex production structure arose whereby black slaves, repartimiento Indians and free workers lived together. This was the consequence of at least three conditions: the mining centres were located in remote areas, some of the more important tasks required a certain level of qualification, and the mine workers found themselves forced to participate in an extremely risky and unhealthy production process. Additionally, these conditions made mine labour unattractive. Mine owners had to resort to various means of obtaining and disciplining labour such as the use of forced labour and corporal punishment, or the practices of paying wages to the work force and the use of incentives in kind. It is incorrect to assume that only one of the above-mentioned means was the basis of the colonial mining production. If one thing is clear, it is that a mine owner was able to best exploit his workers precisely through the use of a combination of all the forms of labour organisation.

After independence, Mexico continued to play an important part in the production of precious metals on a world scale. Silver continued to be Mexico's main export product and, consequently, mining was the prime basis for the relations which the new independent nation established with other countries. Furthermore, independence from Spain not only opened up new possibilities for commerce but also allowed direct foreign investment to take place in the exploitation of the silver mines. Investment capital, machinery and modern ideas came principally from Great Britain. However, the majority of mining companies created with foreign capital faced serious operational problems from the outset due to, among other reasons, the sizeable gap between their return on investment expectations and the real profit-making possibilities offered by the Mexican mineral deposits.

Laying aside the problem of profitability of the British mining enterprises, it is true to say that they did succeed in modifying the technological base on which mining had been conducted until their arrival in the 1820s. On the whole, the investors believed that huge profits could be made in the mining and refining of metals by addressing their presumed Mexican technological and organisational backwardness. In considering these British efforts to modernise the mines, the question that comes to mind is whether or not a general tendency toward the use of salaried labour in the mining centres emerged. In other words, did the opening up of the mines to international commerce, foreign mining investments and technological modification contribute toward giving Mexican mining, and especially mine labour, characteristics of a more capitalist nature?

It is clear that the use of a combination of the different colonial forms of labour continued long after independence and the arrival of the British steam machines. Consequently, it is important to analyse the mining process in order to observe the factors that determined the need to continue the use of the traditional forms of labour. It is rather difficult to analyse this problem from a macro-economic perspective due to the immense proportions of the task and the inadequacy of the available primary sources. In this essay, the problem is attacked through the perspective of a case study. This type of approach provides a closer view of the concrete forms of the evolving labour relations, of the living and labour conditions of the mine worker and of the connection that all of these elements had in labour disputes.

Real del Monte was the location of one of the longer lasting British mining concerns, an investment based on the greatest amount of determination in attempting to overcome the adverse conditions associated with the production process. The British investors and mining administrators arrived in 1824 and undertook remarkable efforts in the rehabilitation of the old and dilapidated mines that had once belonged to the Casa de Regla. After decades of dismal failure, they left, selling their rights and their properties to Mexican investors such as Manuel Escandon and Nicanor Beisteguí, who in 1849 founded the Compañía de Minas de Real del Monte y Pachuca. In contrast to the British, this Mexican enterprise enjoyed noticeable success from the outset, especially since it experienced the benefits of a great bonanza in the Pachuca mines (Randall 1977). From 1850 to 1871, the company earned considerable profits and even though it experienced serious difficulties in the following few years, it managed to maintain a comfortable profit margin from 1877 to 1906.

It is especially interesting to analyse the period from 1824 to 1874 because it was during this time that two experimental enterprises were undertaken with equally different outcomes but whose productive base was substantially the same. The Mexican enterprise based its success on the adaptation of the technological innovations implemented by its predecessor, but in a different type of operation. In this way, one can compare the transforming potential of the investment and organisational strategy by following the evolution of the relations of production through a period of foreign investment and subsequent reorganisation. The constant factors present will help to discover the logic behind the functioning of the labour process as well as to what extent this system was an obstacle to progressive modifications.

In the evaluation of the tendencies present within the productive process two aspects are of particular interest; (i) the degree of dominance and imminence of the capitalist investors in the labour process, that is, over the ways and means in which the mine workers carried out their tasks − this matter can be seen through the ways in which production was organised and in which labour was divided − and, (ii) the degree of development of the process of separation of the worker from the means of production − this question can be observed mainly through an analysis of the forms of payment and obligations involved. This is not an effort simply to superimpose the typical elements attributed to capitalist production on an historical period where these elements were not yet fully developed but rather to analyse those elements which, from the point of view of production itself, furthered or hindered these tendencies. Through exploration of the logic behind the workings of nineteenth century mining production I hope to discover the answers to the above questions.

This case study was made possible through access to the archive of the Compañía Real del Monte y Pachuca. This archive is undoubtedly unique due to the fact that their accounting books covering the entire nineteenth century have been preserved. This allows for a reconstruction of the different ways used by the company in production of the silver-ore and silver.[1]

II Division of labour and organisation of production

Many studies on Mexican mining have noted the fact that since the colonial period, mining developed a most complex division of labour. Undoubtedly, this development was very similar to the division of labour characteristic of the manufacturing sector. Mining, like manufacturing, is a type of work cooperation based on the 'degeneration of a manual task into the diverse partial operations which compose it', 'a production mechanism whose organs are men', where 'the manual task continues to be the basis of everything' and where the entire process takes place under the direction of capital (Marx 1971 I: 274).

It is usually thought, however, that colonial and nineteenth-century mining production was based on the over-exploitation of workers to fulfil the production demands of international commerce. Taking this idea as true, one is led to believe that production activities were based on simple work cooperation with a qualified labour force playing a secondary role to capital, and that when a division of labour did exist, the differentiated tasks involved were quite simple.

This simplistic thesis is in need of revision. In a recent analysis of the Potosí mines in the sixteenth century, for example, Sempat Assadourian is able to show the coexistence of voluntary with forced labour, under the guise of professional mine workers supported by *mitayos* used for the simple tasks (1979: 257). In the Mexican case, various studies on colonial and nineteenth-century mining also show, for their respective regions, the enormous difference in both salary and social background that existed among the various groups of mine workers, particularly between the unskilled (peones) and skilled workers (West 1949; Brading 1975; Hadley 1979; Alatriste 1983; Cross 1976). All these studies show the connection between the way that the workers were hired and the quality of the work produced. With few exceptions, compulsion was not used to obtain skilled workers who were able to perform experienced and able work.

First of all, in order to understand mining's productive process, it is fundamental to note that one of the most common policies carried out by the nineteenth-century mining companies besides extraction of the actual mineral deposits was that of monopolising the greater part of the activities directly related to the extraction of the ore and the refining process. Following this policy, the Compañía Real del Monte was able to add various refining mills (haciendas de beneficio), stables, salt mines, specialised workshops for the manufacture of tools and haciendas to its already considerable number of mines. The company also invested in the construction of roads and paid for a body of armed men for the protection of the properties and silver convoys. The company then divided its personnel throughout the diverse areas of its complex holdings (Herrera 1979: 68–71).

This type of behaviour, however, should not be taken to mean that the various types of activities were incorporated into the same productive process. In general, each type of activity maintained its autonomous nature in relation to the rest of the activities. When there was integration, it responded to the need of ensuring the availability of the various services and supplies while reducing the costs, and not to the need of increasing work productivity.

Despite the various activities involved in the Real del Monte companies, they tended to concentrate the greater part of their personnel and expenses in the extraction and refining process. Between 1864 and 1867, for example, four-fifths of the company's total expenditure was absorbed by these two activities, and of this same total, close to half was absorbed by the workers themselves.[2] Because of this, it is significant to examine the methods of organisation of production and the division of labour present in the mines and refining mills.

An analysis of the occupational categories allows one to differentiate the groups involved in the production process. In the extraction and refining activities, one can distinguish three distinct occupational categories: the employees, the skilled workers and the unskilled workers.

The employees were in charge of the general vigilance and direction of the mine workers, of the accounting departments, and of the general administration of the company. On the whole, the employees were the ones who put into practice the decision-making process of the company and who directed the production process at the cost of the rest of the workers. Because of this, the employees received a better salary, various fringe benefits and good treatment by the company (Herrera *et al* 1981: 20–54).

The skilled workers constituted the middle group of mine workers due to their position and their earnings. The skilled workers carried out all tasks requiring certain manual experience; because of this, they controlled their own process of production. During the period studied, these workers were not displaced by machinery introduced by the company.

The unskilled workers were used for the more simple tasks which were always the most rigorous and the worst paid. Given the nature of their work, these mine workers were usually moved from one small task to another, and consequently were unable to control their own labour process. The accustomed policy was the use of recently hired workers with no previous experience in mining.

There were various differences in the use of the skilled work force; differences based on mining as contrasted to refining activities. The use of skilled labour within the mine was qualitatively and quantitatively more important, particularly when involving underground tasks. The largest category of skilled labour was the barretero, which alone constituted between one-third and one-fourth of the total work force used in the main mining activities. On the other hand, the middle group of skilled workers constituted no more than one-fifth of the total in the refining mills, while poenes made up three-fourths of the work force used in ordinary, daily tasks.

The excavation of ores by the barreteros was what made the mine productive. The technology employed for the extraction of the ore did not substantially change since the end of the colonial period, when the large mining concerns standardised the use of gunpowder (Brading 1975: 184). In 1804, Humboldt noted that '... in most of the Mexican mines the perforation of the mine which is the task that demands the greatest ability on the part of the worker, is done very well ...' (1966: 366). No technological innovation introduced by the British or Mexican companies substantially modified this situation. During the period covered in this study, skilled workers continued to work with traditional technology: picks, wedges and gunpowder. The skills of this type of worker derived from precise knowledge of the use of these tools, and how to obtain adequate productivity from them. Experience was also important in deciding which veins were to be mined and how. The only way to gauge the profitability of a silver deposit was physically to carry out the mining process and to see how much metal was yielded. Experience in the detection of adequate veins was, therefore, fundamental to avoiding unnecessary costs.

Other types of skilled labourers whose tasks were limited to the interior of the mine were the timber workers (ademadores), the whim operators (malacateros), the drainage-pump operators (bomberos) and the blacksmiths (herreros). The first type of worker was in charge of the bracing of the mine shafts and tunnels to avoid collapse, the second type worked the whims, the third type operated the steam machinery and the fourth type sharpened the tools used in the mine.

In the refining mills, productive labour was principally performed by unskilled workers under the direction of the employees. The tasks concerned with the reduction and crushing of the ore, the mixing of the silver ore with quicksilver, and the washing and blasting of the ore absorbed the greatest amount of workers. The same was the case in any of the refining systems used by the company during this period, such as the smelting process, the traditional *patio* amalgamation and the barrel process. In this sense, the employees took part in two different ways. First of all, they were the captains and foremen of the teams of unskilled workers, secondly, they supervised the technological operations involved in the amalgamation and smelting of the silver ore. As in the colonial period, the experience and capability of the supervisor of the amalgamation process (azoguero) continued to be a major element in the profitability of the refining mill.

There were, however, similar tasks involved in the process of extraction and refining of the silver ore. The carpenters and masons were in charge of the construction and maintenance of the exterior buildings and patios. Together with the blacksmiths and timber workers, these mine workers maintained relative independence from the general productive organisation. These work groups had their specific division of labour with a hierarchical structure similar to the one found in the craft workshops. In spite of this, there was a certain tendency throughout the period to utilise these workers for the specific needs imposed by the mining production in general.

Work gangs were a type of permanent work organisation made up of skilled workers. Specifically, barreteros organised groups of one or more pairs of skilled mine workers (paradas), as well as a variable number of unskilled workers in order to work on a specific point of a vein. Other types of skilled workers made similar use of unskilled workers for the simpler yet heavier tasks.

The technical administration of the mines was carried out by specifically trained personnel. It is important to note that between 1824 and 1874, the greater part of those employed in technical posts dealing with the extraction and refining processes were British. Furthermore, long after the closure of the British company, many of the top administrative positions were held by Englishmen who originated from the mining district of Cornwall. This factor aided the company in the degree to which it allowed the company to control the lives of the foreigners and, through them, the lives of the Mexican workers (Herrera *et al.* 1981: 65–7).

It was very common to use child and female labour in the lighter and simpler tasks of the refining mill, for which these workers were paid an inferior day-wage. (The Hacienda de Velasco's paymaster received a communication from the Board of Elementary Education asking him not to employ under fourteen years of age without proof that they are attending night school.)[3] Inside the mines, use of child labour was more limited. Children were used only in loading, and in returning the ore cars. Only an adult would take on the more skilled jobs underground (Cumplido 1845).

Simply through participating in the production process from an early age, the possibilities for the promotion of the work assistants in the interior of the mine were great. There are also reasons to believe that the children employed in the labour process as apprentices were the offspring of the barreteros.

Within the mines, the work of the unskilled labourer was more exacting. The main tasks involved were the clearing of the work area, the carrying of the loads of the silver ore (tenateros) and the picking through and selection of the richest bits of the silver ore. Obviously, possibilities of promotion through these types of tasks were most limited.

There were employees both in the mines and in the refining mills who had almost identical tasks involving general administration, the overseeing of workers and the general security of company property. These administrative employees were in charge of seeing to the proper functioning of the company as a whole as well as the handling of the company funds for the payment of salaries and other expenses.

The overseeing of the work was principally in hands of the captain or foreman, who directly controlled labour productivity. This activity was especially important in the control of unskilled workers. In the case of the skilled worker, control was exercised largely within the work groups themselves, through the hierarchical structure. The barreteros, for example, who were paid according to work accomplished, exercised a relative degree of control over their own work as well as over the unskilled workers who they themselves hired to form part of a work gang.

The guards, watchmen and gate-keepers were in charge of looking after the mines and refining mills as a precaution against theft and vandalism. On many occasions, these employees had the task of searching the mine workers at the end of their turn.

Advances in the technology employed in the extraction and refining of the silver ore were not sufficiently significant to modify the structure of mining production. Because of this, the traditional forms of labour continued. Inside the mines, technological modifications consisted of a more functional type of planning of the interior mine works, the introduction of water pumps for drainage, and the partial substitution of the whims by steam machines. With these innovations, the work of the porters who carried the silver ore up to the surface of the mine was simplified. Steam machines and whims were now used to facilitate the interconnection of the different shafts through which the silver ore was extracted. Thus, there developed a new type of mine worker who operated the new extraction and drainage machinery. As a result, the mine workers who dealt in the now obsolete forms of manual drainage, the norieros, disappeared.

A new type of refining process known as the barrels system was introduced in the refining-mills. It was introduced in an effort to replace the traditional patio method. The British experienced only partial success with the new method due to the high initial investment cost, even though, in the long term, this new process meant considerable savings in the current costs. On the other hand, the new method was chemically very similar to the traditional patio method, the main difference being that the chemical reaction took place within the barrels. Because of this similarity, throughout the period studied both systems were used alternately. Another interesting modification was the use of hydraulic wheels to move the mortars and mills used to crush the metal (Randall 1977: 101–42).

As far as the labour process was concerned, the barrels method meant the elimination of certain unskilled labourers and muleteers whose task it had been to remix the metal in the founts of the patio. The hydraulic wheels also reduced the number of unskilled labourers needed in the mills and mortars. The technological innovations affected the amount of unskilled labour in the mines and refining mills, rather than modifying the character of skilled labour.

One finds little basic difference when comparing the different forms of labour organisation geared towards production employed by the British and Mexican companies. Perhaps one difference worth mentioning, however, is that the Mexican company had a preference for a more efficient administration of the labour process and, as such, used simple labour within the mines to a lesser degree, while simultaneously increasing this same type of labour in the refining-mills. The Mexican company also increased the amount of personnel in charge of working the pumps and drainage systems. None of these modifications, however, signified a radical change or a meaningful innovation in the labour process.

The difficult conditions under which the mine workers operated are one of the most fundamental aspects of the mining process. Particularly noticeable in the case of Pachuca and Real del Monte are the descriptions of the various risks and illnesses involved in mining during the mid-nineteenth century and afterwards, which curiously are the same type of descriptions made during the latter half of the colonial period. Most of the risks involved took the form of frequent accidents, cave-ins and fires. The most common types of illness referred to were derived from the inhaling of dust and gases and from constant exposure to the humidity and suffocating heat within the mine (Velasco 1982: 129–33, Santibañez 1876: 9–13). As late as 1872, a representative of the barreteros stated in one of the labour disputes: 'All of us understand that our life is short and miserable ... in ten or twelve years those of us that are not dead will be left to begging for the few days left of our lives.'[4] The similarity between the colonial and nineteenth century descriptions shows that not only did the introduction of steam machinery have no appreciable impact on the labour process, but also that the exploitation of the mine workers largely depended on the brevity of the productive life of the individual mine worker. In other words, no betterment of the labour conditions took place nor was there any indication that the companies took this factor into consideration at all.

III Forms of payment

As has already been mentioned, the existence of forced labour in the productive activities connected with the exportation of goods has been exaggerated, leading one to believe that this was the dominant feature of the process. In this respect, Marcello Carmagnani, making reference to eighteenth-century New Spain, states '... an analysis concerning the indebtedness of workers in the agricultural and artisan fields can also be applied to the mining field ...', even though he recognises the fact that this field takes on specific characteristics (Carmagnani 1978: 54).

The use of this analysis supposes that work is derived, at least generally, from indebted, unskilled workers. As we have seen above, it is erroneous to consider the unskilled labourer the basis of the mining activity, and, furthermore, it is clear that the majority of mine workers were not indebted. It becomes necessary then to distinguish the various ways in which the actual work force was either paid or forced to work, and relate these to the way in which the mine workers fit into the productive process.

One can clearly distinguish four types of payment made to free workers: salary, *jornal* (day-wage), *destajo* (payment according to task accomplished), and *partido* (a share of the silver ore mined).

According to the sources, salary was the usual means of remuneration for the employees, who were paid weekly. There were, however, exceptions where British technicians and administrators were paid according to a monthly, termly or yearly salary. This form of payment corresponded to the highest salaries paid out since it corresponded to the highest categories of employees. In 1863, for example, an administrator received a salary of 40 pesos per week, as opposed to the 35 pesos salary of a mine engineer and the 12 pesos of a sotaminero (the engineer's assistant). On the other hand, the wage paid to the unskilled labourer fluctuated between 3−5 reales a day; in other words, the wage varied between 2.25−3.75 pesos per week. Privileged remunerated workers were few, and in 1863 represented a mere 5 per cent of the total number of workers employed in the mines being worked at the time.

The high salaries paid to the employees were directly related to the services which each employee rendered to the company and to the nature of the task involved. Although this was the general rule, the British employees were always paid the highest salaries, even when the type of task performed was similar to that performed by their Mexican counterparts.

The payment of a salary and the sense of belonging to a group of privileged employees was seen as a status symbol to be used against the rest of the workers. The weekly wage for a guard, for example, was, in certain cases, less than the wage received by a timber worker. However, the fact that the guard belonged to the employees category placed him on a higher scale within the hierarchy of mine workers. The guards and captains could be selected from the ranks of the unskilled workers to perform the task of overseeing, as a means of remunerating their services to the company. In this way, salaries were not only important wage payments made to the more qualified employees, but also reflected the control mechanism that these employees exercised over the rest of the mine workers.

The day-wages were money payments made to the mine workers on the basis of a daily quota paid by the week. It is safe to say that this was the most generalised type of payment among the artisans and unskilled workers. In 1863, 45.7 per cent of the mine workers employed in the more important mines were paid a day-wage while 85.9 per cent of the workers in the refining mill of the Hacienda de Regla were similarly paid. The relative importance that this type of payment acquired in the refining mills was determined by the large numbers of barreteros within the mines, who constituted almost one-half of the total amount of mine workers and who were paid by destajo or by partido.

The jornal differed from the salary both quantitatively and in that it was payment made on a daily work basis. This meant, naturally, that the amounts paid suffered acute variations from week to week. This variation was further accentuated when one takes into account that the jornal was a way of attracting regional agricultural workers in need of combining their agricultural tasks with other types of work in order to increase their earnings. Therefore, the degree of permanence within this type of work was limited. In order to avoid this, the enterprise utilised the *tienda de raya* (company store) on the haciendas as a means of tying the work-force to the company. In this way, the use of the company store transformed what was an original money payment into a retribution in kind introducing the mechanism of indebted labour as shall be examined below.

Perhaps the most significant form of payment in the analysis of labour relations within the mining enterprise was the partido, a legacy of the colonial period. This system developed out of the need of mine owners with insufficient capital to cover the wages of the mine workers. The substitution of partial payment with a share of the silver-ore mined of the wages to be paid to be barreteros was seen as a solution to the lack of liquid capital. In the smaller mines, for example, it was common practice to divide the silver ore in two equal parts; one part went to the mine owner and the other to the barretero. In this sense, the partido can be considered to be a rent in kind.

The partido was a way of attracting free workers for work in the mines in spite of the poor working conditions involved. The mechanism implied that the more the workers laboured and the better they chose their mining areas, the greater their earnings.

Toward the mid-eighteenth century, this system had developed to the point of obliging the barretero to extract a certain amount of ore in exchange for the payment of four reales (tarea or tequio). Once this obligation was satisfied, the rest of the silver ore was divided into two halves; from his half, the pick-man gave a certain proportion to the porters, timber-men and to the other mine workers who had helped him.

For more than a century, this form of payment caused most of the disputes between mine workers and the company. The attempts made by the Conde de Regla to diminish the participation of workers were the origin of the well-known 1766 uprisings in Real del Monte. Although this conflict went on for years and put the use of the partido into question, it did end with the meeting of the mine workers' demands (Chavez Orozco: 1960; Danks: 1979). In the nineteenth century, the British also attempted to abolish the partido with the idea of retaining all the silver ore for the use of the company itself. The mine workers, however, actively protested, and the only gain that the company was able to obtain was the partial reduction of the silver ore retained by the mine workers to one-eighth of the total amount extracted (Randall 1977: 155–71). When one takes into account that the barretero could originally retain one-quarter of the total silver ore mined, this was a considerable reduction.

The barretero did not always defend the partido. They defended their right to the partido only when they knew, through experience, that a nearby mine was suitable, in other words, that the mine would provide a bonanza. Thus the partido and destajo were used alternately during the British administration of Real del Monte. The use of either of these types of remuneration, however, depended less on the financial decisions of the company and more on the direct relation between the amount of silver ore extracted and the constant and urgent needs of the mine workers regardless of whether they were hired by jornal, partido or destajo.

The companies sought to do away with the partido for two reasons. In the first place, the barretero worked only if the vein contained high quality silver ore, thus leaving the mines with little ore in constant need of mine workers. Secondly, there was always the risk that the mine worker would work hard for only a few days and then abandon the mine to spend the money earned from the sale of his share of the silver ore. This situation made the hiring of a regular work force impossible and, what was worse, it meant that the daily number of mine workers entering the mine became more unstable with the discovery of a bonanza. Finally, the partido was unpopular with the companies because it absorbed a large part of the silver ore without leaving any gain for the owner of the mine.

In order to eliminate the partido, however, the day-wage would have had to be increased greatly and so the destajo was used as a final resort to provide an incentive to work a mine regardless of whether there was a bonanza present. Furthermore, this new mechanism could attract the barretero to the poorer areas of a mine as opposed to the partido system which concentrated too much attention in the productive work process with too little attention given to the drainage systems and the connection between the various shafts. This situation made communication and transportation within the mine difficult, and could lead to possible flooding and cave-ins. The Mexican company was successful in totally substituting the partido with a more attractive destajo. One of the factors which contributed to the company's healthier operations was its ownership of the greater part of the silver-ove extracted from the mines.

For the barretero, the destajo consisted of payment for each length of mine covered. On the whole, gangs of barreteros and of unskilled workers were hired by the head of the group of mine workers. An agreement was then reached with each mine worker over the price to be paid for the length mined in the particular area in which he chose to work. The price to be paid depended on the wealth of the ore to be mined and on the hardness of the rock, with the amount of lengths mined measured by the week. Once the total amount to be paid was agreed upon, the cost of gunpowder, candles and reed was deducted. It was common practice to grant a contract with a fixed payment stipulated regardless of the amount of lengths mined. Once this fixed sum was in the hands of the captain, it was divided up according to the work done by each of the workers in regard to their individual positions within the hierarchy of the group.

Through the use of destajo, the heads of the barreteros who acted as their intermediaries with the mining company were able to subcontract larger work gangs than those already agreed upon. This mechanism converted these group-heads into the actual work supervisors while providing them with supplementary earnings. Because of this, the subcontracted mine workers known as valientes marginally participated in the income of the work-gangs. Their relative importance within the group of barreteros would increase in relation to the growing unemployment which forced potential mine workers into accepting unfavourable contracts.

The companies also used the destajo as a form of remunerating other types of mine workers such as the pepenadores, who were paid by the load of silver ore picked out, and the aserradores, or sawers who were paid according to the vara (approximately 2.8 feet) of wood sawed. In any case, between 1853 and 1874, the barreteros represented almost all of the mine workers paid by the destajo.

The financial problems which the Mexican company began to suffer from 1896 onward were reflected in its labour relations. A drop in the company's annual net profits forced it to implement a policy of general cut-backs. Adjustments in the total amount of workers were made as early as 1872, along with a reduction of the jornal and destajo which caused a direct confrontation between the barreteros and the mining company. During the following three years the area suffered various labour conflicts which caused sporadic shutdowns of the mines. In spite of all the efforts by the mine workers, company policy remained inflexible. Ultimately, the company closed the Real del Monte mines, reduced salaries, jornales and destajos and, in 1874, proposed the re-establishment of the partido, which the barreteros were forced to accept (Flores and Cuauhtemoc 1980: 21–5; Randall 1977: 50–5). This last measure proved to be quite interesting. The company proposed returning to the partido system, arguing the need to cut back on costs. On the one hand, the company would save 400,000 pesos on the payroll annually or the equivalent of 50 per cent of the total amount paid out for the work done in the mines.[5] On the other hand, by returning to the old system and eliminating the subcontracted valientes, the company would be able to exercise a greater degree of control over the fluctuations in company expenditure. The condition under which the company operated at the moment, however, made the return to the partido unfavourable to the barreteros. With the low quality of the silver ore and the uncertainty of the silver market, the re-establishment of the partido meant that the real earnings of the mine workers would be reduced, given the low prices at which the silver ore was being sold at that time.

IV Forced labour

Given the absence of a labour market capable of satisfying the needs of mining, forced or draft labour had been considered since the colonial period to be an adequate substitute for unskilled labour. On countless occasions, Manuel Romero de Torreros, Conde de Regla, solicited Indians from the neighbouring towns for work in the mines and haciendas. Although some of his petitions were seen to by the higher authorities, he found a great deal of difficulty in actually drafting the Indians. The village judges were reluctant to give up any member of the community for fear of draining its labour force, and the Indians themselves preferred to stay in their native communities to work on the various agricultural tasks rather than to go to work in the mines and refining mills. Seeing the many difficulties involved in obtaining Indian labour, Romero de Terreros obtained a royal decree in 1764 that distributed labour to haciendas needing help; to be included were mulattos, free blacks, mestizos, vagrant Spaniards and Indians. The Visitor-General, José de Galvez, argued in favour of the decree saying it was necessary for those criminals and delinquents whose crimes were not capital to be sent to work in the mines for the time that the court decides is necessary to work out their crimes. This decree also affected the Spanish population not involved in an honest occupation. Despite the decree and the many efforts made by the Conde de Regla to retain his work force, he continued to experience problems in the recruitment of workers (Santibañez 1934: 35–42).

Although greater analysis of this problem is required, given the available data, one can suppose that the participation of repartimento Indians in the mines and haciendas of the Real del Monte region toward the second half of the eighteenth century was negligible. The consequence was the use of various methods to retain the available Indian and black slave labour. In 1791, for example, there were fifty slaves working in the La Palma mine (thirty-one men, sixteen women and three boys). In any case, however, it is clear that forced labour was not significant in the case of the mines, especially considering that a few years later the Vizcaína vein of Real del Monte employed over 2,700 mine workers.[6]

Indebtedness is even more difficult to evaluate, especially since the advancement of goods under certain conditions could mean yet another way of attracting labour rather than a form of coercion. An inventory of the Zimapan mines belonging to the same Conde de Regla mentioned in 1784, for example, shows that the previous administration had transferred a few books where the debts of the mine workers were noted and of which 'none had been recalled'. Afterwards, there is mention of a new book of debts owed by people working in the mine Lomo del Toro; from advances given to keep them working in the mine in order to pay the debt.[7] One can then suppose that, at least in the mines, both the direct forms of forced labour and the various means of attracting labour had as their major aim the maintenance of a constant work gang.

On the haciendas, there were more viable conditions for the development of the hacienda store as a retention mechanism. Since 1757, the repartimento Indians complained that in the refining hacienda de Regla, in addition to the poor working conditions, the jornal was but two and a half reales. This bought only four biscuits and the rest was taken for firewood, corn, for Mass and for the dead (Zavala and Castello 1946: XXVII). In 1784 an inventory of the same hacienda store reported that the debt incurred by its workers amounted to over 1,800 pesos, almost one-third of the total inventory.[8]

The British also made use of the hacienda store. In 1830 they gave the concession to administer the Hacienda de Regla's store to Ramon Borbolla in exchange for a rent payment, and the contract was renewed on three occasions. In 1847, a conflict arose between the concessionary and the mining company which made public the way in which the enterprise had been run. Despite the fact that Borbolla was bound by the terms of the contract to exercise 'prudence', his conduct was the main cause of the scandal. He sold alcoholic beverages during the entire day, even though he was prohibited from doing so, causing drunken brawls among the workers both inside and outside the hacienda and absenteeism on Mondays. He also provided shelter to strangers and vagabonds and bought stolen objects including silver; he resorted to usurous practices, lending money at the interest rate of ½ – 1 real per peso loaned per week, and received objects as pledges for a loan. Borbolla was also guilty of selling the hacienda store goods at inflated prices and so maintained the workers in a state of constant indebtedness, having cornered the region's consumer market. Furthermore, Borbolla deducted far more than the maximum one-quarter of the workers' total salaries allowed by contract, and on some occasions even deducted the entire amount. According to his accusers, this situation provoked hostility in the neighbouring village of Huasca toward the mining company. Furthermore, it caused the workers to resort to uprisings, hoping to increase the 'price of their labour'. Although the available sources are scarce in reference to the type of contract signed between the mining company and the concessionary, there is access to correspondence concerning the conflict. It shows that altogether, the hacienda workers owed 7,000 pesos during that time which, considering their number and salaries, meant that they owed between three and six months of their wages.

Ironically, the hacienda administrator complained that the store had everyone indebted and that when the hacienda was in need of workers there were none to be found since the inhabitants of the region understood that they would be working for the hacienda store and not for themselves. The lawyer, for his part, argued that charging the workers more than one-quarter of their earnings was causing their 'departure from the hacienda' which, in turn, placed the mining company 'in the need of recruiting and introducing workers from afar'. These words suggest that the exaggerated methods of labour retention made the proper functioning of the hacienda difficult. Still, the permanence of the hacienda store shows that its use as a retention mechanism was of utmost importance for the normal operation of the enterprise. As far as it is known, the efforts made by the company did not include the closing down of the hacienda store or the termination of the contract. For 1874 there is correspondence referring to the hacienda store which obviously continued to operate through concessionaries, but unfortunately this information is very scarce.[9]

As far as the mines were concerned, the Mexican company revived the colonial custom of using delinquents as mine workers. In December of 1850, the company signed a contract with the state of Mexico calling for the establishment of a prison on the mine site. The company would be responsible for the maintenance of the prisoners in exchange for the use of their labour in the mines. The contract took effect in 1850 and was terminated when the company no longer considered it to be profitable.[10]

The company maintained a varying number of prisoners. In 1853, the Real del Monte prison had an average of 150 prisoners, each costing the company approximately 2⅜ reales per day. On the other hand, the unskilled workers who performed similar tasks received, for the same year, an average day wage of 2⅞ reales. In 1866, about 163 prisoners were maintained at an average cost of 59 cents daily; according to 1868 data, the amount of the day-wage paid to unskilled workers reached an average of 52 cents a day. In 1874, the mining company sought for the first time the termination of the prison contract, arguing that '... según las cuentas de la negociación resulta que es de diez reales diarios [el costo] por cada sentenciado, siendo así que los trabajos que desempeñan, los podrían hacer hombres libres, cuyo jornal no pasaría de tres y medio o cuatro reales'.[11]

This statement clearly shows that the maintenance cost of the prisoners was not as meagre as one would imagine. Salaries paid to the guards who watched over the prisoners were especially high, and consequently, the cost of the maintenance of the prisoners in the 1870s became noticably greater than the cost allotted for the payment of unskilled workers who performed the same tasks. The company was able to maintain the profitability of the prison during its first year because it received support from the State of Mexico, which had promised to pay the guards and reduce the payment of dues on the silver from 3 per cent to 1½ per cent. Years later, however, this support was withdrawn, leaving the company to bear the total cost of the prisoners' maintenance.

The main reason for keeping a prison in the mining zone, as in the previous century, was to have a constant source of workers upon which to draw for hard labour in the interior of the mine. Real del Monte maintained 200 prisoners, all selected by and under the complete control of the company (Burkart 1861: 108).

The prisoners never constituted the essential core of mine workers within the mines. In 1863, they constituted only 3.7 per cent of the total work force employed in the productive mines and, in the best of cases, for the same year, they never constituted over 15 per cent of the entire work force of a single mine.

The prisoners were the ones who most directly suffered the terrible working conditions within the mine. Another argument used by the company to do away with prison labour was the high mortality rate which increased during the final years. Of the 254 prisoners employed between 1869 and 1873, 116 died and only thirty-six were released after completing their sentence.[12] Everything seems to indicate that the condition of the prisoners tended to worsen at the same time that the company found it more difficult to maintain them. The existence and duration of the prison partially explains the poor working conditions and the many dangers confronted in the mine, since it was always difficult to recruit workers willing to expose themselves to illness or death.

V Final considerations

During the entire eighteenth century, and at least until the mid-nineteenth century, the owners and financial backers of the mines of Real del Monte and Pachuca were hindered by the scarcity of available labour for the simplest or more difficult of tasks. As we have seen, this situation worsened, in the case of the mine workers employed in the interior of the mine, because of the difficult working conditions involved.

If the barreteros were able to defend the use of the partido, it was precisely because of the lack of available workers combined with the poor working conditions involved. Under the Mexican company's administration, the barreteros lost their right to the partido but continued to enjoy relatively high earnings through the use of the destajo. However, in the 1870s, the behaviour of both the company and the mine workers had clearly changed. In spite of the strikes, the barreteros were unable to persuade the company to meet their demands, and the company re-installed the partido even though it would be of no use to the mine workers.

One must take into account two complementary elements in order to explain these modifications. In the first place, there was a clear expansion of the labour market during the second half of the nineteenth century. This was the product of a slow and generalised process which has been attributed by many to the dissolution of Indian community land, in other words, the development of a mercantile society at the cost of one dedicated to autoconsumption. Even though the mining sector lay well within the context of a mercantile society, it is not clear whether its development in the Real del Monte area significantly triggered the process of despoilment and advancement. The reality of the situation was that the mining process used, and adapted to, the conditions under which it was able to obtain its work force.

The second element is directly linked to the first. The conditions of boom or depression that the companies experienced at different times, and which were potentially determined by fluctuations in the richness of the silver ore deposits, decisively influenced the type of option used to obtain their labour. The bonanza enjoyed by the Mexican company in its early years opened up the possibility of substituting the partido with high destajo rates and contracted prison labour, which could assure it a permanent source of labour rather than, as one has been led to believe, a savings in costs. On the contrary, the difficulties which came up after 1870 and the need to reduce expenditure led the company to cancel the contract involving prison labour and to reinstate the partido while simultaneously curtailing the scope of its activities.

As far as mining was concerned, it is clear that forced labour did not constitute the fundamental basis of production and that it was the legally free workers who constituted the larger group. What was even more significant was the creation of a combination of both of these types or workers in which each type played a specific role.

It is much more problematic to evaluate the case of the hacienda/refining mills. It is possible to surmise that coercive mechanisms took on a relative importance given the geographic isolation of the hacienda and the preference for the use of unskilled labour. At any rate, the British company was not convinced in 1847 of the benefits which the hacienda store could bring and, in general, the lack of adequate correspondence and other similar documents suggest that there was little interest on the part of the company to use the mechanism of indebtedness.

Seen in its totality, the mining productive process in Pachuca and Real del Monte did not show a tendency to develop capitalist forms. It is, however, clear that in comparing it to the minor mining enterprises, the division of labour functioned as a way of advancing its productive capacity. On the other hand, one needs to keep in mind that this type of organisation was characteristic of mining since the colonial period. Moreover, the introduction of steam machinery did not modify the basic manual character of skilled work and as such did not contribute to the domination of the capitalist mode in the mining productive process.

The separation of the worker from his means of production in the extraction and refining processes was not totally clear. The partido had brought with it a participation of the mine worker in the destiny of the mining process. This is why one can not consider the barreteros to have been totally divorced from their means of production. All the different types of forced labour also imply that the worker was in the same way linked to the said methods, and consequently, to the productive work force of the unit of production. So, while the barreteros were reduced to accepting a type of salary – the destajo, an institution was simultaneously created which provided forced labour – the prison; and when the use of prison labour was deemed unprofitable, the partido was re-installed. It is clear that in returning to the partido, the relation which it had represented for the barreteros of previous times had been modified. However, the spirit of using the partido as a means of reducing monetary costs remained unaltered.

In order to understand the reduced capacity of transformation of the silver mining production, it remains significant that, in this case, it was precisely during the time in which the Mexican company had high profits that the company was less concerned with developing long-term ways to increase productivity. Paradoxically, it was decline which created the need for innovation in the productive process or in the means of providing supplies. The investors preferred to use the returns of the bonanza for accumulating capital or for investment in less risky activities.

The production of the surplus appropriated by the investors derived from the exploitation of a heterogeneous collection of workers and of labour relations. This collection rationally integrated itself according to the available means and the type of task done. The resulting complex unit poses a definition problem. It is not possible to equate this combination of types to any of the traditional modes of production models, especially since its internal logic did not show any tendencies toward a transition.

Notes

This essay was made possible thanks to the collective academic work done in the 'Condiciones de trabajo y situación de las clases trabajadores en México en el siglo XIX' seminar. I am particularly grateful for the help provided by Inés Herrera, William Meyers, Eduardo Flores and Clara García Ayluardo to whom I wish to express my deepest appreciation.

1 The books of expenditures of this archive were used as the principal quantitative source. A sample was constructed from the following series: Minas de Regla (in Real del Monte), 1833–49; Estractos de Memorias (of the mines of Pachuca and Real del Monte) 1853–74; Hacienda de Regla, 1828–67. For more complete information concerning the content of this collection see Flores *et al.* 1981.

2 The exact proportions are the following: 82.9 per cent of the total costs were absorbed by both the refining and the mining sectors, 32.4 per cent corresponded to the former and 50.5 per cent to the latter. The work force represented 44 per cent of the company's total expenditure. In the mines, two-thirds of the costs were absorbed by the work force while only one-third was alloted to the refining process. 'Gastos mensuales en las negociaciones del mineral del Real del Monte y Pachuca según ramos.' July 1864 – December 1867 (Archivo Histórico de la Compañía de Real del Monte y Pachuca (AHRMP)).

3 Letter from the Committee for the Protection of Primary Education to the Hacienda de Velasco paymaster (Omitlán, 1 June 1868, AHRMP).

4 Letter from the barreteros and other mine workers of the Real del Monte mines to the state government (Pachuca, 2 July 1872, AHRMP).

5 Records of the Pachuca and Real del Monte mines for the week of 8 August 1874 when four projects to reduce the operation of the enterprise were proposed (Mexico, 11 September 1874, AHRMP).

6 'Entrega que ha hecho Don Juan Bars a Don Gregorio Lope de las minas, casas y adornos de ella y demás enseres pertenecientes a la dependencia denominada la Veta Vizcaína del Señor Conde de Regla' (Real del Monte, 29 August 1791, AHRMP). 'Estado que manifiesta por menor el numero de dependientes y operarios trabajadores que en el presente mantiene el Señor Conde de Regla en sus minas de la Veta Vizcaína ...' (Real del Monte, 15 March 1979, AHRMP).

7 'Inventario de las minas y haciendas que pertenecientas al Senor Conde de Regla se hayan en el Real de Zimapán ...' (Zimapán, 19 May 1784, AHRMP).

8 'Balance de la tienda, sita en la hacienda de la Nuestra Señora de Regla, que ... practicó Don Joseph de Murrugat, apoderado general del Señor Conde de dicho Regla ...' (Regla, 16 July 1784, AHRMP).

9 The Hacienda de Regla store file. Documents and correspondence referring to the period between June 1847 and September 1848 (AHRMP: vault).

10 Notarised contract signed by Alfonso Fernández, prefect of the district of Tulancingo, and Juan H. Buchan, director of the Real del Monte and Pachuca Mining Company (Mineral del Monte, 4 December 1850, AHRMP). Draft of a letter written by the board of directors of the company to the governor of the state of Hidalgo concerning the renewal of the contract involving the prison in Real del Monte (no date or place (1874?), AHRMP). Letter written by Justino Fernández to Antonio Mier y Celis concerning the renewal of the prison contract (Pachuca, 24 March 1874, AHRMP).

11 Report of the prison's maintenance costs for 1853 (no date or place, AHRMP).

12 Study of the prison mortality rate (Pachuca, 21 February 1874, AHRMP).

References   * *See annotated bibliography for full details of all asterisked titles*

Alatriste, Oscar
1983 *Desarrollo de la Industria y la Comunidad Minera de Hidalgo del Parral Durante la Segunda Mitad del Siglo XVIII (1765–1810).* Mexico: Universidad Nacional Autónoma de Mexico.

Assadourian, Carlos Sempat
1979 *La Producción de la Mercancia Dinero en la Formacion del Mercado Interno Colonial. In Ensayos Sobre el Desarrollo Económico de México y América Latina 1500–1975.* Enrique Florescano ed. pp. 223–92, Mexico: Fondo de Cultura Económica.

Brading, David A.
1975 *Mineros y Comerciantes en en Mexico Borbónico (1763–1810).* Mexico City: Fondo de Cultura Ecónomica.

Burkart, Juan
1861 'Memoria sobre le explotación de minas en los distritos de Pachuca y Real del Monte de Mexico'. In *Anales de la Minería Mexicana* I. Mexico: Siglo XXI Editores.

Carmagnani, Marcelo
1976 *Formación y Crisis de un Sistema Feudal, América Latina del Siglo XVI a Nuestros Dias.* Mexico: Siglo XXI Editores.

Chavez Orosco, Luis
1960 *Conflicto de Trabajo con los Mineros de Real del Monte, Año de 1766.* Mexico: Biblioteca del Instituto de Estudios Históricos de la Revolución Mexicana.

Cross, Harry Edward
1976*

Cumplido, Ignacio
1845 *El Barretero. In Decimo Calendario de Cumplido para 1845.* Mexico: Imprenta de Ignacio Cumplido.

Danks, Noblet Barry
1979 'Revolts of 1766 and 1767 in mining communities in New Spain'. Boulder: University of Colorado (doctoral thesis).

Flores, Eduardo, and Cuauhtémoc Velasco Avila
1980 'Doscientos años de luchas mineras en Real del Monte y Pachuca'. Mexico: Dirección de Estudios Históricos Instituto Nacional de Antropología (unedited manuscript).

Flores, Eduardo, Inés Herrera and Cuauhtémoc Velasco Avila
1981 *Guía del Archivo Histórico de la Compañía de Minas de Real del Monte y Pachuca.* Mexico: Archivo General de la Nación.

Hadley, Phillip Lance
1979 *Minería y Sociedad en el Centro Minero de Santa Eulalia, Chihuahua (1709–1750).* Mexico: Fondo de Cultura Económica.

Herrera Canales, Inés
1979 'La racionalidad económica de la empresa minera Real del Monte y Pachuca, 1849–1874'. In *Organización de la Producción y Relaciones de Trabajo en el Siglo XIX en México.* pp. 68–83. Mexico: Instituto Nacional de Antropología e Historia.

Herrera Canales, Ines, Cuauhtémoc Velasco Avila and Eduardo Flores Clair
1981*

Humboldt, Alejandro de
1966 *Ensayo Político Sobre el Reino de la Nueva España*. Mexico: Editorial Porrúa.

Marx, Carlos
1971 *El Capital, Crítica de la Economia Política*. Mexico: Fondo de Cultura Económica.

Randall, Robert W.
1977*

Santibañez, Abraham E.
1876 *Leucocitemia Relativa de los Mineros*. Mexico: Imprenta de Villanueva y Villageliu. Secretaría de la Economía Nacional, comp.
1934 'Los salarios y el trabajo en el siglo XVIII, legislación y nominas de salarios'. In *Documentos para la Historia económica de México* VIII. Mexico: Publicaciones de la Secretaría de la Economia Nacional.

Velasco Avila, Cuauhtémoc José
1982 'La mineria Novohispana: transición al capitalismo y relaciones de producción'. Mexico: Escuela Nacional de Antropología e Historia (thesis).

West, Robert
1949 *The Mining Community in Northern New Spain: The Parral Mining District*. Berkeley and Los Angeles: University of California Press.

Zavala, Silvio and María Castello
1946 *Fuentes Para la Historia del Trabajo en Nueva España*, vol. 8. Mexico: Fondo de Cultura Económica.

CHAPTER 5

The decline of a mining region and mining policy: Chilean copper in the nineteenth century

WILLIAM W. CULVER and CORNEL J. REINHART

Culver and Reinhart argue that the decline of the nineteenth-century copper industry can be laid at the feet of the political weakness of the industry. Through comparison with the copper industry of the United States, they show how mining policy takes on critical importance in the long-term success of a mining region. This argument runs counter to the often held view that Chilean mining entrepreneurs were technically incompetent.

William Culver and Cornel Reinhart both teach at the State University of New York at Plattsburgh, Culver in political science and Reinhart in history. They are presently working on a history of the politics of the world copper industry from 1800 to 1930. They have recently published an article, 'Chile en el mercado mundial de cobre, 1810–1927', in Minería Chilena (1984).   — The editors

I Mining policy and mining regions

Copper, the great enduring problem of Chilean public life, has persisted as an issue for almost two centuries. No political faction in Chile has ever obtained sufficient power to form a timely and lasting national mining policy. The political struggle to nationalise copper during the early 1970s (Moran), and the ensuing political turmoil over the mining property reforms of 1982–4 (Correa Amunategui),[1] are no more than continuations of a policy debate left unresolved from the nineteenth century. The most intense period of that debate was during the 1850s, a debate which slowly intensified until the 1880s when Chilean copper production itself began a geologically unwarranted decline. While this copper crisis did lead to the organisation of the copper industry for political action, the policy reforms the industry obtained came too late to have sufficient impact to save the national industry. And in the long run the reforms only served to pave the way for the investment of capital from the United States, and thus introduce new complexities to the national debate over the ownership of copper. The unsatisfactory resolution of the 1880s copper crisis has obsessed Chilean politics ever since.

Surprisingly, analysts of this decline have ignored the intense political effort aimed at preventing that very outcome (Prezworski; Bravo; Pederson;

Reynolds; Mamalakis).[2] The political problem facing those working to promote Chilean copper was the same problem basic to all mining regions, namely: the ownership and distribution of mineral wealth. Beginning with a simple tax rate argument during the 1850s, the copper industry's difficulties in facing the declining world price for copper gradually intensified until copper mining leaders wanted and succeeded in obtaining major changes in the entire structure of mining property and taxation in the 1880s. Their 1880s goal was to transform copper mines into private property which could be held, sold, mortgaged or developed in any way the owner saw fit. Their apparent legislative success of 1887, however, was an illusion. Placed in the context of the world copper trade, which by the 1880s had left Chile far behind, these pro-capitalist measures came decades too late.

This chapter argues that the solution to the search for the causes of the precipitous decline of copper output during the 1880s can be found in the relationship which evolves between a mining region and the policy which governs it. A complete answer would, of course, require a broad review of Chilean political history, a task much beyond the scope of this chapter. Instead, the relationship between mining policy and the success of a mining region is explored through a comparison of the competitive relationship between Chilean copper and copper from Michigan and Montana.

In order to appreciate the importance of mining policy for the prosperity and development of a mining region the natural evolution of the mines in a mineralised zone needs examination. Mining regions pass through cycles of discovery, development, exploitation and abandonment; their end comes predictably as ore is exhausted. High mineral prices may speed up exploitation, just as low prices may prolong it, but sooner or later the ore in all mining regions 'gives out'. No matter how great the investment, no matter how ingenious the metallurgical processes invented to utilise lower grades of ore, when a region matures and its ore is depleted, the mining cycle stands complete. Throughout the mining cycle policy influences the way geological theory is incorporated into daily practice, the way technology is applied, the way labour is organised, and the attractiveness of the mining region to new investors.

There are instances, however, where the mining cycle is cut short, and a mining region experiences a premature decline. Such is the case of Chilean copper mining in the late nineteenth century. A premature decline is simply the closing of mines in a mineralised region while ore remains untouched in the ground. In such cases, social not geological factors are at play. The stagnation of an otherwise productive mining region unable to 'meet the competition' is ultimately a problem of mining policy. No matter how rich a mine is believed to be, if it is running at a loss it will be forced to close at some point.

Why is this so? Why does a mining region with reserves of ore find itself unable to match the production costs of its competition? Certainly, relative ore quality and related mining costs are important, but ownership of subsurface mineral wealth is a critical factor for modern capitalist production. Is such a resource the property of the surface landowner? Or does the discoverer own it? Or those with the knowledge to extract it? Or the state itself? Each society answers these questions differently. Certainly Herbert Hoover had considerable insight when he noted that in any given society, the most powerful class can be located by finding out which sector obtains the greatest share of the mineral wealth (Agricola: 82–6). Once the conditions Chileans faced in producing for the world copper market are understood, then the seriousness of the political problems facing the copper industry becomes clear. More than anything else the lack of political power by the copper mining interests delayed for too long the reforms they required to compete with the emerging copper regions in the United States: Michigan, Montana and Arizona.

The objection can be made, that if successful mines are established under an existing mining policy, and that legislation is not changed, how can the same policy precipitate the region's decline? The answer is found in the recognition of mining as a constantly evolving industry where, decade to decade, conditions of ore quality, technology and labour, and thus competition, change. As the organisation of the mining and smelting of a mineral evolves to follow the market, that evolution is shaped by mining policy. If mining policy in a country producing for world markets is out of tune with that proven to enhance mining competitiveness in other countries, the key problem facing mining in the former country is not technical but political. And it is in the political arena that mine operators, always a numerical minority, and usually a minority in power terms, face their greatest challenge.

II Early Chilean copper production

Chile's importance as a copper exporter emerged in the copper trade of the early nineteenth century; a trade which had as a centre the smelters of Swansea (Roberts 1969). Swansea smelters had hopes in the 1820s of displacing Russia as the supplier of copper to France and continental Europe, but the supplies of copper ore from Cornwall, Anglesey and Ireland were barely sufficient to supply domestic British needs. This supply problem, coupled with the opening of official trade with Latin America, led to a much noted boom in mining investment in Chile by British capital in 1824–5 (Jenks 1927: 23; Vicuna MacKenna 1883: 148–55). This interest in Chilean mines was a part of a generalised infatuation with mining investment in Latin America, and while the boom everywhere collapsed in 1827, it did establish that Chile had copper ore and that merchants with trade items and capital for loan to miners could obtain high quality copper. At the time of the boom, the Swansea smelting interests optimistically obtained a slight reduction in copper import duties in the expectation of trade with Chile and, of greater significance, they obtained an import policy change which allowed them to import foreign ore, under bond, if it was for subsequent export. By 1831 Chilean copper, along with copper from Cuba, increasingly arrived in Swansea to provide the basis for British predominance in the European copper market (Roberts 1969: 145).

The smelting interests of Swansea went on to advocate free trade; they wanted to use Chilean and Cuban ore to supply domestic needs. Opposing this were the powerful Cornish mining interests, and their ally the *Mining Journal*, which editorially attacked the 'monopolistic' Swansea Smelters Association (Roberts 1969:145). In 1842 a compromise between the Swansea and Cornish interests was worked out whereby the customs duty on foreign (Chilean) copper for domestic use was reduced by over one half, but the lowered duty was also to apply for the first time to copper ore intended for subsequent export, the smelting in bond privilege being eliminated (Roberts 1969: 145). This compromise was a grave commercial error for the smelters. Suddenly smelting outside of Great Britain became commercially attractive. By 1844 new projects to smelt in Chile were under way (Great Britain 1847: 62). And as the new duties on copper for export effectively raised the price of copper exported across the Atlantic, entrepreneurs in the United States responded by setting up their own smelters, buying ore directly from Chile and Cuba. Significantly, the British customs duty was removed in 1847, but the policy reversal was too late (Great Britain 1847: 71). The copper trade had been permanently decentralised and entrepreneurs in Chile were on the way to building the world's strongest copper mining and smelting industry.

Whatever the short-term benefits the Cornish copper industry obtained from the 1842 compromise on copper customs duty, in the long run the duty fed the decline of copper mining in Cornwall by stimulating competition in Chile and the United States. But if the stimulus to mining in Chile and the United States was shared, their reponse was not. In Chile increased production was based on the extensive expansion of a well-established existing industry. Since the United States had no traditional mining industry to mobilise, a search for copper properties upon which to base an industry had to take place. Lake Superior or Michigan copper was discovered and first shipped in 1845. Its subsequent expansion was rapid (Murdoch 1964; Levy 1968: 8). But for many years the United States had a larger smelting capacity than the Michigan, and the smaller and minor Tennessee, mines could supply. Chile in contrast could not smelt all the ore it produced. The new Eastern Seaboard smelters in Boston and Baltimore thus came to depend on Chile, as well as Cuba, for ore supplies (Egleston 1886). The United States became an important part of Chile's copper marketing.

The early success of Chilean mining capitalists in production of copper in industrial quantities took place under what was basically an eighteenth-century mercantilist code. The Chilean mining policy conceived of minerals as a prime source of state revenue and consequently had regulations to prevent tax fraud by miners (Mendez Beltran 1979: 16–26; Vicuna MacKenna (1883: 604). That Chilean copper thrived under a considerable tax burden is a credit to its mining capitalists; still as late as the 1860s mining in Chile was technologically ahead of Michigan (Davis 1961: 262). During mid-century in Chile, as well as in Britain and the United States, copper production was expanded extensively rather than intensively. Competition was based on exploiting resources at more points, not by increased sophistication of technology at given points. But Chile was the dominant producer and in Chile there were mines which were world leaders in the introduction of machinery. Jose Urmeneta's Tamaya mine stands out as one of the great mines of the nineteenth century (Vicuna MacKenna 1883: 15–51; Douglas 1872).

III   Copper production in the United States

It is only in the 1870s, as intense production began in the United States' copper regions, that mining and smelting methods can be differentiated between Chile and the United States. Technological change, especially the application of gigantic power machinery financed by capital from outside the mining industry, set the stage for further market-dominating United States expansion. But economic expansion in copper was not simply a matter of machinery, however inventive. If that was all that intensive copper expansion required, Chile would have stayed with Michigan, Montana and Arizona stride for stride (Douglas 1872; Chouteau 1887: Introduction). Each region possessed extensive reserves. Entrepreneurs in each region were aware of new large-scale applications of steam and electricity. They all understood the role of railways, increased scales of production, new drills and lighting systems. Capital was readily available for all; skilled labour was at hand. The crucial catalyst for the factors of production to combine in an efficient intensive mine operation was national agreement on the importance of industrial progress, with debate over the best means for that progress, as opposed to debate over the ends of industrial growth itself. Debate over means of mining progress in the United States led to supportive mining policy. In Chile the same debate stalled and the search for a broadly supported, coherent mining policy continues to this day.

It is interesting to note that both Chile and the United States had mid-century civil wars which affected the search for mining policy (Whiteman 1971: 207–8). In the United States the civil war helped to provide a consensus by eliminating a significant conservative perspective from the debate, and in copper industry this translated into several important developments. Among these was the creation of a tariff on smelted copper in 1861, which was raised again in 1864 (US Revenue Commission: 11–19; *Congressional Globe* 1869: 416). Michigan copper spokesmen argued that by ensuring higher copper profits through a tariff, and thus supporting higher mine wages, the development of Michigan and the west was certain. In its hearings on the copper industry in 1865, the United States Revenue Commission heard testimony that only a tariff of 6 to 8 cents a pound, at a time when the price for smelted copper was hovering above 20 cents, could reverse the low profits caused by stiff

Chilean competition (US Revenue Commission: 12). Michigan copper spokesmen complained of low-cost Chilean copper supplying upwards of one third of the United States market and thus depressing Atlantic Seaboard prices.[3] Additional arguments were made that Chileans hurt the United States by their refusal to buy New England manufactures, preferring instead to spend their copper dollars in Britain. And the issue of national security was raised with the assertion that the United States was dependent on foreign supply for vital war materials (US Revenue Commission: 15). These debates led to the Copper Act of 1869, which placed a 5 cent a pound tariff on refined copper. Chilean copper quickly disappeared from the Atlantic smelters, which were themselves soon forced to close.

The dramatic shift in structure of the world copper market favouring United States producers during the 1880s, while helped by the tariff, was primarily based on the 1866 revision of the United States mining code. This crucial reform allowed mineral claims to be made and held with a minimum of bureaucracy and offered complete title security to the claim holder (*Congressional Globe* 1867). The 1866 Mineral Lands Act also constituted a formal rejection of minerals as a source of public revenue either through their sale or by taxation; and secure full possession of a mining property was made possible through a simple claim process (*Congressional Globe* 1867). This policy was reaffirmed in the Mineral Development Act of 1872, when minerals policy was further refined to meet the needs of new corporate mining companies (Doherty 1971: 542–55). The 1872 law made investments even more secure, and thus suitable for securing large loans, as well as making possible the consolidation of many small claims into one large property. Consolidation was obligatory to ensure that a large processing plant could be fed sufficient ore over a long period of time at a planned rate and at a predictable price in order to justify the huge investments new, efficient plants required. Planning and predictability were absolutely critical if financing was to be available for ever larger installations. It was widely believed that 'nothing but steady, unavoidable losses will force a mine into suspension, and therefore, the period leading to the survival of the strongest is usually much longer than is generally believed' (*Engineering and Mining Journal* 1884: 456). Consequently, easy access to financial backing was critical for survival in the face of early years of low earnings or even losses.

As copper prices fell sharply during the 1880s because of overproduction and inter-regional price wars in the United States, United States corporations were continually interested in 'how our competitors are bearing up under the strain' (*Engineering and Mining Journal* 1884: 456). In 1881 a correspondent of the *San Francisco Bulletin* wrote that with the transportation problem of Arizona solved, Arizona could now 'compete successfully with Chile in supplying the English demand ... The future of the Arizona copper mines is bright and full of promise, and now that avenues of transportation are open by which products can reach tidewater, we ought to be able to undersell every other copper country' (*EMJ* 1881: 22). Warming to the copper contest, the

Engineering and Mining Journal in 1884 observed that: 'It seems to be a mighty struggle now between the great producers in all parts of the world' (*EMJ* 1881: 22). The same editorial analysed the capacity for Chilean copper miners to continue producing, noting that as the mines were all in the hands of native families, 'they are probably capable and willing, with that faith which is at the same time the blessing and curse of mine ownership, to struggle at least for years to come' (*EMJ* 1881: 22).

As the 1880s ended, New York and Boston based copper corporations, working deposits in Montana, Arizona and Michigan held a dominant place in world markets. This dominance was founded upon coordinated political action by governors, state legislatures, senators, and representatives. They all joined in legislating mining policy favourable to capitalist needs at the state and national level (Toole 1954; Tausig 1915: 161–70). More than any other single factor, these policies made possible the financing of the technology necessary to change copper mining from a family-run, small-scale, cottage enterprise to a highly developed, large-scale, mechanised industry (Voskuil 1930: 226–7; Chandler 1959). Copper in the United States was able to become an aggressive, efficient and competitive industry, because the state provided the legal, physical and economic infrastructure the corporate copper industry required.

IV Chilean copper in decline

The story of copper in Chile now can be seen more clearly with this comparative case in mind. As the 1870s began, Chile was the undisputed world leader in copper, with the lowest price and the highest output; yet the future still worried leaders in the Chilean copper industry (Chouteau 1887: Ch. 1). They actively sought legislation to rationalise Chilean mining along lines taken by the United States (Vicuna MacKenna 1883: 604). Mining policy reforms in 1874, however, actually made the Chilean situation worse, for surface agricultural land rights were strengthened against those of mining entrepreneurs trying to discover, claim and operate mines. Chilean mines in the 1870s still had the basically colonial/mercantile mining policy, while North American mining was operating under pro-capitalist policies. There is no evidence that the world mining community in the 1870s appreciated the full implications of these differences; how bad the Chilean copper situation was, nor how good was that of the United States. The 'Chili bar' was the standard for world pricing as late as 1882; ten years later it was irrelevant. Chile had a mining policy of state revenue, the United States one of economic growth. Whatever these policies meant to society in general, they had specific and critical implications for copper regions involved in a Darwinian struggle to survive.

The Chilean method of taxation of copper had specific unfortunate consequences. Reforming the colonial tax, which was set at a flat 20 per cent, the post-independence tax policy of 1810 established a rate of two pesos per quintal of copper when the market price was eight to ten pesos (Vicuna MacKenna 1883: 513–17). By mid-century, as prices fell, the tax became proportionally higher and harder to pay. It had to be paid even during periods when the market price fell below the cost of production. Mining resentment of this difficult tax levy, based on volume of production rather than earnings, lay behind the unsuccessful civil wars of 1851 and 1859. These civil wars have traditionally been viewed in Chile as anarchist rebellions but were in fact policy based in the north. Mining operators claimed their taxes supported government projects only beneficial to urban and agricultural interests, while mining had unmet needs for roads, port facilities, and other infrastructure. In the 1859 civil war a rebel government was formed in the north in Copiapo. One of its first acts was to suspend the copper tax and replace it with a lower export tax (Figueroa 1889: 318). The civil war was lost by the copper interests, yet production continued, and even expanded in the 1860s and 1870s. Despite antagonistic policies, so strong was the North Atlantic demand, that Chilean mines, strained as they were by shortages of high grade ore, continued to thrive.

In 1883 the Michigan and western copper producers were jubilant. With detailed production statistics in the headlines, the *Engineering and Mining Journal* proclaimed United States copper superiority (*EMJ* 1883: 27–8).[4] Then what the Chilean miners had feared actually occurred. The United States began to sell copper surpluses in Europe. The relatively superior position of Chilean copper in terms of skills and abundant ore had postponed the crisis – now it was acute. Chile was straining to its utmost to produce what copper it extracted in 1883, yet all the while most mines lost money. Only the hope of better prices and the fear of losing a mining property under the existing code kept more mines from suspension. The crisis intensified demands for mining code reforms.

V The political response in Chile

The creation of the Sociedad Nacional de Mineria (SNM), was the focus for just such a reform movement. It consisted of copper and silver industry leaders from the Norte Chico; all with Liberal Party connections to both the chamber of deputies and the senate (*Boletín* 1883–4).

The SNM was organised and officially recognised by the government in 1883 (*Boletín* 1884: 94). It was not a lobbying group in a 'pluralist' sense, rather it was born through presidential sponsorship and was financed by public monies. The SNM lead an intense effort for congressional legislation favourable to mining. The SNM argued copper prosperity was the only possible foundation for the country's general economic prosperity and future well-being (*Boletín* 1883–4).

The SNM journal, the *Boletín de la Sociedad Nacional de Minería*, argued strenuously that the existing mine code was the principle cause of the copper industry's stagnation; a stagnation resulting from a lack of new investment on a scale required to compete with the United States (*Boletín* 1883: 1–2; 1884: 9–10). They believed that the antiquated mining code sought to increase state revenue, not promote economic growth through productive mines. The code's logic presumed miners would not work a mineral property unless forced to; the code allowed anyone to claim an existing mineral property if work at that property were suspended, even for a short time. But United States competition called for ever larger scales of production, as did falling ore grades. Both conditions required mineral properties to be consolidated, and financed with outside capital. This consolidation was particularly difficult for Chilean producers as the code mandated the working of each small claim individually. When consolidation was attempted, it required a team of lawyers to fend off claim challenges. Such insecurity was not conducive to obtaining the necessary capital for large-scale mines complete with steam machinery, and vast smelting capacity. Thus the existing claim system was criticised as the root cause for Chilean investment funds going elsewhere, particularly into Bolivia and Peruvian mining which had regions with high grade ores and more favourable codes (Bravo 1980: 53).

Thus the SNM's primary goals were to change the mine claim system to that of the 'Patent', the system used in the United States and Spain, and to abolish the export tax in favour of an annual fee on each claim, or a patent fee (Bravo 1980: 53). SNM leadership did not believe that Chilean copper could both compete worldwide and further capitalise itself if it was also expected to be the prime contributor to the state treasury (*Boletín* 1884: 12, 101–2; 1885: 389).

Not until 1888 was a new code approved by the Chilean Congress (*Boletín* 1888: 15; 1889: 293–308). The reform emerged after five years of debate and compromise. The main opposition came from agricultural and coal/nitrate interests. Under the 'colonial' code, only mineral properties containing certain metals could be claimed by anyone, on any property. That right was resented by landowners, and they feared reform would only strengthen and expand it to all minerals, not just metals. This possibility also led the coal and nitrate interests to resist reform; they were unregulated by the 'colonial' code which only conceived of metal mines. So far as the old code was concerned nitrate and coal did not exist. Surface landowners owned subsurface coal and nitrate, free from outsider's claims, thus allowing the minerals to be held in reserve. They could already exercise the very latitude being sought by the copper industry. In the 1888 code reform, coal and nitrates were given exceptional status and continued unregulated (*Boletín* 1889: 194).

With that legislative success behind them the SNM began to push its proposals for rail consolidation, a mining college, mine schools and an international mine exposition (*Boletín* 1889 series 2: 9, 237). Liberal President José Manuel Balmaceda, who provided essential pressure for passage of the 1888 reform, initiated efforts to rationalise the railway situation in the copper region and thus to reduce the rates for ore transportation. His programme called for rail nationalisation. But all efforts were cut short by the revolution of 1891. The SNM had been supportive of Balmaceda, never expressing any concern for his methods. Once the revolution broke out the SNM suspended meetings, and after Balmaceda lost power they were conciliatory toward the new agrarian-supported government (*Boletín* 1889 series 2: 37, 190–2).

The legislative successes of 1888 stemmed from support for mining from both the executive and legislative centres of government. Certainly there were initiatives the SNM could work out with the presidency alone, but fundamental change required legislative approval. After 1891, the willingness of both the chamber of deputies and the senate to go along with reforms stopped. Ministerial use of budget monies on hand could allow for such activities as the International Mine Exposition of 1894 in Santiago, but no legislative support for mining can be found until the end of parliamentary dominance in 1925.

VI Conclusion

During the 1880s Chilean copper interests tried to create reforms that were already twenty years old in the United States. Miners in Michigan and the west had worked out their own relationship with the governments which regulated their mining, writing legislation themselves to legalise existing practices. United States mineral policy was crafted by the mining industry and was aimed at rapid economic growth based on huge, quick profits to individuals. One 'good' mine fulfilled the career ambition of a prospector, an engineer or an investor. North American experience was in stark contrast to Chilean mining developed under a code based on a Spanish colonial experience which sought to create state revenues in the mining of precious, not industrial base metals. The Chilean reforms, born out of crisis, had the intent of increasing efficiency by allowing for large-scale, highly capitalised mining as practiced in the United States. But even had these reforms been achieved earlier, there is no guarantee that Chilean copper would have prevailed, given the predatory international behaviour of the New York-based mining corporations, particularly the dumping of surplus production below cost, while using domestic, tariff-protected earnings to reap a profit.

The parliamentary republic in Chile (1891–1925), from a copper point of view, complicated an already bad economic situation. The reforms of 1888 were too little, too late. And the revolution of 1891 postponed for almost thirty-five years any further attempts to rationalise and modernise Chilean mining policy. This policy, during the years of parliamentary dominance, consisted of measures antagonistic to economic development. It is ironic that the 1888 reforms, while too late to revitalise the Chilean copper, provided the basis for United States capital to enter Chile. Without them there would have been no large-scale foreign investment. Mine holdings would have been too insecure.

In the 1890s the United States copper tariffs were dropped, with the copper mines of Michigan, Montana and Arizona in unquestionable pre-eminence. Changing technology also brought forth new possibilities at the turn of the century; copper expansion now focused on porphyry or dispersed low-grade copper. Scouts from the United States copper giants searched the world for porphyry deposits. Chile had many. So United States capital moved into Chile – the country which forty years earlier had provided the excuse to build a copper tariff around the United States. But the Michigan interests that pushed for the copper tariff were themselves now being pressed 'to the wall' by copper from Montana and Arizona.

It may be that any industry structured around a given set of regulations becomes, in a sense, comfortable with those regulations. Change means uncertainty, and is acceptable only when a crisis is unavoidable. In Chile the mid-century civil wars were policy based in the north, but their focus was limited to the rate of taxation, not a total reform of mining policy. The prosperity encountered by extensive working of Chilean copper postponed the crisis necessary to overcome entrenched anti-copper interests. In Chile the power of agrarian and non-metal mining interests was such that industrial copper needs were not taken seriously by a congressional majority until very late. Even the reforms of 1888 meant little in the face of congressional indifference to copper after the overthrow of Balmaceda in 1891, when interests hostile to mining settled in for another thirty-five years of governing.

As the nineteenth century closed, with their great productive capacity in place and the Chilean competition demoralised, United States producers began to fear the dangers of excessive competition. An era of combinations followed, 'intended to lessen the rigors of the struggle for existance by uniting the strength and interest of many individuals in a company or association or "trust"' (*EMJ* 1889: 452). There was even an occasional expression of concern for the foreign competitors in the industry press: 'The sooner the foreign producers appreciate the ability of our mines to supply the world with very cheap copper, the sooner will they cooperate with ours in effecting some reasonable "working basis" for marketing the large stocks, now in the country' (*EMJ* 1889: 473). Predatory marketing, supported by tariffs, tax, labour and transportation laws favourable to capital coupled with mining codes written by mining industry representatives led inevitably to a United States copper success never dreamed of by Chilean mining capitalists.

Industrial mining arose in the nineteenth century not because of free enterprise, but precisely the opposite, because of the extent and quality of government intervention. Had government not regulated the distribution of subsurface minerals, had mining not been freed of the obligation to contribute to general state revenues and social well-being, had transportation not been subsidised, had limited liability corporations not been legislated into existence, had tariffs not been passed, the United States copper industry would still have emerged – the ore was there. But it would have been a very different industry, and perhaps not the world's aggressive dominant copper producer.

Notes

1 The public mining property debate of August 1983 until late summer 1984 produced a great outpouring of positions on the proper 'ownership' of copper properties. Very quickly the debate became historical as in the cited article by Professor Correa Amunategui.

2 The most influential of these works has been Reynolds, because he is so often cited by others. He draws a number of observations from secondary sources about nineteenth-century Chilean copper. From these observations he concluded that '[T]here were few men in Chile at the turn of the century who had contacts with the large investment consortia and who were aware of the developments in copper technology and their potential application in Chilean mining' (1965: 212). As others subsequently studied Chilean copper, Reynolds was cited to support the argument that Chile was incapable of the transition to becoming an efficient, large-scale and low-cost supplier of copper due to the failings of Chilean miners as entrepreneurs. Less prominent, but also significant is Pederson's study of Chilean mining which expresses a similar view: 'The vast majority of Chile's mining entrepreneurs were incapable of adopting more than the simplest of the new techniques, such as oil lamps as a replacement for tallow candles.' However valid these opinions for the 1910s, there is no evidence that they hold true for the 1880s.

3 Congressional Globe, 40th Congress, Third Session: 416. Senator Zachariah Chandler, of Michigan, stated on 18 January 1869, that copper is 'an interest that is absolutely being crushed by foreign competition – the competition of convict raised ores in Chile'. There is no evidence that his charge was based on anything other than his imagination.

4 This claim was made despite the fact that it took three United States copper regions to match Chile's single region.

References   * *See annotated bibliography for full details of all asterisked titles*

Agricola, Georgius
1556 *De re metallica*, with annotations by Herbert C. and Lou H. Hoover, eds. and
    trans. London: *The Mining Magazine*, 1912; reprinted 1950, Dover Publications,
    New York.

Boletín de la Sociedad Nacional del Minería
1882–91 All issues.

Bravo Juan Alfonso
1980*

Chandler, Alfred D.
1959 'The beginning of "big business" in American industry'. *Business Historical
    Review* 33: 28–31.

Chouteau, Eugenio
1887 *Informe sobre la provincia de Coquimbo*, Santiago.

Congressional Globe
1867 First session 39th Congress, Senate Bill 157.
1869 3rd session 40th Congress.

Crapol, Edward D., and Howard Schonberger
1972 'The shift to global expansion, 1865–1900'. In *From Colony to Empire: Essays
    in the History of American Foreign Relations*. Williams Appleman Williams,
    ed. New York.

Davis, John D.
1961 *Corporations*, New York: Capricorn Books.

Doherty, William T.
1971*

Douglas, James
1872 'The copper mines of Chile'. *Quarterly Journal of Science* 9: 159–82.

Egleston, T.
1886 'The Port Shirley copper works'. *School of Mines Quarterly* 7(4): 1: 25.
1845 'Commerce and progress of Chile'. *Merchants' Magazine and Commercial Review* 13: 325–6.

Engineering and Mining Journal
1881 'Copper – Arizona's place among the producing regions', XXXI.
1883 'Annual review of the metal markets for 1882', XXXV.
1884 'Foreign copper and lead mines', XXXVII.
1889 'The copper question', XLVII.

Figueroa, Pedro Pablo
1889 *Historia de la Revolución constituyente, 1858–1859*, Santiago.

Great Britain, House of Commons
1847 Copies of all Memorials and Petitions Respecting the Duty on Copper One.

Hewitt, Abram S.
1876 'A century of mining and metallurgy in the U.S.' *Engineering and Mining Journal*, XXI.

Jenks, Leland H.
1927 *Migration of British Capital to 1875*, New York.

Levy, Yvonne
1968 *Copper: Red Metal in Flux*, San Francisco.

Mendez Beltran, Luz Maria
1979*

Murdoch, Angus
1964 *Boom Copper: The Story of the First U.S. Mining Boom*. Calumet, Michigan.

Reynolds, Clark Winton
1965 'Development problems of an export economy. The case of Chile and copper'. In *Essays on the Chilean Economy*, Markos Mamalakis and Clark Winton Reynolds, eds., Homewood, Illinois.

Roberts, R.O.
1969 'The development and decline of the non-ferrous metal smelting industries of South Wales'. In *Industrial South Wales: 1750–1914: Essays in Welsh Economic History*. W.E. Minchton, ed. London.

Tausig, Frank William
1915 *Some Aspects of the Tariff Question*. Cambridge, Mass.: Harvard University Press.

Toole, K. Ross
1954 'A history of Anaconda Copper Mining Company: a study in the relationships between a state and its peoples and a corporation, 1850–1950. PhD dissertation. University of California at Los Angeles.

US Revenue Commission
N.D. 'The copper crisis in the United States – 1866'.

Vicuna MacKenna, Benjamin
1883 *El Libro del Cobre y del Carbon de Piedra en Chile*. Santiago.

Voskuil, Walter H.
1930 *Copper in Minerals in World Industry*. Port Washington, New York.

Whiteman, Maxwell
1971 *Copper for America: the Hendricks Family and a National Industry, 1755–1939*.
     New Brunswick, New Jersey: Rutgers University Press.

CHAPTER 6

Capitalist development and labour organisation: hard-rock miners in Ontario[1]

JULIAN LAITE

The Canadian Ontario shield is a harsh environment from which to wrest a living. The following essay recounts and analyses successive efforts by timber interests, farming and mining to extract wealth from the land. Knitted together by the supply needs of the railway, the towns it created, and the transport it rendered, each industry came to interact with the others, eventually to be dominated by the nickel mines.

The author argues that the proletarianisation process was delayed on the shield by ethnicity divisions, the mixture of local economic interests, the readiness of labour to move individually rather than fight collectively (encouraged by the tendency toward regional labour scarcity), and the anti-union strategies of mining management. This chapter describes how these forces were eventually displaced by industry and management policies that promoted a homogenisation of the labour force and its subsequent proletarianisation. Models in vogue regarding when and how proletarianisation occurs are shown here to capture only a fraction of the forces active in Ontario nickel.

Julian Laite (Department of Sociology, University of Manchester, Manchester M13 9PL, UK) is currently studying the political economy of mining and labour organisation in a Canadian region, and the changes in the formal and domestic division of labour associated with shifting patterns of work and employment in north-west England. A recent publication is *Industrial Development and Migrant Labour*, Manchester University Press, 1981.   — *The editors*

Capitalist development may occur with different forms of labour associated with the accumulation of capital (Fröbel 1977; Laite 1981). During the development process the social relations and political activities of these different types of labour will also vary. Analysis of the establishment of capitalism in a Canadian region reveals the extent and importance of such variations. Within the region, different forms of production can be identified, using different types of labour. Over time, the nature of, and relations between, these productive forms changed until one of them − mining − became the dominant capitalist industry of the region. Over time, the social relations and political activities of labouring groups in the region, and particularly of miners, also changed. It is these developments which form the main focus of this paper.

The relation between industrialisation and social and political action has been much investigated. On the one hand, it is hypothesised that industrialisation initially produces social changes which lead to a rapid increase in political activity. Then, as industrial and social changes are assimilated, political activity declines, becomes less confrontational and more institutionalised. On the other hand, challenging such hypotheses, Tilly (1974) has argued that industrial change occurs within particular social and political milieux which affect the level and form of political action. Rather than the unilinear decline of political activity, there are variations as circumstances change (Laite 1980). It is just such variations and circumstances which are of interest here.

Industrial development in northern Ontario from the late nineteenth century to the present day did not produce a burst of initial social and political unrest which has subsequently declined. On the contrary, political unrest prior to the second world war varied throughout the region, with brief moments of conflict, whilst after the war it has been at a much higher level, and often protracted. It is this low level of political activity during the first phases of capitalist development which is investigated here.

The social and political structures of Ontario occupational communities have been previously analysed. Innis (1936, 1956) propounded originally the 'staples' theory, that the extraction of natural resources in marginal societies determined regional development processes through requirements for finance, capital, labour and transport. Clement (1980, 1983) has rightly criticised Innis for being too deterministic and not locating primary production within a context of capitalist development and state regulation. Clement argues for a class analysis of the social relations of production of capitalist mining. Clark (1962) has emphasised the concept of community for explaining social stratification in mining and lumber camps. Workers were offered a stake in the community which overcame social class differences. The tensions and differences which developed were not between social strata within the community, but between the industrial workers of the communities and the rural migrants and outsiders.

The political economy perspective of this paper is similar to that of Clement and complements his analysis which deals mainly with the period after the second world war, and the important technical changes occurring then. However, for the first half of the twentieth century, the focus is broadened from mining to the region of northern Ontario and so embraces the relations between mining and the industries of lumbering and agriculture, also important in the region. In so doing, the roles of community and ethnicity stressed by Clark are reconsidered.

Six major factors explain social and political action in this mining region up to the second world war. The first is the relations of production in the region's major industries, including the forms of labour. In mining, there was an important transition from capitalist petty-commodity production to mechanised and proletarianised production. Secondly, the role of the state is important for the provision of legal regulation, taxation and the supply of labour. Complementing government policy was the third factor of managerial policy, seeking to establish control over the labour force. Fourthly, there were the attempts by the Canadian labour movement to establish an organised response to managerial control, although these were not very successful during the first half of the twentieth century. Fifthly, there was the nature of the labour supply, which was differentiated into several ethnic groups, with various levels of skill and distinct political traditions. Finally, there were the urbanisation of the region and the improvements in transport which increasingly unified the region and broke down the isolation of the camps.

For the first half of the twentieth century, the capitalist development of the region proceeded with a segmented labour market. Labour was divided between industries with different productive forms and these divisions coincided with ethnic differences. Mine management and the state reinforced such segmentation as a means of labour control. Faced with such policies, labour found it difficult to organise and overcome these divisions. As mining expanded and mechanised, and as more workers lived in towns rather than camps, the labour market became more homogeneous, and labour was able to seize on the changes in government policy during the second world war to organise political opposition. The history of capitalist development in the region shows how these events occurred, while the roles of these six major factors is discussed in the conclusion.

I Settlement and industries 1850–80

The mining area of Northern Ontario lies to the north and east of Lake Superior: from Thunder Bay in the west, through Kenora, to the nickel and copper deposits around Sudbury in the southeast, and then north through Porcupine and Kirkland Lake. The region is shown in Figure 6.1. It is the Canadian shield: bare rock, close-standing timber, small lakes, clay pockets, frozen in winter and with only a short summer. Within this area is the clearly identifiable region of the Sudbury basin, a belt of clay lying on the edge of the shield, dotted with both farms and rivers, with the refinery town of Sudbury lying at its centre (LeBourdais 1953). The farms produce mainly potatoes, sod, oats and hay for sale, while the mines produce mainly nickel, with some copper and zinc. Around 100,000 people live in Sudbury itself, with 50,000 living in the basin.

In the mid-nineteenth century (Coleman 1913, Kelley 1971, Lower 1936) the area around northern Lake Superior was populated by the Ojibwe Indians who raised some crops and traded furs with the Hudson Bay Company. There were also isolated mines, access to which was by the lake. Such mines were purely extractive ventures, declining and closing when the ores ran out. There was also lumber cutting around the shores, the logs being floated down the lake, but at this time the main lumbering penetration was from the east as French-Canadian settlers and lumberjacks moved along and out of the Ottawa Valley.

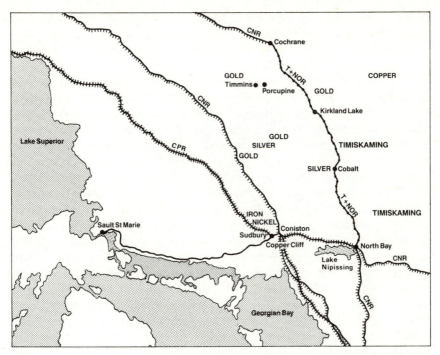

Based on a map in Lower, and Innes, H. (1936)

├─────────┤ 50 miles

┼┼┼┼┼┼┼┼┼┼┼ Canadian Pacific Railway

▬▬▬▬▬▬▬ Canadian National Railway

·⊣·⊣·⊣·⊣· Timiskaming and Northern Ontario Railway

───────── Hudson Bay Railway

Fig. 6.1 Northern Ontario Mining.

Although there were short-lived mining booms, the major changes in the area came with the expansion of the railways. After confederation in 1867, Macdonald's government supported railway construction, and in 1882 the Canadian Pacific Railway (CPR) workers cleared a camp some eighty miles west of North Bay. Prospectors and railway workers found chippings of copper near the camp, and the railway reached the spot in 1883. The railway brought into the region the railway workers, immigrants from Scandinavia and central Europe, prospectors from Britain and America, lumberjacks and farmers.

These last were mainly French-Canadians, migrating down from Quebec (Brandt 1977, Hamelin 1971, Skelton 1913). The French had begun the settlement of Quebec in the seventeenth century, and the French Crown gave the bulk of colonisation to commercial companies. In return for such trading privileges as the fur monopoly, they took responsibility for establishing settlers in New France. The companies granted land free to *seigneurs*: quasi-manorial lords who in turn rented it to settlers. Two centuries of population growth put pressure on the limited land available along the shores of the St Lawrence.

The form of settlement adopted on the *seigneuries* was the parish; a church settlement under the direction of a colonising missionary. The congregation built the church, roads, houses and schools. The *curé* became the teacher, local arbiter and parish representative. Roman Catholicism emphasised the virtues of cohesion in an agricultural community. The immigrants' forebears were French peasants. Indeed, whole French villages were on occasion transported to Quebec and established there. The parishes were peasant communities, partly commercialised through involvement in capitalist production and exchange relations due to farmers working seasonally in lumbering and fishing, generating income with which they bought some farming necessities.

Increasing population led many peasants to move to America. Alarmed at the dissolution of community and faith, Roman Catholic priests combated this exodus by a colonisation drive into the Canadian interior. In 1883 a group of French-Canadian colonists came to Sudbury and established themselves as subsistence farmers in and around the cleared railway sites and encampments. Supportive of settlement, the British Crown allocated large tracts of land to the French-Canadians who built a church and a school in Sudbury itself. During 1883–4 there were fifty French-Canadian families in Sudbury.

The French-Canadian agriculturalists began to provide wood to the CPR, participating in the second major industry in the region – lumbering. There was a symbiosis between establishing a subsistence farm and lumbering. The land was stocked with timber which needed to be cleared before farming could begin, the logs being used to build cabins and out-buildings. During the winter months when agriculture on the shield is impossible, farmers would either cut lumber themselves and haul it to be sawn by nearby lumber mills, or the CPR mill in Sudbury, or they would take their teams of horses and work for the lumber companies as elite hauliers. In both cases they were contract workers, not wage labourers. During the short summer the farmers would take hay, oats, pork and butter up to the lumbering and railway camps, where they would get high prices. Although lumbering was later to become subordinate to the development of mining, at the turn of the century lumbering was undoubtedly a very important industry. Sudbury was both a jump-off and recreation point for migrant lumberjacks. During the spring Sudbury's population swelled by thousands of these peasant farming lumberjacks, French-Canadians who were ethnically, religiously and occupationally different from the miners who were also in the town. There was little fraternisation between these groups of workers (Baine 1952).

Alongside farming and lumbering the third industry in the region was railway construction. Building the railways posed a massive problem for labour recruitment and so labour was sought abroad. The subcontracting of labour by labour agents was adopted and men were recruited at the docks, on the ships, and in Europe itself. The conditions of work were dreadful and many labourers ran off to work in the mines or cities. Ethnic groups emerged amongst the railway crews, with East Europeans doing unskilled work whilst French-Canadians were trestle builders and axe men (Bradwin 1972).

The fourth regional industry was mining. Ojibwe Indian miners were displaced by prospecting immigrant miners. The immigrants reproduced in the mines the same labour systems which had prevailed in Europe: tribute and tut-work. In eighteenth and nineteenth century Devon and Cornwall, hard-rock tributers were small masters who contracted ore faces and hired labourers to work them. Occasionally, workers superceded employers and took over the tribute for themselves at the auction for the contracts which w.:s held each month. The tributer was paid in proportion to the ore mined. Tribute in British mines was giving way to tut-work, a form of piecework. Employers paid workers fixed rates plus a small percentage of the ore mined. In Canada, in the mid-nineteenth century, mine owners were using both tribute and tut-work to get Cornish and Canadian miners to work in them. Between 1861 and 1872 nearly 60,000 miners left the United Kingdom to work in America, Australia and Canada (Todd 1967).

By the mid-1880s the miners coming into the Sudbury region were pro-spectors, usually 'grubstaked' with a forward payment from a third party. Such miners would receive a wage plus a percentage of mined ore. They were looking for deposits of copper and iron, or even precious metals. During the 1880s their mines were small surface diggings, difficult to expand. Develop-ment was hampered by the ecology of the shield, the depressed price of copper, and the $.05 per pound duty on copper entering the United States. The only thing to do with a good deposit was to sell it, for clearly it was to require large amounts of capital and labour to make mining in the region worthwhile.

II Capitalist expansion 1880–1902

Mining companies and the state

The major capital investment in the region was to be American (Main 1955; Buck and Toombes 1967; Pye 1967; Dorian 1955). S. J. Ritchie, an American manufacturer prospecting for hickory and freight for his Central Ontario Railroad, learned of the cliff of copper near Sudbury and formed the Canadian Copper Company (CCC) in 1886 to extract the ore. The CCC bought lots in the Sudbury Basin and sank three mines in 1886: Evans, Stobie and the Copper Cliff mine itself, just to the west of Sudbury. By the end of the year there were sixty-five miners on CCC's payroll. The ore was shipped to the Orford Refinery Company (ORC) in New Jersey, where it proved to be not copper, but cupro-nickel. Nickel was a virtually unknown metal, but now CCC and ORC had a lot of money tied up in it. So, whilst Ritchie looked for markets, smelting was begun at the Copper Cliff site to minimise transport costs and to stock-pile matte (smelted but unrefined ore). The smelting consisted of roasting the ore in huge open fires which continually smouldered and gave off a sulphuric smoke, destroying all vegetation for miles around (Cox 1970). In 1888, a furnace was installed which produced twice as much as world production in the previous year. The marketing of the metal was to the armaments industry and particularly the American navy, for experiments had shown that nickel-steel was tougher than just steel. The contract with the American navy under-wrote the industry's expansion in the 1890s and established nickel as a major Canadian resource.

This capitalist expansion in the mining industry was supported by the Canadian state. However, since the capital was American, the dilemma arose of how to control foreign investment without alienating it. Since the mid-nine-teenth century the state had played an important role in capitalist mining expansion. Prior to 1845 the Crown laid claim only to the gold and silver found by miners, but as settlement and prospecting expanded, a series of parliamen-tary orders built up to form a set of mining regulations. From 1850 to 1864 the minimum claim area was reduced and claim land prices dropped to $1.00 per acre. In 1864 came the Gold Mining Act, in 1868 the Gold and Silver Mining Act and in 1869 the General Mining Act. These acts withdrew the Crown's reservations on gold and silver, reduced the taxes on mined ores, and favoured exploitation by large companies.

The state looked to the mining industry as a source of revenue and develop-ment, but by 1890 more than half the mining capital investment in Ontario was American, and it was the Americans who supported new ventures. One compensatory move was to increase the price of land to $2.00 per acre, but the prize was on a bigger scale: from 1870 to 1900 Ontario mining was to earn $33 million in gold, silver, copper and nickel, yet pay not a penny in tax. So, in 1886 the Canadian parliament instructed CCC to refine in Canada, but CCC demurred. Again in 1890 the Royal Ontario Nickel Commission recommended imposition of a 3 per cent sales tax on all mined ore. This was agreed to by the Ontario legislature who in return supported further railway development to mining centres, offered bounties and tax-free concessions to industrialists, and set up a Bureau of Mines to help miners. However, the wrath of ORC was not assuaged. It informed Prime Minister Laurier that if the Act were implemented ORC would buy nickel from New Caledonia. Laurier backed down and the Act did not come into force.

Not only were the refining companies strong *vis-à-vis* the state, but also *vis-à-vis* their own suppliers. The refineries had sufficient supplies of matte and restricted the activities of the mines. Internecine war between producers and refiners broke out, resulting in Ritchie being ousted and Thompson of the ORC taking control of both companies. With matte still piling up, Thompson pursued the marketing strategy to Europe, competing with the French Rothschilds and selling to the German Krupp. A price war broke out.

In the Sudbury district the price war meant the rationalization of the industry at the turn of the century, for the victims of the war were the smaller producers. In 1892 there were four companies in the region, all of whom, due to the low-grade ores, had erected smelters of a sort. Mining capacity was far in excess of smelting and refinery capacity, so employment in mining and smelting fluctuated sharply. In 1893 CCC cut its work force to 495 men, only to raise it to 650 men in the following year. However, whereas CCC could survive the price war due to its marketing outlet through ORC, the other companies could not and by 1895 CCC was the only producer. Yet, so important was the refining bottleneck that once it was broken by technological advance, producers could once more set up. Thus, in 1898 Ludwig Mond, having developed a new refining process, bought mines in the region and became the second largest producer. The success of nickel-steel in the Spanish–American war of 1898 had boosted the metals' prospects considerably.

Labour and communities

By the turn of the century regional capitalist development was occurring with different forms of production drawing labour from a segmented labour market in which the proletarianisation of labour was limited. In the smelter yards and lumber mills there was large-scale capitalist production. Many contract miners, lumberjacks and commercial farmers were petty-commodity agricultural suppliers. Settlers farmed on a peasant household basis. These productive forms were interlinked. Capitalist merchants used petty-commodity agricultural suppliers, mining companies used wage labour and prospectors, lumber companies used wage labour and contracted peasant farmers. Ethnicity reinforced labour market segmentation. British and American miners, European railway workers and French-Canadian lumberjacks and farmers formed separate occupational and social communities.

Until the first decades of the twentieth century there was no concerted effort by capitalist enterprises and the state to homogenise the regional labour market through large-scale proletarianisation. On the one hand, the state was not embarking on a programme to release labour from the French-Canadian farming communities. On the other hand the labour camps in mining, railways and lumbering were not free markets for labour. In fact, these camps were *per se* a means of labour control, constraining labour and directing it to work as a particular industry developed. As production fluctuated, which it did markedly during the 1890s, labour was switched from one mine and camp to another: the mines of Copper Cliff and Creighton expanded, but by 1901 the Stobie mine was already closed.

These controls on labour were both economic and social. In the camps there were wage labourers, grubstaked prospectors, and contracted workers. Some miners were like the immigrant railway workers in that their employment contract was a debt relation, for they were obliged to work to pay off a sum of money previously advanced.[2] Other miners were like lumberjacks in that they contracted either ore faces or work. Workers were penalised for leaving before the contract was complete. Proletarian wage labour, with payment by the day, was clearly only one form of contract.

In the mining camps the miners and the smelter-yard labourers worked for a mixture of contracts, bonuses and minimum day rates. They worked twelve hour days or fourteen hour nights, six days per week. Mine wages ranged from $145 per month for a mine captain to $1.40 per day for surface workers. Contract miners earned up to $3.00 per day sinking the shafts and drifts. Despite the long hours, overtime was frequent and there were two holidays, on Christmas Day and Dominion Day.

Socially, the camps were small ethnic groupings housed in collections of huts and barracks around a shaft. Although some labour was available in the region it was unskilled and mining companies preferred to recruit Cornish and Welsh hard-rock miners. At the Copper Cliff mine, two miles from Sudbury, there were about three dozen log cabins, with kitchens and two bedrooms. In 1885 CCC built a boarding house there for bachelors. The wooden buildings were charred by the hot sulphurous fumes from the roasting yards. During the summer, pigs and a few cows roamed around the houses, to be slaughtered during the winter. In some camps the miners paid the mining company for their lodgings and board, in others the company supplied food and essential items. At Copper Cliff the CCC provided stage-coaches to bring miners from Sudbury, at $.25 for the round trip. So, for many months of the year, many miners lived in geographically isolated camps and were dependent upon mining companies for work, sustenance and access: it was a bunk-house life (Bradwin 1972). Different ethnic groups controlled particular occupations, restricting access to them. Thus, in various ways, control was exercised both through and in the camps.

The town of Sudbury itself had prospered from 1880 onwards, not as a mining town but as a commercial centre serving the region (Stelter 1971, 1974).

In 1885 the CPR had moved on and the town's population had fallen from 1,500 to 300. Usually, this meant the end of a railway camp but Sudbury was supported by the presence of the French-Canadian settlers, the peasant lumber-jacks and the increasing mining activities. By 1890 population had risen to 1,400 and the largest demographic group in the town was the British ex-patriates, accounting for nearly half the population. However, they were divided on religious grounds between the Protestants and the Irish Catholics. The largest single homogeneous group was the French-Canadians, accounting for one third of the population (Brandt 1977). Increasing residential stability heightened the importance of their ethnic differences, correlated as they were with occupational differences. British and French established separate schools, separate churches, and lived in different areas of the town. The French were either merchants, serving the local farmers or channelling some of their pro-duce to the mines, or they were unskilled labourers, farm migrants perhaps, working in Sudbury. The British were professionals or employed by the mines and railways. The French established their own institutions, most notably the Sudbury District Agricultural Society. Sudbury was incorporated in 1893 and by 1901 was a brisk town of some 2,500 people.

By the turn of the century the French-Canadian peasant farmers in the Sudbury Basin had consolidated themselves. The farms were scattered on the pockets of workable soil but some villages/towns had emerged: Blezard, Chelmsford, Hanmer, which performed several functions. They were impor-tant meeting and marketing places for the farmers, were distribution points for supplies to them, were lumber mill sites and log-driving centres, and were supply towns to the lumber industry still in the Basin. In 1903 lumbering was still probably the largest employer in the region, with seventeen lumber com-panies employing some 11,000 men in mills and bush at peak periods.

During this period of capitalist development, and the following period until the end of the second world war, labour was relatively quiescent. Industrial-isation of itself did not produce social unrest (Canadian Official Publications, 1882–1979). The only recorded strike of the period in the Sudbury region occurred in 1899 when employees at four of CCC's mines struck for higher wages. The strike lasted one day, and the ringleaders were fired. The main reasons for this lack of labour disturbances derive from the nature of this period of capitalist development. However, although this quiescence continued during the second industrialisation phase until 1944, the reasons for it altered with the changes in industrial development.

Firstly, at the turn of the century many miners were still prospectors or subcontracting tut-workers, negotiating through work rates and contracts rather than strikes. Secondly, miners were not surplus labour, not a pauperised peasantry having no alternative but to negotiate in the industrial milieu. Rather, they were migrant workers, adopting a mobility solution to poor wages and conditions, rather than a solidaristic one. Within the camps company control was such that miners preferred to leave rather than confront the company. Thirdly, the shortage of skilled labour at the end of the nineteenth century meant that some workers, whose support would have been crucial during strikes, could find employment elsewhere. Biographies of miners, derived from accident reports, show that they worked across the length and breadth of North America.

The fourth reason for the small number of strikes was the lack of trade union organisation amongst the miners. Although a Canadian miners' union, the Provincial Workmen's Association, had emerged in Nova Scotia in 1879, it was mainly confined to that province. United States unionism was beset by debilitating divisions due to conflicts over mining unions. In 1894 the conservatives in the American Federation of Labour, led by Samuel Gompers, turned against the radical Western Federation of Miners, formed in Idaho in 1893. In 1897 the WFM left the AFL, supported the American Socialist Party, and was attacked consistently by the AFL. Although the WFM made some headway in the west, establishing locals in British Columbia, it made little ground in the east, unsupported by the AFL and out-manoeuvred by determined employers.

When the CCC fired strike leaders, it claimed that they were WFM militants. On the one hand this may have been the case. By 1899 the WFM was a single-industry union which aimed to recruit across the whole of North America. As such, it would have been an important organisation to which mineworkers in Canada, themselves Americans and Europeans, could turn. On the other hand, it may have been that CCC was using an occasion of local unrest to warn militant union organisations to stay clear.

III Industrialisation and proletarianisation, 1902–44

The first half of the twentieth century witnessed a major change in the capitalist development of the region. No longer were there to be several important industries using different forms of labour, but rather there emerged one dominant industry, mining, which sought to industrialise rapidly and proletarianise its work force concomitantly. Again, through fiscal and immigration policies, the state was a major force in helping shape these new developments (Nelles 1974).

By 1902 CCC had a dominant position in Sudbury mining. Generally the problem facing the mining industry was the shortage of capital. However, by 1902 American investors were looking for foreign outlets. During the American Depression of 1893–7 there had been rationalisation of major companies that had begun exporting to unload products that could not be sold at home. One such company was US Steel forged in 1901 by J. P. Morgan, and the largest company in America. To safeguard the monopolistic position of US Steel, Morgan had to guarantee his sources of supply. Nickel was to be used in armaments and cars and US Steel was ORC's and CCC's biggest customer. In their turn the two companies were the world's primary suppliers of nickel.[3]

In 1902 the nickel syndicates of US Steel founded the International Nickel Company as the holding company for CCC and ORC. In effect US Steel had bought them out (INCO 1920; Collins 1933; Thompson and Beasley 1960; Bleach 1974).

Having established itself in the Sudbury region as the major producer, refiner and marketer, INCO was establishing external control over competition and markets, and internal control over production and labour. Externally, Morgan and INCO engaged firms in competition over finance and prices. F. H. Clerque, a paper magnate who had built a nickel smelter in the region, competed with Morgan for finance, but soon went bankrupt. In 1917 the British American Nickel Corporation was formed but INCO immediately engaged it in a price war and it, too, went bankrupt in 1924. INCO bought its Sudbury assets the next year. The only lasting competition of sorts came from Mond, but this was more a policy of 'live-and-let-live' in an expanding market, with INCO as market leader and price setter. When investment plans required in 1928, Mond was eventually absorbed into INCO. Market control was ensured by setting the price of nickel just low enough so that new entrants were discouraged and large consumers would not seek other sources, but would engage in long terms contracts with INCO.

Internally, INCO sought to establish control over production and labour (Clement 1980). The turn of the century and the arrival by INCO marks the transition of Ontario mining from capitalist petty commodity and small-scale production to full capitalist industrialisation. Changes in management policies matched changes in technology. INCO geared its production to monthly changes in demand. When demand levelled off or fell, refinery production was adjusted. When demand picked up again, stocks were used up, rather than increasing mining production. The level of stocks, rather than employment, became a central company concern.

In terms of technology there was a sustained drive for capital intensive production. This was linked to the proletarianisation of labour and the strict regulation of the labour force. Prior to 1900 technology had been labour-intensive, with hand drills in the mines and hand stoking of the roaster fires on the surface. After 1900 this changed, requiring proportionately less labour and different types of labour. Around 1900 steam drills were introduced and then in 1906 pneumatic drills using a piston and hammer action were brought in, which were also much lighter. These mechanical drills increased production eight to ten times over hand drilling. In 1914 hollow steel drills, water-cooled drills and hammer drills were introduced. Hoists were improved and in 1913 single point loading at the shafts was introduced, saving labour time massively. Electricity supplies were rationalised and the mines and smelter of Copper Cliff were electrified in 1903. Mines would either share hydroelectric power plants or compete for power, so the mining companies established their own plants to provide them with electric power which would replace labour. Electric cars came to replace horse-drawn cars.

This influx of American capital brought sharp increases in production. From 1900 to 1910 production tripled, then doubled again in 1916. These increases were due both to improved production and marketing. INCO built new smelters and expanded mines. The Ontario government surveyed the whole ore body, renewing its recommendation that refining be done in Canada. Under pressure from the Borden government, INCO commenced a refinery at Port Colborne in 1916 and completed in 1918. Mond also expanded mines and plant. From 1914 to 1918 the production of Sudbury nickel doubled. Due to the European re-armament drive and the first world war, sales of nickel expanded. In 1913 INCO sent two-fifths of its output to the German re-armament drive. Keeping prices moderate, INCO secured long-term contracts; the British Admiralty contracted, for example, to buy all the nickel produced. These changes in production were matched both by an expansion in mining employment, and, again, by marked fluctuations in that employment. From 1902 to 1904 the number of miners employed in the Sudbury region actually declined and, in 1904, their numbers were only half the 1901 level. However, after 1904 employment expanded steadily to 2,156 by 1910 and 3,512 by 1913.

With the problem of capital investment resolved, the issue facing the Canadian mining industry became that of labour − its supply, quality and control. Concomitant with technological changes, the structure of the labour force altered. The division of labour increased markedly, entailing the simplification of tasks, so that low-paid labour or machines could perform them. This was matched by the specialisation of tasks, so that no longer were miners familiar with all aspects of mining, and therefore essential to production. Rather, specialist workers were created who remained strategically important but whose role in the labour process could be circumscribed. As Braverman (1974) has observed, job hierarchies are in themselves forms of labour control. By 1917 the Royal Ontario Nickel Commission observed that there was a growing hierarchy in the mines and was able to identify fifteen different job categories. The mine workers were moved over to wage payment systems, varying according to the nature of the ore bodies. Wage payment systems for all workers were introduced in 1917. Usually, INCO established its rates through comparison with rates paid by other companies. To attract labour, INCO paid somewhat more.

As well as stratifying the labour force INCO also paternalised them through improvements in working, living and financial conditions. The mining camps became company towns, developed to attract and keep a stable work force and to exercise socio-economic control over them. There were to be four main company towns: Copper Cliff and Creighton, owned by INCO, and Levack and Coniston, owned by Mond, with the first being the most important. At Copper Cliff the number of workers grew from 2,500 in 1901 to 3,000 in 1911, although by 1921 it was back to 2,600. Within this population the number of unskilled workers dropped from 1,800 in 1903 to 1,300 in 1911, as Copper Cliff became a management centre. The number of unskilled then rose to 1,600 in following decades as the technology in use required this type of labour. In the other towns, numbers fluctuated around 2,000.

In the company towns INCO and Mond at first owned and controlled land, buildings and utilities. Yet divisions of functions soon emerged. The company became directly responsible for the hospital, the recreation club, sewage disposal and water supply. The municipality was responsible for the streets, lights, health, police and schools. However, the municipal officers were themselves company managers, elected by acclaim at a mass meeting. These officers raised municipal revenue by taxes on the company and rates on the populace. At the smaller mines there were company settlements but these were really just labour barracks, not towns in themselves. As the four towns reached a population of 10,000, then, under Canadian law they became independent municipalities. In the towns, hospitals and schools were regularly improved and clubhouses and recreation facilities were built and turned over to the men. More and better houses for workers were built.

The extent of the attempts at social control is revealed by the example of alcohol sales. In 1906 the Copper Cliff council (management) refused to allow the building of an electric street railway from the supposedly alcohol-free Copper Cliff to Sudbury, arguing that it would give the men access to hotels and drink in Sudbury. Moreover, it would enable the men to spend money in Sudbury when there were some stores in Copper Cliff. Of course, bootlegging between the two towns was rife and Sudbury as a commercial and shopping centre always competed with the facilities available in the company towns.

As well as their relatively superior living conditions on the Canadian shield, the workers of INCO and Mond were paid higher wages than those received by other miners. They also worked with more advanced tools and equipment, making work somewhat less arduous for them and more productive for the companies. Further, the larger companies brought in retirement and insurance schemes, INCO introducing them in 1928.

Immigration and ethnicity

The problem still remained of guaranteeing a supply of labour to fuel this industrial expansion, and this the state resolved through an immigration policy. This in turn, in the short run reinforced the ethnic differences already extant in the industrial labour force. The state had always played an important role in immigration, particularly since Canada always had to compete with America for labour (Avery 1975). In the 1860s a system of free grants and easier sales of land was introduced. In the 1870s Canadian government agencies were established in Britain and Europe, sea passages and rail fares were subsidised, and a bounty was paid for each adult immigrant recruited. Yet the policies were not really successful and for most of the second half of the nineteenth century Canada was a net exporter of labour to America. Many of those first immigrants were agriculturalists, offered free homesteads and land and forming a 'stalwart peasantry'.

When the immigration policy was revitalised in 1897, however, it was because both Canadian industry and the west were clamouring for labour. This time the immigrants were East Europeans searching for industrial work and were absorbed by the railways and the mines. From 1901 to 1911, 1,700,000 immigrants arrived in Canada, increasing Canada's population by 34 per cent and net migration to Canada became positive for the first time for four decades. It was industrialists who pressured premiers Laurier and Borden to encourage immigration. Due to their political connections the railway companies in 1906 were themselves recruiting illegally from Europe and America. The move from settler-labourer immigrants to industrial proletarian immigrants is shown by the facts that while in 1907 31 per cent of immigrants were unskilled labourers, by 1913 they were 43 per cent, and that while in 1907 29 per cent of immigrants were from Central and Southern Europe, by 1913 they were 48 per cent.

These immigrants became the unskilled workers in mining. In 1911, whereas only 17 per cent of drillers in Ontario mining were immigrants, and only 30 per cent of the borers, of the operatives, 48 per cent were immigrants (83 per cent in the nickel industry) and of the labourers, 70 per cent were immigrants (91 per cent in the nickel industry). The immigrant groups formed different gangs: Italians and Russians were labourers and shovellers, underground drillers were Finns, and so on. Orders and instructions to each group were in their own language.

Immigrants provided nearly half the population of INCO's Copper Cliff mine. Not only did they go into particular occupations, so reinforcing the usefulness of ethnic identification, but they also formed different residential groups. A residential section for British and American executives was built, while barracks were erected for labourers and were filled by East Europeans and Scandinavians. Land at the very foot of the smokestacks was rented out and Italians erected their own houses on it, creating 'Little Italy'. There was no French-Canadian area.

On the one hand, ethnicity provided workers with a sense of common identity and a basis for trust relations. More than this, Finns and East Europeans brought to Canada their communitarian organisation and socialist ideologies from their industrial communities in Europe. The clubs formed by immigrants were not only places of recreation where mutual trust societies for accidents could be organised, but also political pressure groups. In 1908 the Finnish Social Democratic Party in Canada was founded and by 1914 Sudbury was its centre. On the other hand, this ethnicity provided a means through which mine managers could divide and control their labour forces. As one manager of the War Eagle mine in British Columbia observed:

In all the lower grades of labour and especially in smelter labour it is necessary to have a mixture of races which includes a number of illiterates who are first class workers. They are the strength of the employer and the weakness of the union (Scott 1977).

Prior to the outbreak of the first war an economic depression in Canada had put immigrants out of work and, due to the uncertainty generated by the declaration of war, INCO shut down its Sudbury mines and smelters for three months. Many men joined the army. As production picked up, so employment increased, and in 1916 was the highest yet recorded in INCO. But manpower was bled away by the war and employment dropped, leading INCO to introduce labour-saving technology.

Fortunes for the mining industry during the inter-war years were mixed, although of course the basic trend was one of large increases in production. Record production and smelting during the first war, which polluted and devastated the surrounding countryside (Cox 1970), was followed by crisis as metal prices fell at the end of the war. In 1920 there was a 15 per cent wage cut by INCO. Then in 1921 the mines and smelters were closed and two-thirds of the 3,000-strong labour force was laid off. Research and development produced new alloys and, when the automobile industry picked up in the mid-1920s, so did the sale of nickel. By 1929 output of Sudbury nickel was seven times the output for 1922 and 90 per cent of world output. These improved prospects for the industry brought a new mining company – Falconbridge – into the region in 1928. In that year Mond and INCO merged.

The beginning of the 1930s brought the Depression and again men were laid off. INCO's total labour force declined from 12,000 in 1929 to 4,000 in 1932. There was no workmen's relief. INCO joined the world copper cartel to reduce copper stocks. Salaries were cut again. But the mining industry (the world over) climbed quickly out of the Depression. The first signs of revival came in 1932 and by 1933 mines and refineries were reopened and men were taken back on. Employment was back to 12,000 in INCO by 1935. Throughout the period INCO invested heavily in new smelters, acid plants, mines and equipment.

Due to the fluctuating growth of mine employment, more and more workers came to reside in Sudbury itself. In the company towns were to be found the British-Canadian management and skilled workers, and some ethnic groups, such as the Italians in Copper Cliff. In the labour camps were mainly immigrants, with few resident British. To Sudbury came those workers who could not find a place in the company towns or who could commute to them. INCO, at least, recognised the costs as well as the benefits generated by the towns. When, in 1925, it planned the development of the Frood mine, it requested that Sudbury house the required labour, rather than bear the cost itself. Sudbury readily acquiesced. Thus, Sudbury became the garrison for the reserve army of labour required or fired by INCO when opening or closing pits. INCO could not construct housing in Sudbury itself since CPR owned most of the land and would probably have asked a very high price for it.

During the twentieth century the population of Sudbury expanded drastically from 2,000 in 1901 to around 100,000 in 1980. As in mining, however, alongside this expansion there were also important fluctuations in Sudbury's growth, and changes in the composition of the town. It is important to locate the town's growth within the settlement pattern of the region (Stelter and Artibise 1978). Prior to the first war, that settlement pattern crystallised, with British and immigrants in mining; British, French-Canadians and immigrants in Sudbury, and French-Canadians in the agricultural settlements of the Basin. Indeed, by the outbreak of war the most noticeable changes were the rise in immigrants in mining and the fall of British descendants in agriculture, presumably because they found the shield too harsh to farm. All three locations grew in this pre-first world war period: Sudbury doubled, Copper Cliff grew by one quarter, and the agricultural settlements by one quarter.

In the inter-war years, the pattern of growth was quite different, and more variable. Sudbury continued as it had begun: a commercial town servicing the regions, mines and agriculture. Its population doubled again in the decade after the first world war. Mining expansion and a construction boom contributed to this. The rise in population consisted mainly of increases in French-Canadians and Europeans, with French-Canadian middle and labouring classes accounting for over one third of the population, and Catholics for about half. Indeed, during the 1920s, the percentage of British residents declined from one half to one third, whilst that of Europeans rose from one tenth to one quarter. Meanwhile, the size of the agricultural settlements of the Basin remained exactly the same, but the proportion of French-Canadians there declined sharply.

Clearly, the 1920s were a period of French-Canadian agricultural migration into urban employment in Sudbury, rather than into the mines. The growth in mining population was provided by Europeans, who lived both in Sudbury and at the mines. These patterns changed again in the 1930s. Sudbury continued to grow, this time by 75 per cent and the increase in British outpaced that of the French-Canadians. The agricultural settlements grew slightly, but the percentage of French-Canadians there increased sharply.

Evidently, French-Canadian migrant labourers were returning to the settlements during the Depression years. Yet they returned not only to the farms. Population growth now meant that off-farm work was essential and consequently there was an increase in French-Canadian representation in mining. In fact, more French-Canadians than British were recruited into mining during this period. Many went into INCO's Coniston mine, which expanded during the 1930s. By the 1940s population expansion by the major groups in the region was not being contained by the productive forms in which they had previously worked and lived.

The pattern of coincidental occupational and ethnic segmentation of the regional labour market had been reinforced by the form of agricultural development in the region. That is, French-Canadian household farming did not become capital intensive and shed labour, but rather retained it. Establishing themselves as peasant farmers in the second half of the nineteenth century, the French-Canadian settlers in the Sudbury Basin were contained in that role by the opening up of the West. In the 1880s the prices of meat and wheat were halved (Canadian Official Publications 1915). Unable to compete, the Quebec and Ontario farmers shifted over to mixed farming or tried to specialise in fruit or dairy produce. At the end of the nineteenth century the area under crops in the eastern provinces actually diminished. This move to mixed farming was, in fact, reinforced by the growth of local urban and mining markets during the twentieth century. Prairie beef, pork and wheat swamped the mining centres and the mining companies placed their contracts with western merchants. Excluded from those products, local farmers provided mainly hay and oats to the mines, which they originally supplied to the lumber camps. Potatoes and vegetables for the miners were also grown by local farmers.

Partly commercialised subsistence farming thus continued in the region and did not give way rapidly to capitalist farming specialising in cash crops and shedding labour. Household farming using family labour endured. Some cattle were raised for the Toronto market, but this activity was controlled by middle-men and French Canadians did not take it up on a large scale. Agricultural reports by inspectors in the 1920s mention disinterest by peasant farmers in cattle rearing (Kay 1921). Rather, the peasants reared horses and gained income through working as teamsters, and hauliers in the lumber camps. By the 1930s however, the lumbering alternative for French-Canadians had all but disappeared from the region. The timber frontier had been pushed well to the north and west and employment in lumbering was in the camps there. Often such camps were around the lakeside since American cities had emerged as a major market, due to improved shipping and expansion in construction. After fluctuations in its fortunes during the 1920s the timber industry went into decline during the 1930s (Goltz 1974).

Labour organisation and unrest

The lack of labour organisation during the late nineteenth century continued throughout the first half of the twentieth century. Trade unionism was not established on a permanent footing in the region until 1942. Of course, there was always local conflict and haggling, with men stopping work during disputes, as well as absenteeism and industrial sabotage. Further, there were two attempts to form unions, in 1913 and 1936, and there were two recorded strikes by INCO workers in the region: at Copper Cliff in 1904 and Port Colborne in 1908. However, both the attempts to unionise and the associated strikes were short-lived. Such a situation is in dramatic contrast to the period after 1942 (Lang 1970). Since that date the INCO refinery workers and miners have been continually unionised and have struck in 1943, 1944, 1950, 1954, 1956, 1958, 1966, 1969, 1975 and in 1979, when they engaged in strike action for eight and a half months over a new three-year contract. The reasons for the low level of labour unrest during a period of industrialisation, and the differences in pre- and post-second world war activity are discussed in the conclusion.

The two strikes amongst INCO workers were by non-unionised men taking direct action. The first was at Copper Cliff in January 1904, when mine labourers struck for twenty-five days over their methods of payment. This strike was resolved in favour of the employers. The second 'strike' was by brass workers in Port Colborne and was in fact a 'lock-out' by INCO management, penalising workers for absenteeism. Again, this strike was resolved in favour of the employers. Thus, strikes by INCO workers were about economic issues, quickly resolved by management, and did not lead to widespread labour organisation.

Instead, labour organisation originated amongst non-INCO precious metal miners to the north (Baldwin 1977). In 1903 silver was discovered in Cobalt and miners flocked in, many from Nova Scotia. The high profits and labour shortages led to mine owners competing for labour and wages rose. To counter this the mine owners formed a Mine Manager's Association in 1906 and declared a common, lower wage. In response the miners established a Western Federation of Miners (WFM) local affiliated to the new labour federation, the International Workers of the World (IWW), and went on strike. The management reaction to unionisation and strike action was to fire unionised miners and replace them with immigrants. One thousand Nova Scotia miners were replaced, many by Italians recruited in Montreal. The strike was broken, although bonuses improved.

However, labour organisation had been initiated so that when, in 1912, employers in the new gold fields of Kirkland Lake tried to replace workers with labour from Cobalt, the World Federation of Miners established a local in nearby Porcupine and demanded recognition and an eight-hour day. The mine owners resisted, and there was a strike which, again, the miners lost. Following the persecution of the WFM in the gold and silver fields, the WFM began to organise in Sudbury. In 1914 Sudbury became the WFM headquarters for the region. In 1908 the WFM had left the International Workers of the World, finding them too radical, and had affiliated to the American Federation of Labour. However, after its defeats in Canada and in the Michigan Copper strikes it went into decline and in 1916 changed its name to the International Union of Mine, Mill and Smelter Workers: Mine Mill. Although a Mine Mill local continued for a few years in Sudbury it was small, having only some fifty to sixty members, and by the end of the first war it had disappeared.

Clearly, the potential for union organisation was being deliberately thwarted by management at INCO and in other mines. These were the decades following the defeats of American labour at the end of the nineteenth century and during which American management was seizing the initiative to keep labour on the defensive. At INCO, for example, Pinkertons infiltrated the Sudbury labour force and in 1916 200 workers were fired with 'We don't want troublemakers' on their pay-slips. Sudbury became notorious for its abundance of company informers.

Moreover, the Canadian state was also hostile to non-British immigrant communities and ideologies. During the war the shortage of workers improved labour's potential bargaining position. In 1917 the Russian revolution threw a long shadow over East European immigrants. In 1918 there was labour unrest in British Columbia and Northern Ontario. In 1919 came the Winnipeg general strike, seen to be organised by immigrant agitators. The Borden government responded to these perceived threats first by preventing Finns from organising or publishing in Finnish, and then in 1919 suppressing the foreign language press completely and outlawing socialist and anarchist organisations.

As the war drew to a close both industrialists and government feared that there would be falls in production and employment and that returning soldiers would join with radicalising labour forces. One solution to these problems was the dismissal and deportation of aliens. Supported by government, industrialists offered to take on returning soldiers by dismissing aliens.

Since it took three years' residence in Canada to gain Canadian citizenship, many redundant immigrants were without either work or citizenship and so, having no claim to remain in Canada, were liable for deportation. One estimate is that by 1919 some 150,000 Central and Southern European migrants returned to Europe. In that year INCO dismissed 2,200 of its 3,200 employees, the vast majority of those fired being European immigrants. Of course, not all workers left employment involuntarily. On a continent where some opportunities were available elsewhere, mobility was a constant labourer's strategy in the face of poor conditions. Many immigrants would use their wages to pay for either a piece of land or a business. Also, remittance of cash to Italian and German families in Europe was very high and immigrants voluntarily used their wages to pay for their return fare to Europe. So, the policies of industrialists and government combined to reduce the perceived threat from unemployed militant immigrants.

Mining employers, therefore, used various techniques in the inter-war years to maintain labour control. The high supply and induced turnover of unskilled European labour were advantages for employers in that they militated against union organisation and reduced pressure for higher wages. When skilled labour was required for mining development, the company town was established. When new mines were sunk or old ones expanded, company towns were planned and implemented, such as those by Mond at Coniston and Levack, and by INCO at Creighton.

Large amounts of capital were required to exploit the mineral potential of this region and industrialists used their ownership and control of this capital to institute effective policy mixtures of paternalism and threat. After the failure of Mine Mill in 1920, the next attempt to organise the INCO workers was by the Mine Workers Union of Canada, formed in 1925 by a left-wing splinter group from the United Mine Workers of America. The Mine Workers appeared in Sudbury in 1933, supported by the Finnish community, but the group lasted less than one year. Some ex-Mine Workers joined a new Mine Mill local which opened in Sudbury in 1936 and, despite being attacked by thugs, the local survived a little longer. However, by 1938 it had only twenty-nine members and by 1939 it had disappeared.

Elsewhere, however, events had taken a different turn. In 1937 the United Steelworkers of America finally signed a contract with INCO at Huntington in the USA. INCO combated this in Canada with a 10 per cent wage increase and plans for a company union, but the union foot was in the door. The outbreak of war in Europe necessitated the mobilisation and cooperation of working classes for the defence of Britain and the empire. Once these were secured in Britain, the Dominion Labour Act was passed in 1942 to provide a framework for the cooptation of imperial labour. Wages were fixed during the war years but workers were free to organise and unions were allowed. In Kirkland Lake, Mine Mill struck over recognition, as did the Auto Workers in Windsor, and in 1942 Prime Minister King recognised the validity of their grievances. The Kirkland Lake leader of Mine Mill moved to Sudbury and organised a new local there, which was recognised by INCO in 1944. For two years the Sudbury Mine Mill local was the largest in Canada. It was also one of the most radical, leading to allegations that it was communist controlled, and its suspension from the Canadian Congress of Labour. In 1962 Mine Mill was replaced by the United Steel Workers of America, which represents the metal workers today, and has led them on four major strikes.

IV Conclusion

Capitalist development in northern Ontario during the first half of the twentieth century was characterised by a segmented labour market, divided between productive forms and ethnic groups. Changes were under way in this period, such as urbanisation and large-scale industrial investment, which were bringing about occupational and residential homogeneity in the regional labour market, but it is after the second world war that a freer, more fully proletarianised, regional labour market emerges. This transition in the nature of the labour market was thus concomitant with the transition in the form of capitalist production in mining, the dominant regional industry.

Despite the steady industrialisation of the region, which brought with it important social and economic changes, the first half of the twentieth century did not witness a high level of political unrest on the part of labour. Earlier it was argued that political action is not an automatic consequence of industrialisation, but depends upon the social and political milieux in which industrial development occurs. Six factors influencing political activites were outlined, which may now be discussed in more detail.

The social relations of production influence the division of authority, perceptions of interest, and the generation of conflict. In the mining industry there was a general move from capitalist, petty-commodity production to mechanised and proletarianised production. Clement (1980) has analysed the consequence of this transition in particular instances of technical change and the labour process. In so doing he was established the links between the organisation of production and social action, missing in Innis's work. The emergence of a large class of wage labourers competing with management for control of both revenue and working conditions both distinguishes between these forms of capitalist production and helps explain later political activity in mining.

For the earlier period, however, it is necessary also to consider the relations of production *between* mining and other industries, and within those other industries, to complement the analysis. For a period, mining coexisted with farming and lumbering in the region. Although there was the circulation of commodities between all three industries, as food, pit-props and building materials were exchanged, yet there was not an open circulation of labour. Mining did not rely on labour from the farming sector, which rather provided skilled and semi-skilled labour to lumbering. These limits to the supply of labour were partly due to farming being household based, with the peasant farmers preferring seasonal skilled work in lumbering to permanent proletarianisation in the mines. The limited nature of mining operations reinforced this.

Secondly, the role of the state influences political activity – in this case, in several ways. Initially, the state provided a support to mining, selling land, doing surveys and so on. It then helped solve the industry's labour supply problem through its immigration policies, resulting in ethnic diversity in Canada's basic industries. The state intervened to control the political activities of those immigrants, but did little to prevent the domination of the Canadian mining industry by American capital. It was really only when state regulations regarding labour were changed, during the second world war, that a large-scale organised labour movement was established in the mines, although, as outlined here, other regional changes had created the conditions for this. The mining unions did not suddenly spring into existence.

Managerial policy affected the political activities of the miners both directly and indirectly. It is evident that mine management directly resisted the formation of labour unions until the second world war through the use of detectives, worker layoffs and ethnic mixing. Indirectly, labour organisation was affected through labour camp and investment policies. The camps, and towns such as Copper Cliff, were controlled through managerial paternalism. The resident workers were dependent on the companies both for basic necessities and for the many benefits of leisure, hospitals and pensions which the companies afforded them. At the same time, management steadily introduced capital into the mines, markedly so after the second world war, which, as Clement (1980) demonstrates, placed many workers in similar proletarianised occupations, and some in crucial locations in these companies' operations.

Fourthly, it must have been increasingly difficult to manipulate labour in a region whose population was growing and urbanising in centres outside of company control. Indeed, given the expansion of the labour supply in the towns, labour camps became unnecessary. Moreover, the Canadian economy was opening up, with improvements in transport, as was Canadian society. Imperial control was in decline and democratic institutions were growing. Such changes meant that large employers needed to negotiate new relations with their work forces.

Immigration and managerial policies are both linked to the next element influencing labour organisation — the structure of community including the extent of ethnic diversity. For Clark (1962) the community of the northern Ontario labour camp overcame social class differences. The social and occupational core of the camps were British skilled workers and management, benefitting from company paternalism. These identified with each other against the unskilled workers, and the French-Canadian farmers surrounding them. The anti-communist Catholic church discouraged the Irish and French workers from unionising. Consequently, unionisation occurred amongst the culturally independent European immigrant workers, who organised their libraries, schools and work demands along nationalist lines.

The evidence offered here supports Clark's view, and that of Davis (1980) for the American working class in general, that ethnicity was an obstacle to widespread labour organisation. Certainly, ethnic identity provides a basis for the establishment of trust relations among strangers and can lead to political action (Laite 1974, 1981). However, as this Canadian case shows, such ethnicity can be divisive and manipulated by management. This is particularly so when the regional economic structure receiving ethnic groups is already divided along national and cultural lines. Occupational and ethnic segregation then reinforce one another. Yet, important as the concepts of community and ethnic identity are in the short-run, they cannot be placed at the heart of the analysis of labour organisation, for changes in the other regional and national elements outlined here resulted in the decline of the labour communities and the emergence of a widespread labour movement in mining.

Finally, labour organisation in this region was influenced by wider North American labour politics. During the first decades of the twentieth century the divided labour unions of North America could offer little support to the miners of Northern Ontario, confronted by intransigent management. Even indigenous Canadian unions found great difficulty in organising in the region's mines. And although it was a breakthrough in America which opened the door to unionisation in INCO's Canadian operations, yet the divisions in the labour movement continued after the second world war, resulting in the demise of Mine Mill and affiliation with the United Steelworkers of America.

Thus, the natures of labour organisation and political activity are not direct consequences of industrial change, peaking during its initial stages and then declining. Rather, capitalist development involves several processes and elements, from which emerge political activities and strategies. In the region of Northern Ontario, capitalist development proceeded from differing forms of production and a segmented labour market, to a dominant productive form with a more homogeneous labour market, providing a changing framework for changing labour organisation.

Notes

1 This research was carried out whilst I was Visiting Associate Professor at the University of Guelph, Ontario, during 1978–9. Guelph University is the old Ontario Agricultural College which was part of the University of Toronto. As such, it has a magnificent collection of nineteenth-century manuscripts as does the library of the University of Toronto. The data collected was thus historical, documentary material, comprising census information, rural reports, accident reports, mining analyses and autobiographies, as listed in the bibliography.

I would like to thank Gilbert Stelter and Wallace Clement for the discussions I had with them whilst I was at Guelph University, and also S. D. Clark for comments on a version of this paper which I gave at the Institute for Canadian Studies at the University of Edinburgh in 1981.

2 Such debt contracting was also being used in South American mining during this period, where it was called *enganche*: 'the hook' (Laite 1981).

3 Nickel was not the only raw material Morgan moved to control. In that same year, 1902, he helped form the Cerro de Pasco Copper Company, which was to produce most of the copper, lead and zinc from Peru over the next seventy years (Laite 1981).

References   * See annotated bibliography for full details of all asterisked titles*

Avery, D.
1975 'Continental European immigrant workers in Canada 1896–1919: from stalwart peasants to radical proletariat', *Canadian Review of Sociology and Anthropology* 12(1).

Baine, R.P.
1952 'Settlement of the Sudbury region'. Master's thesis, Toronto University, Toronto.

Baldwin, D.
1977 'A study in social control: the life of the silver miner in Northern Ontario'. *Labour/Le Travailleur* 2: 79–106.

Bleach, N.
1974 'Nickel capital: Sudbury and the nickel industry, 1905–1925'. *Laurentian University Review* 6(3).

Bradwin, E.W.
1972 *The Bunk-house Man*. Toronto: Toronto University Press.

Brandt, Gail C.
1977 'The French-Canadians of Sudbury: 1883–1913'. PhD dissertation, York University, Toronto.

Braverman, Harry
1974 *Labour and Monopoly Capital*. New York: Monthly Review Press.

Buck, W.K. and Toombes, R.B.
1967 'One hundred years'. *Canadian Mining Journal* 88(4).

Canadian Official Publications
1882–1902 Ontario Sessional Papers: 'Report of the Commissioner of Crown Lands'. Toronto.

1885–1902 'Annual Report on Agriculture and Arts'. Ottawa.
1889 'Report of the Royal Commission on the Mineral Resources of Ontario'. Ottawa.
1915 'The rise in prices and the cost of living in Canada, 1900–1914'. Ottawa: Department of Labour.
1891–1979 'Annual Report of the Bureau of Mines'. Ottawa.
1901–79 *Labour Gazette*. Ottawa.
1978 'Strikes and lockouts in Canada'. Ottawa: Department of Labour.

Clark, S.D.
1962 *The Developing Canadian Community*. Toronto: Toronto University Press.

Clement, Wallace
1980 *Hardrock Mining: Industrial Relations and Technological Change at INCO*. Toronto: McClelland and Stewart.
1983 *Class, Power and Property*. Agincourt: Methuen.

Coleman, A.P.
1913 *The Nickel Industry*. Sudbury: Canadian Government Press.

Collins, E.A.
1933 'Early days of the International Nickel Company'. *Canadian Institute of Mining and Metallurgy Bulletin* (July).

Cox, G.L.
1970 'The effects of smelter emissions on the soils of the Sudbury area'. Thesis, master's of science, Guelph University, Guelph, Ontario.

Davis, M.
1980 'Why the U.S. working class is different'. *New Left Review* 123 (September–October): 3–44.

Dorian, C.
1955 *The first 75 years*. Devon: A.H. Stockwell.

Fröbel, F.
1977 'The tendency towards a new international division of labour'. *Review* 1(1).

Goltz, E.
1974 'Española: the history of a pulp and paper town'. *Laurentian University Review* 6(3).

Hamelin, J.
1971 *Histoire Économique du Quebec*. Montreal: Fides.

INCO (International Nickel Co.)
1920 *The Mining and Smelting Operations of INCO*. New York.

Innis, Harold A.
1936 'Settlement and the mining frontier'. In *Canadian Frontiers of Settlement*. W.A. MacIntosh and W.L.G. Sperg, eds. Toronto: Macmillan.
1956 *Essays in Canadian Economic History*. Toronto: University of Toronto Press.

Kay, W.
1921 'Field supervision of settlers under the Soldier-Settlement Act in the Sudbury district'. Master's thesis, Guelph University, Guelph, Ontario.

Kelley, G.T.
1971 'Early Canadian industrialisation'. PhD dissertation, Clark University, Wooster, Massachusetts.

Laite, Julian
1974 'Paisanos'. *Journal of Peasant Studies* 1(4): 509–11.
1980*
1981*

Lang, J.
1970 'A lion in a den of Daniels: a history of Mine Mill 1942–1962'. Master's thesis, Guelph University, Guelph, Ontario.

LeBourdais, D. M.
1953 *Sudbury Basin*. Toronto: Ryerson Press.

Lower, A. R. M.
1936 *Settlement and the Forest Frontier in Eastern Canada*. Toronto: Macmillan.

Main, D. W.
1955 *The Canadian Nickel Industry*. Toronto: Toronto University Press.

Nelles, H. V.
1974 *The Politics of Development*. Toronto: Macmillan.

Pye, E. G.
1967 'One hundred years of mining in Ontario'. *Canadian Mining Journal* 88(4).

Scott, S.
1977 'A profusion of issues: immigrant labour, the world war, and the Conminco strike of 1917'. *Labour/Le Travailleur* 2: 58–74.

Skelton, O. D.
1913 *General Economic History of the Dominion*, Toronto: Publishers' Association.

Stelter, G. A.
1971 'The origins of a company town: Sudbury in the 19th century'. *Laurentian University Review* 3: 1–37.
1974 'Community development in Toronto's commercial empires'. *Laurentian University Review* 6(3): 1–53.

Stelter, G. A. and Artibise, A.
1978 'Canadian resource towns in historical perspective'. *Plan Canada* 18(1): 7–16.

Thompson, J. F. and Beasley, N.
1960 *For the years to come*. New York: Putnam.

Todd, Arthur C.
1967 *The Cornish Miner in America*. D. B. Barton, Truro, England.

CHAPTER 7

Bolivian mining, a crisis in the making[1]

DAVID J. FOX

The Andean massif holds both the curse and the lodestone of Bolivian national viability. A zone where any economically valuable resource − even arable land − is scarce, the lure of fabulous mineral wealth is a singular Bolivian hope. Bolivia's colonial beginnings arose on one of history's greatest silver bonanzas. Three centuries later great fortunes were made in tin. But lodes are few and the mountains tenacious in their resistance to the miner's pick.

The essay that follows examines the condition and prospects − which are bleak − of Bolivian mining as a national industry. Taken as a totality the economic costs of extracting the likely remaining deposits of metaliferous ores are increasingly unfavourable relative to the costs of other producer nations. The political and moral dilemmas this deepening crisis poses for the community of nations are compelling and urgent.

David Fox (School of Geography, University of Manchester, M13 9PL, UK) has a monograph on the Bolivian mining industry in the twentieth century under preparation. He is the author of articles on Bolivia each year since 1979 in *Mining Annual Review* (London).
— *The editors*

The world knows Bolivia, if it knows it at all, as a mining country.[2] The silver mines of Potosi were the richest in colonial Latin America; more tin has been won from the Catavi mine during this century than from any other tin mine in the world. Bolivia is the largest producer of tin in the western hemisphere (although Brazil is beginning to challenge Bolivia's eighty-year-long hegemony) and until five years ago was second in the world league of tin producers. She remains one of the major producers of tungsten (wolfram) and of antimony in the world. Bolivia is a member of a small group of African and Latin American countries for whom mining is the cornerstone of their economies; no other country of this group, which includes Chile, Peru, Guyana and Jamaica in the Americas, has been so dependent on mining over such a long period as has Bolivia. But nowadays, as in history, great mining wealth has brought only poverty to the average Bolivian; in no other country in South America is the standard of living so low.[3]

In 1983 mining generated about 6 per cent of the gross domestic product of Bolivia and gave employment to about 5 per cent of the country's labour force, or to about 75,000 people. These figures are imprecise but are lower than in recent years: in 1977 7 per cent of the GDP came from mining; in 1973 8.6 per cent and in 1969 10 per cent. In absolute terms the recent contribution of mining has been static or declining while that of the rest of the economy has been growing and has become more diversified in origin. But such figures belittle the importance of mining in Bolivia. Mining remains the only substantial and legitimate source of foreign exchange. It regularly supplies two-thirds of Bolivia's export earnings and between 1920 and 1970 the figure was 90 per cent. During the mid-1970s the value of exports of oil and natural gas was greatly enhanced by OPEC-induced price changes but production subsequently flagged and mineral ores and, increasingly, smelted metal reclaimed their domination of Bolivia's trade with the rest of the world. In 1983 overseas sales of minerals and metals seemingly yielded only US $342 million or 43 per cent of Bolivia's total export trade valued at about $790 million. This is the lowest official figure for a century. But the official statistics ignore the illicit export of gold which in 1983 was variously estimated to be worth between two hundred and five hundred times the official value (of $½ million) and, when incorporated, gives a figure of between 50 per cent and 60 per cent. Consequently the well-being of her mining industry plays a large part in establishing Bolivia's credit-worthiness in the eyes of the overseas banking fraternity. Further, mining supplies a major part of governmental revenues: in the 1940s the fraction was over one-half, today taxes on mining and mineral royalties account for between one-quarter (in 1983) and one-third of such revenue; such economic figures give some measure of the political importance of mining organisations and of miners in Bolivia. For the last half century the miners, in concert with the military and more recently the campesinos (rural agriculturalists), have been one of the major political forces in the country: the miners' union has been the incubator of political parties, philosophers and practitioners and, just as important in a country long characterised by political uncertainty and violence, their refuge in hard times.

The Bolivian mining industry in 1984 is in severe difficulties and the country is facing its most serious social and economic crisis in thirty years. To some extent this situation is not of Bolivia's making. Bolivia's enormous overseas debt (of between $3.5 to $4 billion in mid-1984, imposing debt servicing charges equal to the total value of her official exports) and the belated recognition by her overseas creditors of their exposed position has made it increasingly difficult for Bolivia to obtain the external financing upon which her solvency has depended for the last decade. In 1982 and 1983 she was obliged to obtain agreements to delay payments of interest and principal when they fell due and in mid-1984 shocked the international banking world by unilaterally halting repayments on debts owed to overseas private banks. The collapse in the dollar value of the peso since 1982 has led to a 400 per cent annual inflation rate and a tripling of local prices to the consumer. International demand for Bolivia's minerals has slackened and metal prices weakened as the world

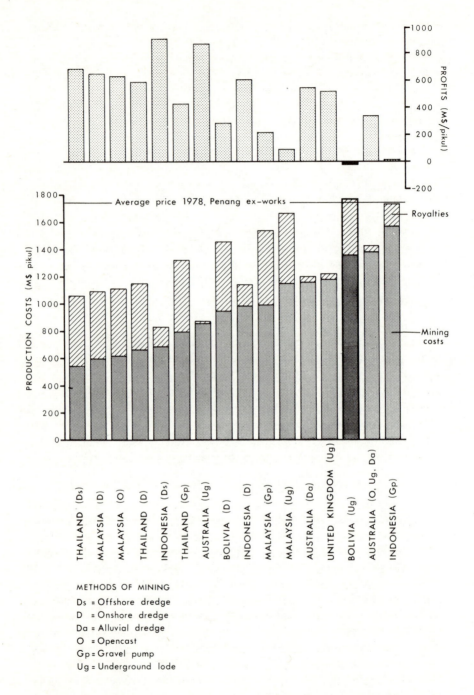

Fig. 7.1 Comparative tin mining costs, royalties and profits, 1978. *Source: Tin production and investment* (International Tin Council), London 1981. 127.

industrial recession exacted a toll. But these unfavourable external circumstances overlie difficulties within the mining industry which are partly of
Bolivia's own making. The current tragedy for the country is that the mining
industry of today is barely capable of generating enough momentum to keep
its own machinery ticking over let alone provide the motive force so badly
needed to transform the national economy. The major problem is that by the
standards of her international competitors the Bolivian mining industry is a
high-cost one.

The explanation of such higher costs lies partly in the nature of the ores,
the size of deposits and their location. It lies partly in matters of an institutional
nature including constraints on the adoption of more efficient mining practices,
the manner in which mining enterprises are organised, and the relationships
between the industry and the government, the smelters and the markets. And
it lies partly in the broader economic and political context in which mining
in Bolivia takes place. This essay now looks further at the different factors
which threaten the competitiveness of Bolivia's mining industry.

I General considerations

Most Bolivian ores are by their geological nature expensive to mine. Many
of the important economic minerals are genetically linked to granitic structures
and to their metamorphic aureoles; they are found in veins deeply impregnat-
ing highly durable country rock. This is particularly true of tin ores. These
accounted for 81 per cent of the material brought to the surface in 1982, a
typical year. It also applies to the ores of lead and zinc with associated silver
which made up a further 16 per cent and to some of the ores of wolfram,
antimony and copper.

Most veins worked today are narrow (no wider than 10 cm in the case of
tin), and mining entails the removal of an unusually large proportion of worth-
less host rock to extract the payable ore. Drilling (the most costly element in
hard rock mining) is slow, wear on bits heavy and consumption of (imported)
explosives high; once mined the broken material is relatively more expensive
to mill than is softer rock. Typically the veins vary substantially in thickness
and tenor and many peter out over short distances. Most are steeply inclined
and the ores they contain often change their character with depth: near the
surface they tend to be oxidised, richer, and easier both to work and concentrate, whereas at depths below, say, 100 m, the harder, leaner sulphide ores
are the norm. The veins are frequently fractured and faults dislocate the
geological structures in which they are contained. It is not easy to predict the
possible extension of metaliferous structures and the problem of proving
reserves and extrapolating likely levels of future mineral production are more
taxing than in many other competing mineral fields. Individual deposits are
small and the potential for economies of scale limited. Mines that are large
by Bolivian standards are only modest by world standards. For example, the

mineralised Salvadora stock at Catavi – which during this century has provided almost one-third of Bolivia's tin production, or one-twentieth of the world's total recorded output – measures only 1.2×1.0 km at the surface and extends for only 0.7 km under the ground. Mines that are small by Bolivian standards are tiny on a world scale and their very existence would be an anachronism in most other countries.

The value of most Bolivian deposits is further reduced by their scattered distribution, inaccessibility and inhospitable setting. Most of the mining activity takes place within the boundaries of an arc, 50–150 km wide, which runs from Lake Titicaca southwards for 800 km before fading away beyond the Argentine border (Fig. 7.2). Within it there are thousands of individually worked deposits and from it comes over 90 per cent of the mine production of the country. This arc is coincident with the great flexure in the axis of the Andes induced during Triassic times by the eastward movement of the stable Brazilian shield. The minerals now mined were segregated in deep-seated magmas associated with the orogeny and subsequently injected upwards along plains of structural weakness to produce a series of mineral fringed batholiths; later erosion, uplift and warping has exposed some of these frozen infills as fractured veins skirting resistant granite bosses. In the vicinity of Lake Titicaca and La Paz the granites form the seemingly indestructible core of the highest sierras of the glacial Andes; further south recent layers of volcanic outpourings blanket part of the mineralised zone, obscure underlying geological relationships, and make veins difficult to locate. It is thus no accident that the most important mining region of Bolivia coincides with the widest and highest part of the Andes thereby creating mining conditions of almost unparalleled difficulty.

No other country has to contend with mining at such high altitudes as does Bolivia. All but 10 per cent of her production is from mines above the 3,700 m contour line or over 12,000 ft above sea level. The entrances to the highest mines are above 5,000 m (almost 17,000 ft) and above the snow line even in these tropical regions. At these altitudes the oxygen content of the air is only between two-thirds and one-half of that at sea level and both machinery and human physiology must adapt to this rarified atmosphere. The only mining activity of interest below 2,000 m, or 7,000 ft, takes place in the Tipuani goldfields; it is quite atypical of Bolivian mining in many other respects beyond that of mere location. As well as thin, the atmosphere over much of the mining area in Bolivia is bone dry. Many mines have only an unreliable water supply for their mills and in some of the higher areas this supply is frozen for part of the year. Trees are a rarity and mine timber expensive; it is only east of the Andean cordillera that the ocean of cloud heralds the different Bolivian worlds of the Amazonian selva and the savanna of the Oriente. The mountainous situation of most mines hampers their access and increases their costs. The journey to many a mine is over narrow and precarious trackways which hairpin their way to their remote destinations. A skeletal railway network, created piecemeal between 1873 and 1917 to meet the requirements of the mines, links Potosi, Catavi and the Quechisla mining regions in the south to Oruro, La Paz, the Pacific ports of Antofagasta and Arica, and the Titicaca terminus at Guaqui. In the last fifteen years the beginning of a metalled road system has emerged. This has drawn such business as there is away from the railways, making their financial position precarious. Large tracts of the mining zone generate almost no traffic except that created by mines: the barren land supports only a very thin scattering of campesinos contributing little or nothing to the commercial life of the country. In practice this means that the miners have to meet much more of the costs of creating and maintaining their lines of communication with the outside world than is the case in most other countries. And what is true for roads and railways applies with equal force to the provision of other elements of the regional infrastructure.

The high costs individual mines have to bear because of their isolated and mountainous local situations are compounded by the remote and land-locked situation of the country at large. Metals or concentrates leaving Bolivia, like mining supplies entering the country, have to bear the cost, say, of a 1,000 km journey overland by truck or rail, perhaps even by lake steamer, between the mine and Antofagasta, Arica, Mollendo or Matarani on the Pacific coast. These ports were lost to Bolivia and gained by Chile or Peru a century ago following the War of the Pacific: at best their foreign ownership is a source of constant irritation to the Bolivians, at worst a potential stranglehold on the economic lifelines of the country. Once at the Pacific coast there is a 10,000 km journey by sea to the major world markets for Bolivia's minerals. In 1915 such transport costs represented one-quarter of the total cost of placing Catavi tin on the market; fifty years later the figure was still a considerable 6 per cent with half the cost being incurred by the journey from the mines to the Pacific.

The methods of mining used in Bolivia reflect the predominance of small and dispersed deposits and of vein ore rather than disseminated ore, the disadvantages of geographical location, and a shortage of investment capital. For example, it has so far proved impossible to use modern methods of bulk mining in Bolivia. Rather, the typical mining practice in Bolivia is that of selective underground mining using methods that were commonplace at the beginning of this century and a technology which is largely of nineteenth-century design. This means both that the costs of mining per unit of production are, in general, higher than those of competitors, and also that the costs of mining form a higher proportion of total realisation costs than in most other mining fields. The driller and his gang remain the key figures in most Bolivian mines and each block of material dislodged underground is handled by several pairs of hands before it reaches the mill. The whole operation is relatively labour intensive. For example, Catavi has as many personnel on its payroll as does Chuquicamata, the great open pit copper mine in Chile where over thirty times as much material as at Catavi is extracted each year. On the positive side it can be claimed that mining generates a larger number of jobs than would be the case if more capital intensive methods were used. Bolivia has no shortage of potential miners and needs employment opportunities for them: the cost of labour is governed by the local labour market and paid for in local currency, unlike almost all the capital equipment and material needs of Bolivia's mines. On the negative side, however, Bolivia is swimming against the modern tide, and in consequence has been unable to keep abreast of some of the more recent innovations in mining and drilling. Every year the country finds herself in deeper and deeper water. Old-fashioned working techniques and outmoded and worn-out equipment help increase unit costs, squeeze profits, reduce the opportunity for re-investment, and lessen the likelihood of external financing. The geological and geographical difficulties of proving reserves, of planning their exploitation and of winning a return on them makes Bolivia a technically higher-risk country for new mining ventures than most others.[4] Mining expertise may be able to counter such risks but it is a very mobile factor and there is a strict limit on the availability on such talent, even in a mining country like Bolivia.

To an adverse physical setting must be added a difficult political environment; it, too, exacts a financial toll on the viability of Bolivia's mines. The state is, understandably, heavily involved with mining in Bolivia and this means that decisions affecting mining may not always be taken with the best commercial interest of the industry uppermost in mind. The state impinges on mining in many ways. First, there are substantial tracts of country in which mining is not permitted. These include the so-called fiscal reserves, a category which includes regions close to national frontiers, much of the Oriente in the eastern half of the country and certain other areas. There are other reserves which have been pre-empted, including those under the jurisdiction of the armed forces. The most interesting are those that lie in the valley of the Tipuani upstream from the current goldworkings. Secondly, the state extracts taxes and royalties on mineral production and upon exports. Since 1982 such levies have pushed many mines beyond their financial limits and large amounts of money have remained unpaid. The indirect fiscal burden borne by the industry was increased following the much stronger currency controls imposed in 1982 on foreign exchange earned abroad from sales of minerals and metals. Thirdly, the state imposes legal obligations on mining enterprises to provide their employees with medical facilities, schooling, housing and other social services thereby passing on to the industry costs which their competitors abroad may not have to bear.

But the involvement of the state goes well beyond regulating concessions, levying taxes and imposing social obligations. Two-thirds of the productive capacity of the industry now operates under the aegis of the state mining corporation, the Corporación Minera de Bolivia (COMIBOL), and has done so since 1952. Further, half the remaining mining capacity is required to sell its production only to the state mining bank, the Banco Minero de Bolivia

(BAMIN), an institution founded in 1933. The degree of state involvement has become heavier during the last decade with the emergence of a national smelting industry. Smelting is a state monopoly in Bolivia and more than half of the minerals produced are now smelted domestically. This level of state involvement has militated against Bolivia enjoying the technical and financial benefits available to other countries through the multinational mining houses. The expropriation of the Gulf Oil assets in 1969, and the withdrawal in 1971 of the concession granted a US consortium to work the Matilde lead and zinc mine, were signals widely read outside Bolivia; in the mid-1970s the Grace interests decided to pull out of mining in Bolivia, and indeed from Latin America at large. There are only a few exceptions to the general rule that overseas mining companies are not interested in risking capital in Bolivian mining ventures. In general the bulk of the industry has to look for its capital requirements to its own resources, which are negligible, or to government-backed concessionary loans or grants-in-aid from the developed world. Tied loans or grants may not necessarily finance the most appropriate developments for the industry. Their availability is, more often than not, determined by the political complexion of the Bolivian government of the day and by the self-interest of the donor country. It is unfortunate that the average lifespan of most Bolivian governments is much shorter than the lead time needed to bring to production new mining prospects or to reap the benefits of many types of desirable mining investment. Where successive governments have been able to influence investment decisions they have preferred the quicker returns available in more speculative ventures elsewhere in the national economy. The mining industry has gained little benefit in recent years from its close association with the state. Between August 1980 and September 1982 Bolivia was governed by military juntas viewed with deep distaste both by the governments of the industrialised world and by those of the eastern block; a moratorium was called on almost all prospective intergovernmental loans and grants. The slide in value of the Bolivian peso, the withdrawal of financial support by Argentina, and strikes in the mining sector and elsewhere, led to a growing loss of confidence, both in the country at large and within the armed forces, in the ability of the military to run the country. The armed forces were forced to return power to the same Congress which had been elected (and suppressed) in 1980.

The first job of the civilian government was to persuade the International Monetary Fund (IMF) to give Bolivia the standby credit facilities denied the previous military dictatorship. The government was obliged by the IMF to institute an austerity programme. This programme bore particularly heavily on wage-earners and strained but did not break the ties binding the mine workers to the new government. The reconstituted miners' union agreed to relatively modest wage rises as part of their contribution to the support of a pro-labour government. The United States, through, for example, the Inter-American Development Bank (IDB) and the western international aid agencies, made some limited money available as a token of their support for a democratically-elected government. And the left-wing cast of the government – one in which in 1984 both the Minister of Mines and the Minister of Labour were Moscow-line communists – brought some tangible concessions to the country and its mining industry from the Soviet Union. On the other hand the new government had little appeal to potential overseas investors from the private sector. The dependence of the mining industry upon domestic political circumstances is very clear.

II Sectors of the mining industry

Within the harsh confines of nature and the limitations of financial and political circumstances there is a wide range of achievement within the three organisational sectors of the Bolivian mining industry. Within the state sector the major achievement has been its ability to survive. It came into being in 1952 with the nationalisation of the holdings of the big three mining barons – Patino, Aramayo and Hochschild – who in popular estimation had bled the country white. Nationalisation was seen at its inception as a radical departure heralding an era of improvement; the last thirty years have shown that in reality COMIBOL has been a highly conservative body. Innovations have been few, and changes have only been accepted with reluctance. The mining landscape of 1984 was very little different in outline from that of 1952. Not a single mining community under COMIBOL care has died – a remarkable achievement in a country littered with the desiccated corpses of abandoned mining camps reminding modern Bolivians of the ultimate fate awaiting most mining communities. Within the COMIBOL communities the material standard of living and the quality of life of the average miner and his family has greatly improved. On the other hand, whatever the social achievements of COMIBOL, conventional economic diagnoses of the state-run mines have many times prescribed wholesale cauterisation. Yet only one mine – Pulacayo in 1955 – has been closed. Unfortunately no completely new mining prospects have been brought into production since nationalisation to help rejuvenate a naturally ageing industry. For examples of innovation one must turn to the private mining sector. Not only has this sector shown a remarkable ability to survive in an environment which is doubly hostile, but it has also demonstrated a willingness to exploit modern advances in mining technology, to take risks, and to invest in new prospects and an ability to attract some limited capital into the industry and to make money. The achievement of the third sector, that of the peasant miners, lies in their inherited ability to wrest an albeit miserable existence from a land as bleak and unyielding as that to be found anywhere in Latin America. Each of these three sectors make their own distinct contribution to the Bolivian mining scene and are now considered further.

The state-run mines

Whatever the achievements of COMIBOL[5] in the past, the auguries for her future are not good. Some argue that the Corporation is a national liability absorbing scarce resources, burdening the country with almost insupportable debts and is long overdue for drastic change. Whatever the merits of such a view the current plight of COMIBOL is plain for all to see. Recent production trends are downwards. The output of tin from COMIBOL mines in 1983 was only 14,527 tonnes. This compares with 23,306 tonnes in 1977, the year in which the generally rising trend since 1961 (14,812 tonnes) reversed; the last time COMIBOL's tin production was as low as in 1983 was in the aberrant year of 1971 when the figure (14,723 tonnes) was the lowest since national-isation in 1952. It should be noted however, that more tin comes out of COMIBOL mines than official statistics might lead one to believe. In 1983 the manager of Huanuni mine, currently Bolivia's most productive tin mine, estimates that theft was costing his mine the loss of about 100 tonnes of tin a month, one-quarter of its potential output, or one-third of its actual out-put. Such tin is not lost to Bolivia: it is 'laundered' through the parallel small mines sector and either sold to COMIBOL or to the regional agency of BAMIN. Tin production is more important in the state mining sector than elsewhere in the mining economy: during the last thirty years COMIBOL has derived three-quarters of its income from sales of tin. Production trends for other significant minerals produced by COMIBOL mirror that of tin; those of lead, zinc and copper are exaggerated versions of the trend in tin. The only mineral of importance in the last few years to have been at variance with tin has been wolfram. The ores reaching the COMIBOL mine heads are getting leaner in quality: the average tin content of COMIBOL ore in 1983 was below 0.5 per cent tin; in 1975 the comparable figure was about 0.7 per cent. Thus although the output measured in metal content of COMIBOL's mines has slumped, the volume of rock displaced to yield that ore has increased year by year. The size of COMIBOL's labour force may reflect that increased volume of work: it rose from 22,695 at the beginning of 1971 to 25,573 five years later and to 26,525 at the beginning of 1981. Over half of the tin contained in ore reaching the surface is lost in the process of concentration; the average recovery rates in COMIBOL mills run between 40 per cent and 55 per cent.

The mills produce a low-grade concentrate averaging overall only 23 per cent tin content (the comparable figure in 1975 was 27 per cent): the richest concentrates are those from Huanuni which averaged 41–2 per cent in 1982, the poorest concentrates came from Catavi assaying 12 to 13 per cent tin content. Low-grade concentrates are obviously more expensive to smelt and are consequently worth less at the smelter than are higher grade concentrates.

For these and other reasons it will be appreciated that COMIBOL's unit costs of recovered metal are rising and productivity is falling. The economic effects of these production trends have been compounded by the fall in metal prices in recent years. In 1982 and 1983 the IMF index of primary metal prices

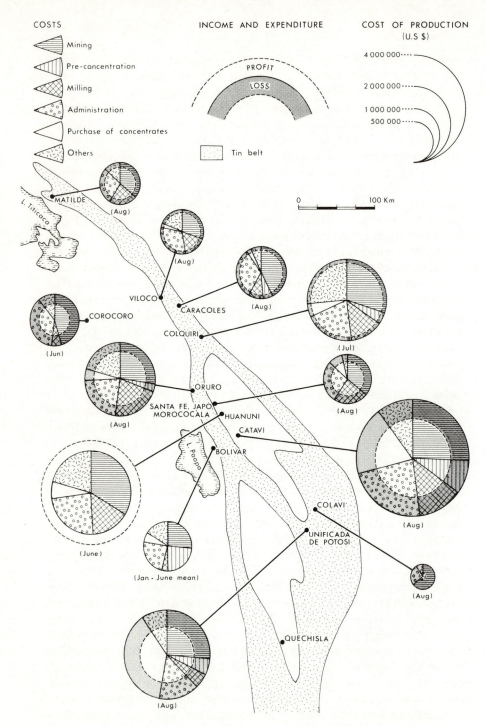

Fig. 7.2 Bolivia: COMIBOL mining divisions: monthly gains and losses, 1981. *Source: Plan de Rehabilitación de la Corporación Minera de Bolivia.* La Paz 1982.

stood at 78 (1980 = 100); the average price of tin on the New York Metal Exchange in 1982 and 1983 was 20 per cent below that of 1980. The fall in the real value of COMIBOL's (and Bolivia's) metal production is even greater when metal prices are deflated by changes in the US wholesale price index, a good indicator of the cost to Bolivia of imported mining supplies. COMIBOL's income from sales in 1983 was a little below $250 million; in 1982 a little over $300 million; in 1981 about $450 million; and in 1980 a record $555 million. In 1971, a poor year for tin production, the figure was only $98 million; when translated into current dollar values this clearly was a less worrying outcome than that for 1983. Whereas for the years immediately prior to 1980 COMIBOL's annual profit and loss statement showed an, albeit small, profit (averaging 1.5 per cent of the income from sales) since 1980 substantial losses have been incurred. In 1981 COMIBOL lost US $44.7 million; in 1982 the figure may have been as high as US $80 million; and in 1983 $160 million. In mid-1982 Catavi alone was reliably reported to be running at an operating loss of US $2.5 million a month. To some extent these figures are only paper losses to the Corporation: in 1981 COMIBOL paid seemingly only US $14 million of the US $47 million due to the government in royalties and taxes. If to these current account losses one adds the costs of capital investments in mining and social projects, in machinery and equipment, and the amortisation of loans less any assets, grants or financial assistance obtained, then a global deficit for COMIBOL can be calculated. In 1973 this deficit was only US $14.5 million; by 1979 it was over ten times this figure (US $155.7 million). A year later the deficit had risen to US $286.3 million, a figure equal to about half the total assets of the Corporation, and at the beginning of 1984, it was around the $500 million mark.

This is not the first time COMIBOL has faced severe difficulties. In 1952 COMIBOL inherited mines which already bore many of the signs of decadence, and nationalisation exacerbated their difficulties. Both capital and mining expertise have remained in short supply and pressure to widen the social responsibilities of the Corporation have proved irresistible and expensive. Such profits as COMIBOL has made during its thirty years of operations have largely accrued to the government or have been transferred to other sectors of the country's economy. COMIBOL has received substantial injections of overseas aid, notably under the Triangular Plan of the 1960s and generous amounts of often sound technical advice, but the problems of an almost senile industry have remained. Over the years many reports[6] have been commissioned on the state of COMIBOL, often written in support of requests for money and assistance from outside bodies and spurred by the desire to help provide a sound basis for the national economy: such reports have been an almost routine response to crisis. These reports tend to emphasise the same recurrent problems and to offer the same general advice. Most are comprehensive surveys made by mining consultants unlikely to be receptive to the concept of a nationalised industry. The detailed report on the machinery requirements of COMIBOL commissioned by the United Kingdom (socialist) government and prepared by the (British) National Coal Board in 1976 is an exception to this rule.

The latest report was commissioned from Price Waterhouse by the military government in 1981 and presented in 1982. The consultants were asked to report on how the corporation could become an 'efficient and financially viable organisation' with a view to seeking financial help from the World Bank. The civilian government of 1983 inherited the report and has reactivated requests for help previously turned down. The proposals in the report give an accurate insight into the current position of COMIBOL.

The report argued the need to devolve decision-making away from the general manager in La Paz and to increase the authority of individual mine managers. This is partly to improve effective day-to-day management, partly to distance the Corporation from the repercussions of political instability. In recent years as political fortunes have ebbed and flowed there have been more general managers than presidents of the Republic; general managers have on occasions been unwilling scapegoats for government mistakes.

At the same time the investigators recognise what many would agree as the almost untenable burden of responsibility which now rests on the shoulders of the mine managers: they are responsible not only for mining operations within their area but, in effect, for almost all the welfare services in their locality. The report argues that the mine managers should be divested of their inherited and state-imposed responsibilities for housing and health services, for education, for transport, for the stocking of the *pulperías* (commissaries) with basic domestic supplies, and for other ancillary activities and be left free to concentrate on mine operations. The report reveals the scale of these service activities. Between 1975 and 1981 COMIBOL spent over US $15 million on building schools, houses, health centres etc., and needed a further $114 million to meet measured demand. There are ten times as many children attending COMIBOL's schools as there are COMIBOL miners working underground. Subsidies paid on goods supplied to the puperias in 1981 were the equivalent of US $120 per employee per month, or US $37 million over the year; in effect miners have no control over half their effective income. In recent years many of the labour troubles at the mining camps were caused as much by the inability of the corporation to keep the commissaries stocked with even the most basic of foodstuffs as by the fall in the purchasing power of the monetary wages of the miners.

Perhaps predictably the report targeted overmanning and overtime for reform. In the eyes of the consultants a quarter of the labour force, say between 4,500 and 6,000 miners, are surplus to requirements. Further, a simplification of the complicated wage structure, re-introduction of a profit-sharing scheme, and an adjustment for the loss of pulpería privileges would, it is argued, make the real economic situation of the mines less obscure to both miners and politicians alike. The replacement of existing antiquated systems of accounting and inventory control would work in the same direction. Interruptions to production caused by the delayed supply of essential materials cost an estimated 5 per cent of COMIBOL's production in 1981; unaccounted losses of concentrates in transit to the smelters, the inefficient routing of shipments, the use of obsolete rolling stock on the railways, among other factors, all reduce profits.

Most fundamental, perhaps, were the criticisms made of mining practices. Repeatedly, at each of the mines, the assessors found obsolete equipment, inadequate supplies of materials, and excessive supplies of surface labour. The lack of spare parts is becoming a more desperate problem as shortages of foreign exchange grow. For example, in May 1984, Huanuni was at a standstill because of the lack of spare parts for the rock drills.

COMIBOL has developed no new deposits outside existing mine precincts for a quarter of a century and there is none which is ready to be developed. The investigators recommended that over a five-year period US $23 million be put towards a systematic exploration of promising areas with attention being particularly focused on the Morococala-Japo area, northeast of Oruro, where it was thought an open pit operation might prove feasible. Mining practises at existing underground mines are old-fashioned and need to be systematised. Some specific proposals were made: for example, the value of continued block-caving at Catavi was questioned. The report underlined the large loss of mineral in the COMIBOL concentrating mills. All these mills were built at least thirty, and many fifty, years ago and in other countries would be classed as obsolete: they were designed for much higher grade ores than they receive today – the ore from Catavi, for example, is currently running at under 0.4 per cent tin content.

The general criticisms levelled at COMIBOL in the report can be detailed for particular mines and regions. The implications for the 100,000 people who depend directly or indirectly upon Catavi mine of the mining data shown on Figure 7.3 will not be lost on readers. The COMIBOL mines at Potosi are in an equally vulnerable state as reserves run out and costs escalate. It has even been proposed that the conical hill looming over Potosi – a hill honeycombed with old workings and which has been the symbol of the Bolivian mining fraternity for four centuries – should be completely excavated and worked as an open pit. To contemplate the closing of Catavi or the desecration of the Cerro Rico is to think what was, until recently, unthinkable in Bolivia. Such is the plight of individual mines that in the whole of COMIBOL only the drab mining community of Huanuni may feel safe from the cold wind of economic bankruptcy. Yet even here technical deficiencies and labour troubles continue to raise questions about the future.

The hurdles to be overcome before COMIBOL could be placed on a sound financial footing are high. The principal hurdle is finding the capital to achieve the changes envisaged by the consultants: the sum involved would be about US $500 million at 1981 values. The World Bank is moving cautiously. The

E

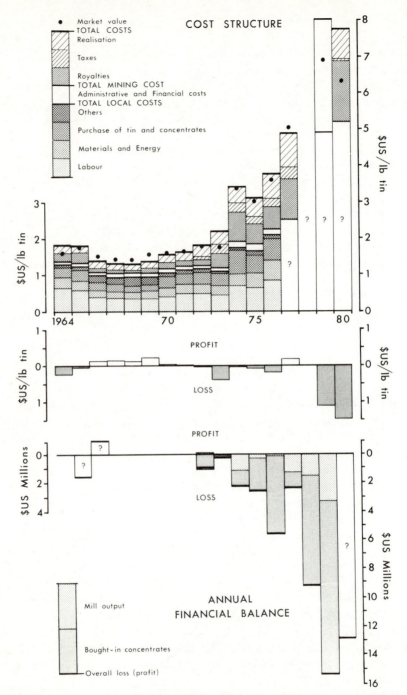

Fig. 7.3 Catavi: cost structure, 1964–80. *Source:* Alfredo Deheza Pérez. Los factores externos que influyen en los gastos de producción de la minera nacionalizada en Bolivia. Unpublished paper, Simposio Internacional del Estaño. La Paz 1977. Amando Canelas Orellana. *¿Quiebra de la Minería Estal Boliviana*? La Paz. 1981. p. 157.

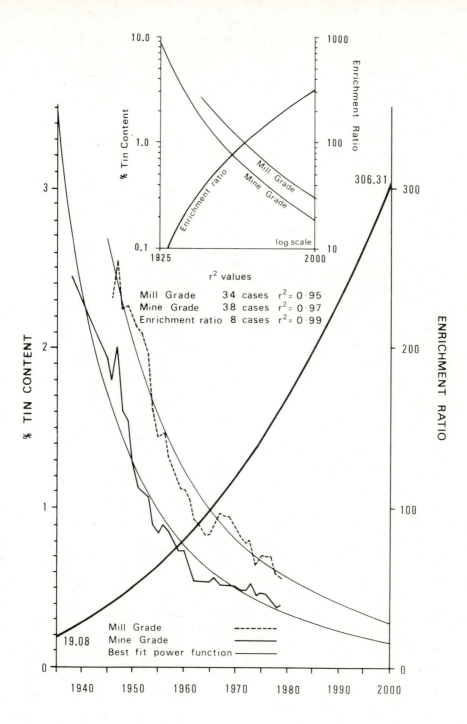

Fig. 7.4 Catavi: quality of tin ore, 1935–2000.

proposed changes would affect deeply embedded vested interests in Bolivia. COMIBOL is important to the state and, whatever the hue of the government in power, that government would be reluctant to be less closely involved than at present in the Corporation's affairs. The mining unions are also part of the Bolivian political scene and have fought many bloody battles to establish their power and to protect the interests of their membership. For example, the Federación Sindical de Trabajadores Mineros de Bolivia (FSTMB) was successful in acquiring in 1983 workers' representation (co-gestion) at all levels in the administration of COMIBOL from the board of directors downwards. But even given the will and the financial means to allow change, there would remain the challenges of acquiring new expertise and skills and the winning of new reserves and markets. COMIBOL will remain in deep crisis for some time to come.

The privately-owned mines

A second commercial mining sector[7] has existed side by side with COMIBOL since 1952. This is a group of twenty-four (thirty-one in 1977) small-to-medium sized privately-owned mining companies which together generate about a quarter of the value of Bolivia's mining output. They make, perhaps, a more interesting contribution to Bolivia's mining economy than their size would suggest. Their production is more diversified than is that of COMIBOL and less dominated by tin: the price of tin plays a marginally less significant role in this sector of the mining industry. In recent years the so-called medium mining sector has been responsible for over three-quarters of the country's production of antimony, half of her tungsten and gold, and one-third of her zinc, one-quarter of her lead and one-eighth of her silver.

Their impact has been more than merely national: it was the private mine owners of Bolivia who spearheaded, so far unsuccessfully, attempts to establish an International Tungsten Council along the lines of the International Tin Council. A strong sense of the market and a more diversified portfolio of minerals meant, for example, that the private miners suffered a smaller reduction in income in the early 1980s than did their neighbours in COMIBOL. These medium mines are predominantly owned by Bolivian companies and the trend since 1952 has been toward increased domestic control. This trend has been encouraged partly by default of foreign-owned mining companies (for example, the Grace trading enterprises decided in 1973 to withdraw from mining in Latin America and sold their Bolivian holdings to local interests), and is one which has suited Bolivians' sensibilities. Because the decisions of the private mine owners are largely taken in a Bolivian context, domestic entrepreneurs have seized opportunities which might not otherwise have been recognised or available to foreign owners.

A combination of good management, a strong political sense, and good luck has led to the emergence of at least two groupings within this sector which have moved strongly against the receding tide of recent mining prosperity in

Bolivia. These two groups have been the leaders in introducing successful innovations into the otherwise largely tradition-bound mining scene in Bolivia. One group, organised within the Estalsa Boliviana holding company, is responsible for about 30 per cent of the overall productive capacity of this sector. It has benefited particularly from the withdrawal of Grace from Bolivia. Within this group is the Chojlla tin and wolfram mine in the Yungas which was the first, fifteen years ago, to use the Bartles-Mozley concentrating unit to recover more of the products held in the fine fractions of the milled ore. The same subsidiary company, International Mining, has used experience gained at Chojlla in their recent development of the new Enramada wolfram-tin mine nearby. Another subsidiary is seriously considering a joint venture with Asarco in New York to work by open pit methods the highly fractured quartzites which outcrop on either side of the 5,000m contour line on Chacaltaya, one of the highest peaks in the Bolivian Andes. This would be the first open pit mine in Bolivia. The group already has experience with unconventional mining methods elsewhere in Bolivia. The Estalsa tin dredge has been working alluvial deposits in the Pazna valley since 1967, moving in the shadow of the group's more conventional underground mine at Avicaya. The dredge has almost exhausted the local reserves in its current concession and is looking for a new area of employment. At one stage a joint venture with COMIBOL to exploit the Centenario reserves near Catavi seemed possible.

The second largest private mining group in Bolivia is the Compania Minera del Sur or COMSUR. It, too, has successfully invested in a tin dredging operation: this one on the Pilcomayo river not far from Potosi. It also has conventional underground mining operations: it acquired the difficult mines of the former British-owned company Fabulosa (difficult technically because of the geological circumstances and difficult managerially because proximity to La Paz has given the work force a more militant posture than most in Bolivia) and, under lease from COMIBOL, mines at Porco. Perhaps COMSUR's most unusual operation is that of the gold dredge which works the Tipuani river where it empties onto the lowlands which flank the eastern Andes. Bought from Grace for a song in the late 1970s, the dredge paid for itself within months as the price of gold rocketed. The main potential of this elderly dredge lies in a negotiated exploitation of the alluvial gold which lies in the reaches of the river upstream from COMSUR's existing concession, gold which now lies in a state reserve. Luck favours the brave and COMSUR has not been afraid to take risks; the management has committed itself to Bolivia and individuals have been politically active during periods of civilian government. A third group, the Empresa Minera Unificada SA or EMUSA, is smaller but sufficiently unusual to merit note. It is the largest producer of antimony in Bolivia (and Bolivia is the leading supplier of antimony in the western world). A slump in demand for antimony in recent years has led EMUSA in 1983 to close her largest mine, Caracota, and to diversify her mining interests: EMUSA has

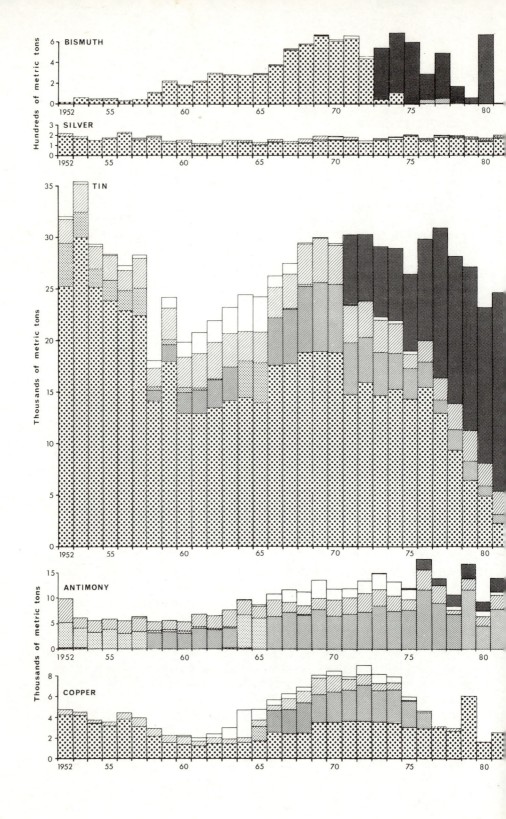

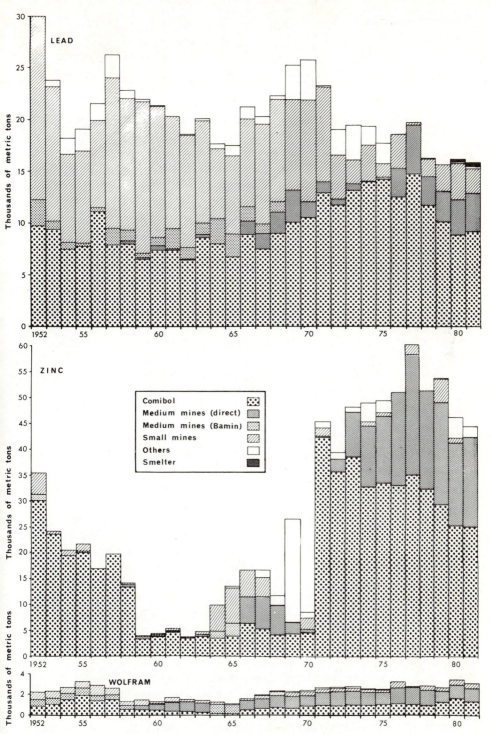

Fig. 7.5 Sources of mineral exports from Bolivia, 1952–81.

announced her intention to develop an open pit gold and silver operation at Inti Raymi in a 50–50 joint venture with Westworld of Texas.

All these mining groups (but not others) have made judicious investments in the Bolivian mining sector, have introduced modern mining practices to the country, and have turned an honest profit. In general the private mining sector has been more successful than has been COMIBOL. The state recognises this success by presuming lower costs of production (*costos presuntos*) than in mines belonging to all other sectors of the mining industry; this suits the government's purpose since it imposes taxes on the difference between such presumed costs and the current retail price. The Asociacion Nacional de Mineros Medianos actively campaigns against this discrimination. It also campaigns against the preclusion of private mining from significant regions of the country – for example, areas close to COMIBOL concessions. In 1982, the private sector was very badly hit by the need to impose restrictions upon their freedom to spend the foreign exchange earned overseas, and by the requirement to change part of those earnings into pesos at a rate well below the free market. The group emphasised, with some justification, that the sector depends largely on self-financing for its survival (the stock exchange in Bolivia hardly exists and the financial institutions have shown more interest in those sectors of the economy more likely to show quicker returns than mining). On the other hand, foreign grant-supplying bodies, like the IDB, have shown themselves more ready to help the private sector than to help COMIBOL.

Mining cooperatives and the peasant miners

The third sector of the Bolivian mining industry is the so-called small mining sector. This is the least well documented[8] sector and the most varied in composition. It includes some mines comparable in their degree of mechanisation, income, and level of output to certain of the mines in the medium mining sector, the largest with up to about 1,000 members and in total involving up to 20,000 people, working concessions which may or may not have been worked previously by private mining companies or COMIBOL. It includes about 1,000 registered concessions being worked by family groups or peasant communities either above ground or underground. And it includes the most primitive of occasional workings attended only outside the agricultural season, or only when metal prices are right, or only when all other means of life support fail. Altogether, this third sector employs about 40,000 people, which is more than the other two mining sectors combined; yet it is responsible for only 10 per cent of the value of the country's total mineral output.

Although the relative contribution to the national economy is small from the mines in this sector, they possess one economic attribute which is not necessarily shared by those in other sectors: their production costs are almost always equal to or below the market price for their product. The small miners are obliged to sell their minerals either to COMIBOL or to the State Mining

Bank (BAMIN) and at a price which varies with the quality of the ores and prevailing world metal prices but is always fixed to ensure a margin of profit to COMIBOL or BAMIN. From that price received the miner has to meet certain unavoidable costs – tools, dynamite, candles etc. – and accept whatever valuation the world places on his labours. The combination in 1983 of depressed metal prices and an extraordinary drought drove many sub-marginal peasant miners and their families to utter destitution; this drive was accelerated by liquidity problems in BAMIN which meant that the small miner had to wait for payment on minerals already delivered. The production of the small miner is of some significance to COMIBOL. In 1980 approximately 8 per cent of COMIBOL's tin, 5 per cent of her silver, and 68 per cent of her tungsten was bought by the Corporation from affiliated cooperatives; unlike the production from her own mines these minerals when sold by the Corporation are sold at a profit to COMIBOL.

Small miners are more important in the more remote parts of Bolivia – for example, in Potosi Province more than in Oruro, and Oruro more than in La Paz. The most successful are undoubtedly the 6,000 or so members of the gold cooperatives who work the Tipuani goldfields. In the last five years the area has been linked to Guanay and thereby with the outside world by trackway, passable by four-wheel-drive vehicles for much of the year. The goldfield is one of the few areas to have benefited from Banco Minero loans in recent years and the traditional wooden pan or batea is now supplemented by scrapers, draglines and washing plants. Today, a high proportion of the mud-floored houses of cooperative members now boast refrigerators paid for in gold dust. Unlike the other metals mined in Bolivia, gold does not need smelting, can easily bear the cost of transport even from the most remote parts, can easily become contraband, does not suffer from a depressed industrial demand, enjoys a relatively high price, and offers the individual the hope, albeit faint, of a life of relative ease. Gold is the only element mined in Bolivia in which the small miners play a dominant role. The value of that production is obscure. An informed estimate suggests that mechanisation allowed the cooperatives to extract about 100,000 ounces of gold worth about $60 million in 1983. The bonanza brought by enhanced activity will not last for long as fossil deposits are exhausted.

The metallurgical industry

No survey of the Bolivian mining scene would be complete without a necess-arily brief mention of the metallurgical industry[9] which has grown around the products of the mines during the last ten years. The smelting and refining of ores in Bolivia is a virtual state monopoly. Until 1970 only token amounts of Bolivia's ores were treated in Bolivia: in general, Bolivia's smelting was done on a toll basis in the United Kingdom, at Williams Harvey in Liverpool (a smelter in which the Patino family had a significant interest) and at the Capper Pass smelter near Hull, in western Europe, and at Texas City in the

United States during the second world war. No other tin-producing country and very few other mining countries were in such a dependent and, to Bolivian eyes, invidious position. Since 1978 the majority of tin-in-concentrates leaving Bolivian mines has been smelted in Bolivia, however, and in 1982 the proportion had risen to 71 per cent. In 1983 the upward trend of the previous decade was reversed when only 62 per cent of a smaller total was treated at home (the residue continues to be shipped to Capper Pass: Williams Harvey has closed). This change has come about with the completion in stages of a 21,000 tonnes per year high-grade tin smelter and, more recently (in 1981), a 6,400 tonnes-per-year low-grade tin smelter, both situated at Vinto near Oruro and both technically sound. Bolivian ore is varied and not easy to smelt, a problem sometimes eased abroad by blending it with concentrates from other sources. Bolivian smelting costs are relatively high for reasons partly unavoidable: energy and other inputs are expensive and the loans incurred in constructing the smelters, even though they may be soft loans, have to be paid off. With declining grades of concentrate, reduced demand, and reduced production the smelters are running at below capacity and unit costs are high and rising; local critics claim that the national smelting body, Empresa Nacional de Fundiciones, or ENAF, employs and pays twice as many people as is justified.

Although proud of having advanced beyond the stage of being a mere producer of minerals, thoughtful Bolivians are increasingly dubious about the advantages that ENAF brings to their country. ENAF has suffered, like other state bodies, from political patronage and managerial incompetence and these have played a significant part in creating what many miners in Bolivia now see as an economic albatross around their necks: ENAF lost $100 million between 1980 and 1982 and in 1983 an investigation was underway to locate 400 tonnes of tin that were unaccounted for. Under the new government the role of ENAF has been reduced. In 1984 the declared policy to set up a separate metal marketing agency was modified when COMIBOL was entrusted with selling tin and antimony smelted by ENAF. The smelting corporation even suffered the indignity of having its headquarters transferred from La Paz, the capital city, to Oruro, an indication of the body's reduced political power.

There were two additional clouds on the Bolivian metallurgical horizon in 1984. The first lay over the new lead and silver smelter at Karachipampa outside Potosí. Construction work began in 1979 but the project took much longer to complete than anticipated: there were several false start-up dates given in 1984 and estimates of the final cost were as high as $130.7 million (although this figure is officially denied). It also became apparent that, if construction work continued, there was not enough ore locally to supply the 50,000 tonnes per year of concentrates upon which the design of the smelter was based: nearby reserves in the María Luisa and Chirona deposits and at Bolivar have proved disappointing and although COMIBOL claimed in 1983 that 87 per cent of annual requirements could be met from domestic sources, in 1984

Minpeco of Peru was contracted to supply concentrates on a toll basis while the export of Bolivian concentrates suitable for Karachipampa was prohibited.

The second cloud was the disappointment following the completion of the tin-volatisation plant built by the USSR at La Palca, also near Potosí. Already long delayed it ran into severe pollution problems during the start-up trials in 1982. It produces tin dust from local low-grade (3.5–4 per cent tin-content) concentrates and when planned and commissioned by COMIBOL in the days of cheap oil and natural gas was to be the forerunner of other plants designed to utilise some of Bolivia's leaner ores. It is operating at well below capacity and appears to be a substantial charge against COMIBOL revenue. Now with energy costs transformed and a shortage of overseas funds, any extension to this programme is clearly in abeyance.

The most serious challenge to Bolivian metallurgists has, however, been present for longer than the last decade. That challenge is the low recovery rates from Bolivian ores in the concentrating mills. The loss of tin in Bolivian mills averaged 47 per cent in 1982 and is, for example, twice that in the Cornish mills. The comparison with Cornwall is a valid one since the tinfields of Cornwall offer the closest geological, mineralogical and mining parallels with Bolivia but have a different geographical, historical and institutional setting. Whereas Cornish tin production is currently running at higher levels than at any other time since before the first world war, Bolivian production has slumped and losses at her mills mount. Most Bolivian mineral ores are still processed by exclusively hydromechanical methods using jigs and tables and differences in specific gravity to concentrate tin, wolfram, gold, lead, zinc, antimony and other metals. Differences in electrical conductivity and magnetic properties are also exploited. But floatation techniques are still very little used in Bolivian mills although as losses rise and ore must be ground ever more finely it is increasingly the most appropriate technology to employ to improve recovery rates. Volatisation is as yet little tried in Bolivia: it is appropriate for complex ores and for those finely disseminated ores rich in sulphides. But the experience at Karachipampa has been a chastening one, reminding Bolivians that the costs of applying modern technology to the difficult circumstances of their country can be a high one and may lead only to disappointment for would-be innovators.

III Concluding remarks

Bolivian mines have always faced difficulties spared their competitors in the world's markets. In 1984 these difficulties seemed even greater than in the past. To the extra costs of an adverse geographical and geological environment must be added those of a political climate which has denied much of the industry the opportunity to employ modern methods of more effective mining. These difficulties are likely to remain. Bolivia has become increasingly belligerent in seeking compensation for higher production costs by promoting higher metal prices. For example, Bolivia was one of the first three countries to sign the First International Tin Agreement in 1955.[10] That, and subsequent agreements, have coincided with a period of greater price stability and with generally higher prices for tin than for other metals. In 1982 Bolivia decided not to sign the Sixth International Tin Agreement however, preferring to place her support behind the new Tin Producers Association, a body narrower in its interests than that of the International Tin Council and designed to strengthen the hand of producers within the Council. She has subsequently agreed to cooperate with the signatories of the Sixth Agreement and has agreed to reduce exports in an attempt, largely successful, to hold up the tin price. A high price is more important to Bolivia, almost the highest-cost producer in the world, than to other tin mining countries. Bolivia may be deemed unfortunate in being so heavily dependent upon the export of tin: it is a metal for which demand is relatively inelastic and global consumption of tin has been rising at a lower rate than has that of almost any other major metal. Nevertheless she can draw some comfort from the most recent projection of commodity prices to emerge from the World Bank, forecasting a growth in the real value of tin of about 20 US cents per pound over each of the next dozen years or so. All Bolivia has to do is to continue to produce the tin if she is to benefit from this trend!

It is possible that in the longer term the clouds that are currently casting deep shadows over the Bolivian mining industry will lift. It may be hoped that the kind of cooperative geological work which teams from the United Kingdom and Bolivia have been quietly pursuing in the Oriente during the last five years, and the extension of the pioneer inventory of natural resources of Oruro using LandSat data begun with United States support, will help create new mining opportunities and restore a degree of confidence to the beleaguered miners of Bolivia. Paradoxically, it may be in the very paucity of systematic mineral exploration over the last several decades and in the very lack of the application of modern exploratory tools to this old mining country that the real hope for the future lies: the opportunities to revive her reduced fortunes may still be awaiting tomorrow's prospector.

Notes

1 This essay is based upon data and impressions gained on a visit to Bolivia in April–May 1982, partly paid for by a grant-in-aid from the University of Manchester.

2 Current information on mining in Bolivia is widely available in, for example, *Mining Annual Review* (London) published in July each year which includes articles on Bolivia, tin and other metals, and technical progress reports; *Mining Journal* (London) weekly; *Tin International* (London) monthly; *Latin America Commodities Report* (London) weekly; as well as in the mining press elsewhere. Within Bolivia newspapers such as *El Diario, Presencia, Ultima Hora, Hoy*, etc. are usually well informed on mining matters. Statistics are collected in, for example, *Bolivia en Cifras 1980* (Instituto Nacional de Estadística, La Paz); *Anuario Estadístico Minero* (Ministerio de Minería y Metalurgia, La Paz) 1976, 1978; and in *Boletín Estadístico Minero* (Ministerio de Minería y Metalurgia, La Paz) monthly.

3 In 1970 the income of the average Bolivian was four-ninths of that of the average South American; in 1983 it was one-third. The purchasing power of the average Bolivian has been falling: by 3.1 per cent in 1981, a further 11.5 per cent in 1982, and a further 8.7 per cent in 1983. In 1984 the average Bolivian was worse off in real terms than at any other time in the previous fifteen years. For insights into the effects of poverty on the daily lives of Bolivian miners the reader may turn to the testimony of Domitila Barrios de Chungara, a miner's wife from Catavi in *Let Me Speak* (London) 1978, to June Nash's scholarly work *We Eat the Mines and the Mines Eat Us* (New York) 1979, and to *Estaño, Sangre y Sudor: Tragedia del Minero Locatorio* (Oruro), 1980, written by Segundino Palaez R. and Marina Vargas with the financial assistance of Oxfam.

4 The failure in 1980 of the Huari-Huari lead-and-zinc project developed by New Jersey Zinc is the most recent Bolivian example.

5 General information on COMIBOL is in, for example, *Memoria Anual* (Corporación Minera de Bolivia, La Paz) which is available for a number of years since 1965–6, and COMIBOL: Consolidated Accounts prepared by Arthur Young Cons. for various years since 1972. Amado Canelas Orellana has been a long-standing critic of COMIBOL and his books include *Historia de una Frustración* (La Paz) 1963, *Mito y Realidad de la Corporación Minera de Bolivia* (La Paz) 1966 and *Quiebra de la Minería Estatal Boliviana?* (La Paz) 1981.

6 The reports on the mining problems of COMIBOL have not normally been published. They include 'Report of the United Nations Mission of Technical Assistance' – the Keenleyside Report (UN) 1951; 'Report on the mining industry of Bolivia' by Ford, Bacon, and Davis, Inc. (New York) 1956; 'Report on the ... mines of COMIBOL' for the Inter-American Development Bank by C. C. Houston and Associates (Toronto) 1960, 1961; 'Bolivia: the Mining and Metallurgical Sector IBRD' (Washington) 1971; 'Technical requirements of Bolivian mines' by PD–NCB Consultants Ltd (London) 1976; 'Plan de rehabilitación de COMIBOL' by Pinnock, Allen and Holt, Inc. (La Paz) 1982.

7 The Asociación Nacional de Mineros Medianos publishes an annual report; it also publishes *Noticias Mineras Internacional* on an approximately weekly basis. *Taxation and Mining* by Gillis *et al.* (Cambridge, Mass.) 1978 includes a survey of the medium mining sector.

8 The Banco Minero publishes irregularly a bi-monthly review *Revista Minera Bamin* and an annual *Memoria*. Information on mining cooperatives is given in *Co-operativas Mineras* by A. Alurralde Anaya (La Paz) 1973.

9 José Miguel Velasco in *Mito y Realidad de las Fundiciones en Bolivia* (La Paz) 1964 provides some historical background on smelting. 'Fundiciones e industrialización – política metalúrgica de Bolivia' in *Temas en la Crisis* 15 (La Paz) 1982 reviews the present situation. The unpublished papers of the Simposio Internacional de Estaño organised by the Ministerio de Minería y Metalurgia in La Paz in 1977 include some papers on Bolivia's metallurgical problems.

10 For a history of the operation of the International Tin Council and a survey of the world's tin industry see W. Fox, *Tin: the Working of a Commodity Agreement* (London) 1974.

PART TWO

Human responses
to mines and mining

CHAPTER 8

Entrepreneurial risk management in peasant mining: the Bolivian experience[1]

RICARDO A. GODOY

Although small-scale mining in the Andes produces a modest percentage of the total regional output of metal ores, large numbers of
individuals are engaged in its extraction. In political, economic,
demographic and ethnic terms, small mining enterprises in the Andean
countries are ubiquitous and highly important. The ways in which
individuals participate in these small mine enterprises are numerous
and intricately interlocked. Until recently, little first-hand information
has been available about this sector of mining; the following chapter
offers a significant beginning toward remedy.

An individual's decision to work in small-scale mining, and an
entrepreneur's to invest in it, depends not only on such standard
economic factors as income maximisation, cost of capital, market price,
and costs of production, but also on a matrix of social obligations,
alternatives of time and interest for both worker and investor, and
tenets of the moral universe in the peasant sector. Such factors are not
often listed in a small investor's handbook. The Andean web of
decision factors is well illustrated here by Ricardo Godoy's essays on
small-scale antimony entrepreneurs in the Jukumani territory of
highland Bolivia.

Ricardo Godoy is associated with the Harvard Institute for International Development (Harvard University, 1737 Cambridge St,
Cambridge, Massachusetts 02138, USA). His current research deals
with small-scale mining in developing countries and in the decision-
making structure of Bolivian miners.   – *The editors*

I Introduction

The most distinguished anthropological studies on risk have one feature in
common: peasant agriculture.[2] Anthropologists delight in pointing to the
'institutional context' (Wolf 1969: xv) and farming techniques used by peasants
to reduce the probabilities of crop failure and the uncertainties of the market.
A partial list of such techniques includes intercropping, plot scattering, and
sowing several species or varieties of the same crop. Social mechanisms to
reduce uncertainty include patron–client relations, sharecropping, redistribution of land and crops, reciprocity in food and labour, and strategic
marriage alliances.

That anthropologists are interested in how peasants hedge agricultural risk is not surprising, for the received wisdom of the discipline holds that farming constitutes the cornerstone of peasant livelihood (Foster 1967: 4–5). For example, Wolf (1955: 453), whose seminal writings on peasants have influenced a generation of anthropologists, defines peasants as 'agricultural producers', and explicitly excludes from his definition 'fishermen, strip miners, rubber gatherers, and livestock keepers'. Firth (1950: 503) takes a broader approach and is willing to include fishermen, craftsmen and other small-scale producers, yet he notes that the 'primary means of livelihood of the peasant is cultivation of the soil'. Redfield (1956: 19) and Shanin (1971: 14–15) likewise regard land husbandry as a key attribute of peasant societies.

It may have been historically accurate to define peasants as cultivators, but such an 'occupational' definition (Silverman 1979: 65) is an anachronism in a rapidly changing world. Land shortage and rural poverty make it increasingly necessary for peasants to supplement agrarian pursuits with outside income. Documentation of this point comes from Africa (*cf.* Eicher and Baker 1982: 235–7), Asia (*cf.* Johnston and Clark 1982: 76–9) and Latin America. Deere and Wasserstrom write,

... in Latin America, today, the majority of rural families obtain a substantial portion of their family income, between 30 and 60 percent, from activities conducted outside their farmsteads, and in almost all cases these activities include some form of wage labor ... (1981: 152, author's translation)

It should not be surprising, therefore, that in richly endowed mineral nations such as Bolivia, Sierra Leone or Malaysia, peasants should do more than farm. In these countries they also explore and produce minerals in areas of marginal value to capitalised enterprises. Not surprisingly they may become the precursors of these enterprises (UN 1978),[3] and, in many countries, as in Bolivia, they account for a substantial and growing share of national output (Godoy 1984a).

Despite the importance of peasant mining operations in developing countries, studies on the subject are scant, anecdotal and narrowly focused on engineering and technical aspects (Neilson 1982; UN 1970, 1972, 1978; AGID 1982; Baxter 1975; TWMC, 2, 1979). In this frankly exploratory exposition I propose to remedy these inbalances by analysing how the mine owners of small-scale ventures in Bolivia share and shift risks with the peasants from the surrounding countryside, the Jukumani Indians of Northern Potosí. I say exploratory because the focus of the article is on small-scale mining and entrepreneurs rather than with the traditional concerns of anthropology, peasants and agriculture. As such, the ultimate aim of this article is largely heuristic: to awaken thought, interest and respect for artisanal mining ventures.

Table 8.1 Status of twenty-two small mines: mineral, output, ore and access

Mine name, (last worked)	Mineral(s)	Average monthly output	Ore quality	Access
Capacirca (working)*	Antimony, wolfram, gold	2–3 kg gold 10 fine tons Sb (5*)	60% 54–8% Sb; low lead and arsenic (1%)	1* (see key)
Cebadillas (working)*	Antimony, gold	40 metric tons	55% Sb;high impurities 8% lead	1*
Santa Rosa (working)*	Antimony, gold	1–1.5 kg gold 5 metric tons Sb	50–60% Sb; high impurities	1*
Irpa Irpa (1979)	Antimony, gold	50 metric tons Sb 7 kg gold	55% Sb; high impurities	1*
Tolani (1973)	Antimony			2*
Wataria (1973)	Antimony	1.5 metric tons Sb	70% Sb; low impurities (.5% Pb & As)	3*
Wayrojo (1973)	Antimony	1.5 metric tons Sb	70%; low impurities	4*
Totora (1973)	Antimony, gold, lead	–	low impurities	4*
Luluni (working)*	Antimony, wolfram	–	high lead	4*
Ch'ikaqawa (1973)	Wolfram, antimony	–	low impurities	4*
Alicia (working)*	Antimony, wolfram, gold	–	high lead	1*
Ch'isllankiri (1955)	Lead, gold antimony, silver	–	–	3*
Colquepampa (1973)	Antimony	–	low impurities	4*
Sakasakani Mayu (1973)	Antimony	–	low impurities; veneros	4*
Jankolayme (working)*	Antimony	100 metric tons Sb	65% Sb; little lead	1*
Mina India (working)*	Antimony	–	–	1*
Jaquerana (working)*	Antimony	3 quintales Sb	low impurities	4*
Walparuna (1972)	Antimony, copper	–	low impurities	4*
San Juan (working)*	Antimony, tin, wolfram, copper	–	–	–
Hallazgo (working)*	Antimony, lead, silver, tin	–	–	–
Condor Wayra (working)*	Antimony, lead, wolfram, silver	–	–	–
Choqo Choqo (working)*	Antimony, lead	–	–	2*

Key   * Working = as of early 1981
     1* Feeder road, accessible all year
     2* No feeder road but accessible all year
     3* Has road, inaccessible in rainy season
     4* No feeder road, inaccessible in rainy season
     5* Sb = Stibnite (ore of antimony)

Table 8.2   Tenancy, ownership and cross-ownership: twenty-two small mines

Owner(s)*/ tenant(s)	Residence of owners or tenants	Complimentary occupation of owners/tenants	Other mines worked by owners/tenants
Capacirca (1)	Cochabamba	Cattle-raising in Beni; 10,000 heads; sells to COMIBOL	Tin mines in Cochabamba and Oruro
Cebadillas			Antimony mine in Oruro
(2)	Cochabamba	Construction	
(3)	USA	Ambassador	
(4)	La Paz	Presidential candidate	
Santa Rosa (*see* Cebadillas)	”	”	”
Irpa Irpa (5)	Oruro	Construction business; Oruro; hotel in La Paz	Antimony and tin mines in Northern Potosí
Tolani (6)	Sucre USA	Physician in USA; owns hospital in St Cruz	–
Wataria (7) (8)	Uncia Cochabamba Llallagua	(7): public employee in Uncia (8): owns drug-stores, hospital and 50 houses in Llallagua	(7): tin rescatiri in Uncia; hires 120 locatarios (20 from Wataria) to COMIBOL sites (8): rents a tin mine in Northern Potosí
Wayrojo (8)	Llallagua	Owns drugstores, houses, which she rents, and hospital in Llallagua; tin rescatiri	Rents tin mine in Northern Potosí
Luluni (9)	Chukiuta	Employee in Cebadillas; antimony rescatiri; store owner; middleman	(*see* Sakasani Mayu)
Ch'ikaqawa (10)	Uncia	Store owner	–
Alicia (10A)	Potosí	None	Owns tin mine in Potosí
Ch'isllankiri (11)	–	–	–
Colquepampa (12)	Cochabamba	Owns drugstore	Owns tin mine in Northern Potosí with F.A., President of the Small Miners Association in Northern Potosí
Sakasakani Mayu (9)	– – *see* Luluni – –		
Jankolayme (13) (14)	La Paz Uncia	Store owner and poultry shop Lawyer, doctor tin rescatiri	– Sb, Sn, Wo mines; Northern Potosí
Mina India (15) (16)	– –	– –	– (16): Owner of Cepadillas and Santa Rosa

Table 8.2   *continued*

Owner(s)*/ tenant(s)	Residence of owners or tenants	Complimentary occupation of owners/tenants	Other mines worked by owners/tenants
Jaquerana			
(17)	Uncia	Tin rescatiri; employee for government	Tin mine in Northern Potosi
Walparuña			
(18)	Uncia	(18): store owner; govt. employee	–
(19)	Uncia	(19): Owns hotel in Uncia; rents 3 houses; teacher	
San Juan			
(20)	Chukiuta	Owns restaurant in Tacopalca and Llallagua; COMIBOL employee	–
(21)	Chukiuta		
Hallazgo			
(22)	–	–	–
Condor Wayra			
(23)	Chukiuta	Rescatiri; retired teacher	–
Choqo Choqo			
(24)	Aymaya	–	–
Totora			
(25)	Llallagua	Owns drugstore in Llallagua; tin rescatiri; bought 2 houses in Cochabamba	Rents tin mine in Northern Potosi

* Numbers for individuals are used to provide anonymity.

II Mining concessions in the Jukumani territory: a brief overview

The Jukumani Indians are an ethnic group or *ayllu*, who live in the central section of the eastern Andes of Bolivia. Like other ethnic groups in northern Potosí, the land of the Jukumanis is spread out along several different and dispersed ecological tiers. Their major territory, which is also their political and ceremonial centre, is a block of land measuring close to 300 square kilometres extending over parts of the counties of Aymaya, Amayapampa, and Chikiuta in the province of Bustillo, Department of Potosí.

The highland nucleus is bisected by a dirt road connecting the city of Sucre, department of Chuquisaca, on the south, to the Llallagua-Uncia-Catavi mining complex on the north. The principal settlement within the Jukumani *ayllu* along this road is the town of Chukiuta, which is 30 km from Uncia, about 130 km from Oruro, and over 220 km from Sucre. Eastward and downward, Jukumanis have many plots interdigitating with the lands of other *ayllus*. The valley lands are located in the canton of Micani, province of Charcas, Department of Potosí. Demographically, only about 5 per cent of the Jukumani population lives in the valley; most of the 9,000 Jukumanis dwell in the upper zones, though many have double domicile and travel to and from the valley to exchange products, visit and celebrate festivities.

Tucked in between steep cliffs, along sinuous rivers, behind jagged hills, several dozen small-scale antimony mines fade imperceptibly into the stark highland landscape of the Jukumani territory. None have track cables, few have feeder roads, camps or mills (see Tables 8.1 – 8.4). Even the omnipresent emblem of Bolivian mines, the slag pile by the mine mouth, is barely noticeable in these self-effacing operations. Labourwise, scattered workers, squatting spalliers, roving panners and lone prospectors add little to visibility. And during seasons of rain, verdure covers narrow foot paths, dumps and hills, concealing the mines even further.

But if the mines are physically inconspicuous, they are not in other realms. Linguistically, mines are Spanish and Quechua oases polka-dotting an essentially Aymara desert. Ritually, the mines are the province of the devil, *supay*, rather than *pachamama*, the mother goddess of agriculture. Even in the style of human interaction, the worlds of agriculture and mining contrast sharply with one another. While life in the village, fields and pasture grounds is generally tranquil (except during festivities) intercourse in the mine is tense and boisterous: coarse manly joking among work crew members showing bravado in the face of danger, anxiety in finding lodes, depression in failures, worry of accidents, ambivalent feelings of fear, awesome expectancy and obsequious humility toward subterranean demons, whose power and capricious will is at once envied and dreaded. Such is the paradoxical world of Andean peasant mining – inconspicuous but different, revered but despised, alluring but frightening.

At first sight a mining claim in Northern Potosí, as in the rest of Bolivia, looks like an ordinary piece of land, but more careful scrutiny reveals concessions are composed of variegated patches, each of which, in turn, consists of a surface area (*canchamina*, lit: mine plot) and an underground work place (*paraje*, lit: wall). Connected to each other by footpaths and dirt roads, canchaminas vary in size from diminutive plots to sizeable expanses. Abode storage rooms, jigs, buddles, meshes, timber, tools and refuse heaps of mineral pockmark the canchamina.

In contrast to the subterranean world, the canchamina has a touch of freshness, congeniality and leisure. Here workers gather before entering the mine to fill carbide lamps with water, to chew coca, to puff on cigarettes, to exchange information on upcoming crops, to comment on the quality of recent festivals and to discuss new discoveries. At mid-morning, noon, and again at midafternoon, if it is not raining or if the end of the tunnel is not too far, workers gather in the canchamina to have lunch and take both a coca break and a breath of fresh air. Here also, twice each year, in August and in February, animal sacrifices and libations (*ch'allas*) to underground deities take place. It is part of an exchange, miners say: we give them food, they give us mineral – and security.

Table 8.3　Acquisition, source and labour type: twenty-one small mines

Name (area)	Mode of acquisition	Labour pool**	Type of work*		
			Contract	Wage	Workers
Capacirca (74 Has)	Bought from MF	Chukiuta	Y	Y	50
Cebadillas (400 Has)	Rented 5 yrs by construction company from Cochabamba, from SA	Pujrallapu, Belen, Chukiuta	Y	Y	60
Santa Rosa (300 Has)	"	Lukata, Chukiuta, Mik'ani	Y	Y	60
Irpa Irpa (462 Has)	Bought from MS for 200,000 pesos	Wakuta, Irpa Irpa, Laymis (Lagunillas) Mik'ani	Y	Y	120
Tolani	Inheritance	Chukiuta	–	–	–
Wataria	Jukumani told 7; 7 inherits from father; 8 denounces and gets half the property	Wataria	Y	N	60
Totora	Bought	Vila Vila	–	Y	20
Luluni	Petitioned	Luluni	Y	N	10
Ch'ikaqawa	1st owner	Ch'ikaqawa	Y	N	10
Alicia	Bought from 12	Pujrallapu, Belen	Y	Y	8
Colquepampa	Bought	Watari, Colquepampa (Pocoata)	Y	N	20
Sakasani Mayu	–	Luluni	Y	N	9
Jankolayme (96 Has)	Denounced by 13	Saroqa, Phutiwana, (Laymis), Pikina, Jankolayme, Condoriri	Y	N	50
Mina India	–	Luluni, Paymis, Pata, Pata Grande and Chico, Tirani, Churichulpa, Patuma	Y	Y	35
Jaquerana	Denounced by E.P.	Vila Vila	Y	N	5
San Juan (150 Has)	Petitioned 29 Oct. 70	–	–	–	–
Hallazgo (50 Has)	Petitioned 28 July 70	Chiruya, Laymis	–	–	–
Condor Wayra (30 Has)	Petitioned 25 April 79	Choqo Choqo	–	–	–
Choqo Choqo	–	Choqo Choqo	–	–	–
Wayrojo	8 denounced former owner, E.P., truck driver, butcher, and store owner in Uncia	Wayrojo, Q'anupacha, Vila Vila	Y	N	60
Walparuña (60 Has)	Petitioned 20 Sept. 68	Pairumani, Chiruya, Laymis	Y	N	5

*　Y = Yes; N = NO
**　Including dispersed settlement

More prosaically, in the canchamina ore is spalled, washed and stored. This is why it is the favourite hunting ground of *jukus* or ore thieves, who inspect it to see if there is a guard and also if the antimony merits stealing. Mario, a doughty teenager, looked disappointedly at a pile of mineral in a canchamina late one night after a long trip from his village and decided to return home empty-handed. '*Mana convienewanchu*' ('it is not worth it for me'). The mineral was too dirty, the buyer's house too far, the night too cold, the price too low.

Beneath the canchamina, inside the mine mouth, lies a maze of narrow, winding tunnels. The main gallery is driven into the hillside at an inclined, slightly upwards angle so water and corves, trammed by gravity from the interior, travel naturally downhill to the surface, saving both effort and drainage equipment. From the adits, drifts and crosscuts traverse the orebody in all imaginable directions. On several occasions miners, oblivious to their overall trajectory, have drifted back to the surface. To the professional geologist such gophering is the epitome of unplanned, irrational exploitation, a regrettable waste of the nation's mineral wealth; to the peasant, a sensible way of exploring orebodies of irregular mineralisation.

Tunnels are short, low and narrow. Hence, workers crawl or walk bent at the waist, holding a carbide lamp in one hand and tools on the other. When walking, the head is kept bent, chin close to chest, gaze fixed on the floor, lest a protuding rock or horizontal beam injure one's head. The trip from the mine entrance to the lode's face (or vice versa) may be exhausting, to an outsider at least, not from the distance, as was true in British coal mines (Orwell 1937), but from the cramped conditions of travelling. As was true among British colliers, children are used to transport ore and tools because their small and supple bodies manoeuvre through pits with ease and speed.

The inside of the mine is cool, slightly clammy. Muddy footwalls are common throughout the year from internal seepage; flooded shafts from poor drainage. On the positive side, dampness conserves stulled stopes. Dank tunnels, workers claim, are safer than dry ones because posts shoring up the latter, lacking humidity for preservation, rot and break faster under the pressure of the hanging wall on top.

At the face of the adit the most important work takes place; here ore is found, broken and thrust into wheelbarrows or tubs. Sandwiched in a mucky floor between narrow walls and a low roof, the peasant, crouching or lying on his stomach, drives the crow bar into the rock by striking it with a mallet, turning the bar with one hand after each pound. A driller cudgels for an uninterrupted minute or two, then rests by prying loose rocks or by pushing with his bare hands miry ore and slime down the incline. When punching holes, drillers examine the rock formation to determine the location of the next ore shoot; when clearing debris they also look at the rocks themselves for spots of gold in the antimony. Lighting being poor inside the mine, potentially rich mineral may be set aside for later inspection on the surface. Gold veins are covered with mud and boulders to prevent others from sharing in the discovery.

Blasting, a subtle operation, requires years of apprenticeship before it can be done well. Depending on the quality of the rock face and direction of the tunnel, various charges of dynamite are inserted in holes of different depths and directions and then tamped with drill-hole sludge. Centre holes go off first, thus creating a cavity which facilitates rock breakage for surrounding detonations. Once dust settles, peasants re-enter the work place to tear out hanging rock, remove overburden, and prepare the rock face for the next shooting. The cargo of rocks travels down the incline to the tipple, where it is dumped in front of women spalliers known as *palliris*. Squatting all day, they crush ore with mallets, handpicking the best nodules from the dross. An old, infirm man or two may help triturate ore, jerk the jigs, feed the buddle or shovel waste rock to the dump.

From the mine mouth concentrates travel as ballast on beasts of burden or on human backs to entrepots and local towns, and thence in trucks to a custom smelter in Tupiza, southern Potosí, or to the state-owned ore purchasing agency, BAMIN (Banco Minero de Bolivia), 130 kilometres to the north, in the city of Oruro. During the rainy season landslides and poor road conditions make transportation difficult, forcing mine owners to stockpile ore.

Productivity varies at once capriciously and rhythmically: erratically because prices move unevenly; rhythmically because the available supply of labourers ebbs and flows around the mines in tune with seasonal agricultural chores. Figure 8.1 shows how small-scale antimony production in Bolivia parallels price changes in the international market. Figure 8.2 highlights the seasonality of productivity, with troughs in January, the rainy and tilling season, and November, the season for planting and rituals for the deceased. Finally, consider Table 8.5 where one can see that productivity (as measured by tunnelling, mill throughput, and propping) rises during the winter months when peasants enjoy freedom from agrarian chores.

III Claiming concessions: peasant and proprietor's perceptions

Mining interest in the Jukumani territory, as in the rest of Bolivia, may be acquired for a lifetime provided holders pay approximately US $1,000 in stamp taxes and licence fees when first demarcating the claim. Thereafter, semiannual payment of nominal property taxes (*patentes*) based on the extension of the concession ensures the rights to keep the mine in perpetuity. The government does not obligate proprietors to make minimal investments, nor does it subject them to a compulsory timetable for undertaking reconnaisance operations. Thus, although the original expense of taking up ground may constitute an entry barrier, once obtained, mining stakes may be kept for modest fees without committing concessionaries to invest, explore or to exploit the land. As Table 8.2 showed, none of the mine owners are Jukumanis; they are all

Table 8.4 Ore processing components, twenty-two small mines

| Name (altitude) | Trucks | Electricity | Camp size* | Concentration | | | |
				Flotation	Crushers	Jigs	Palliris
Capacirca	25	Y	–	Y	Y	Y	Y
Cebadillas (3750–4000 mts)	4	N	8 families	N	N	N	Y
Santa Rosa (3800–4000 mts)	–	Y	5 families	Y	Y	Y	Y
Irpa Irpa (3700–4200 mts)	3	Y	N	Y	Y	Y	Y
Tolani	–	N	N	N	Y	Y	Y
Wataria	1	N	N	N	Y	Y	Y
Wayrojo	1	N	8 families	N	Y	Y	Y
Totora	2	N	N	N	N	N	Y
Luluni	–	N	N	N	N	N	Y
Ch'ikaqawa	–	N	N	N	N	N	Y
Alicia	1	N	6 families	N	Y	Y	Y
Ch'isllankiri	–	N	N	N	N	–	Y
Colquepampa	1	N	N	N	N	N	Y
Sakasakani Mayu	1	N	N	N	N	N	Y
Jankolayme (3700–900 mts)	10	Y	30 families	Y	Y	Y	Y
Jaquerana	1	N	N	N	Y	Y	Y
Walparuña	–	N	N	N	N	N	Y
San Juan	–	N	–	N	N	N	Y
Hallazgo	–	N	–	N	N	N	Y
Condor Wayra	–	N	–	N	N	N	Y
Choqo Choqo	–	N	N	N	N	N	Y
Mina India	–	Y	80 families	Y	Y	Y	Y

* N = no permanent camp

Table 8.5 Measures of monthly productivity in one small-scale antimony mine[a]

| Month | Year | Mill throughput[b] | Tunnelling | Wooden frames | Minerals in tons | |
					Fines	Ballast
Nov.	1979	52.5	50.5	4	3.9	–
Dec.	1979	57.5	47	8	5.8	–
Jan.	1980	160.0	49	12	1.1	8.1
Feb.	1980	32.5	55	15	1.0	5.4
Mar.	1980	55.0	19.5	8	.4	4.4
Apr.	1980	90.0	43	–	.2	2.4
May	1980	206.2	24	10	–	1.2
June	1980	208.7	83	20	5.3	.4
July	1980	240.0	45	13	2.1	.1
Aug.	1980	233.7	12	32	1.1	4.6
Sep.	1980	185.0	18	–	–	6.2
Oct.	1980	206.2	4.5	27	–	4.7
Nov.	1980	58.1	–	4	–	4.7

a. November 1979 – November 1980   b. In metric tons.
c. Refers to any underground digging, regardless of the angle of inclination. It also includes clearing debris from abandoned sites.

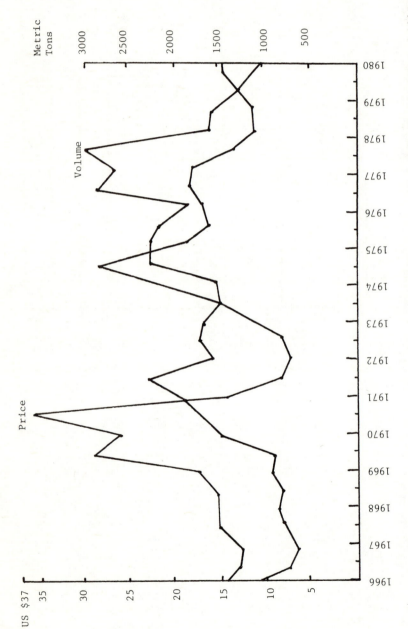

Fig. 8.1 Bolivian small-scale antimony production, 1966–79, and semester average nominal prices for lump sulphide ore, 60 per cent antimony, c.i.f., US $ per metric ton unit (10 kilogrammes). *Source:* BAMIN (1966–75a), *Metal Bulletin* (1966–80).

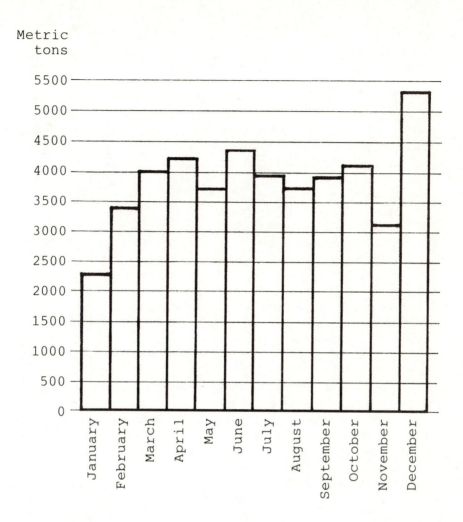

Fig. 8.2 BAMIN total antimony purchases, 1966–79, by month in metric tons.
Source: BAMIN (1966–75 a).

outsiders of diverse social rank and wealth and include a former presidential candidate, a Bolivian ambassador, a Bolivian physician practising in the USA, drugstore owners, ranchers, teachers and tin ore buyers.

But the procedures so far outlined to obtain a mine – payment of stamp taxes, license fees, *patentes* – defines the state's and the *mestizo*'s conception of property rights to natural resources, not the Jukumani's. The right to exploit an exhaustible stock, to the former, is predicated on meeting fiscal duties. These rights are something to be acquired, bequeathed, mortgaged, speculated with.

Jukumanis conceive of mining concessions and mineral wealth differently. It is there for labour to find it, crack it, remove it and process it. At best people have rights of use to galleries and rivulets. Between ethnic groups or villages mineral wealth is shared. For instance, the fluvial placer deposits in Sanoma, Ch'iuta and Sakaroka, known since colonial days for their riches, lie between two villages and are exploited by turns by peasants from the two communities. Two to three weeks before their patron saint feast of San Miguel on 29 September, people from Qhari come down to search for gold in river gravel, benches and in high terraces flanking the river; afterwards, people from Belen poke around for mineral in the same places for their own upcoming fiesta of *El Rosario* in early October.

As with land rights, the claim of villagers to exploit non-renewable common property goods in their own domain may be relaxed or hardened depending on various pressures. During the planting or harvesting seasons, for instance, while families are busy in their fields one may pan in neighbouring villages unworriedly, but during the quieter seasons after harvest, entrance rights are stinted. Similarly, on slim years villagers exercise stricter control over non-villagers moving in to use local mineral deposits.

Underground and placer reserves exist for peasants to exploit as the need for cash arises; they are not there to increase the peasant's stock of wealth *per se*. To the insistent questioning of a jeweller (and an anthropologist) in February 1980, when the price of gold reached an historic pinnacle, as to why they were not panning, an old Jukumani, speaking for an assembled group of men, responded with simplicity that befitted the terseness of his answer: 'because we don't need to'. The deposit was there to satisfy specific, immediate monetary needs; once met, other villagers, villages or ethnic groups (in that order) could use these natural endowments.

Exploration rights and rights of usufruct to ore bodies are thus not legitimised by payments of fees, but rather by the need for cash and by the labour invested. A worker who strips overburden from an abandoned pitch, searches for ore, breaks it, hauls it to the surface, and dresses it to an acceptable level of commercial value is regarded by Jukumanis as the true owner of the site and the mineral, if ownership must be specified at all. Because labour invested, not legal deeds or fiscal payments, crystallises the rights of exploration and exploitation, what from an outsider's perspective is one mine with one legal owner, from a bottom-up perspective is a honeycomb with many men working different cells, each miner the owner of his own site. Having shouldered some of the risks of exploration and the costs of extraction and processing, it follows, in the Jukumanis' understanding, that the worker has a right to enjoy the value added of the discovery.

Who then owns the work site? The answer is contextual and varies depending on the level of discourse: to the government, the owner, the actual worker. According to the share of exploration risks and production costs borne by labour or mine owner, the ownership perimeters differ. If peasants bear all (or most) of the risks of exploration, the systematic marketing risks and the costs of production, the work site, in effect, 'belongs' to labour. Alternatively, if the outside owner bears the risks of exploration and the market, then ownership rights shift away from the peasant.[4]

IV Geological risks

Mineral exploration is generally risky because the location, size and quality of the reserve are unknown and must be found. In uniform veins, the discovery occurs but once, the orebody thereafter being exploited from the top down, from the best to the worst ore. If the lode is of heterogeneous quality and discontinuous formation, geological riskiness increases. Having once struck a shoot, prospecting must start afresh.

Such is the case with the mineralised antimony deposits in the Jukumani territory. Although there are some rich moranic accumulations by river banks, most antimony does not come from placers but from spotty and pockety underground lodes in the shape of rosaries. Like a rosary, deposits consist of a sterile veinlet connecting bunchy concentrations or pockets of stibnite, the principal ore mineral of antimony. But unlike the rosary's string, the veinlet or stringer lead disappears and meanders unpredictably. To complicate matters, the size, location and quality of pockets themselves vary widely.

Jukumanis view the business of winning ore from nature as a desultory activity, dependent on serendipity and experience. Several signs, however, help detect mineralised areas: scrubby growth relative to the surrounding vegetation, smooth host rocks (*lluskas*), and stockworks of white quartz. Furthermore, reddish brown igneous outcroppings by the hillside indicate the presence of a concealed pocket of stibnite mixed with gold. Yet none of these leads ensures certainty of discovery. Success ultimately rests with luck.

In the oreshoot itself, the metallic content, though uneven, may reach 50 to 60 per cent, allowing the mineral to go 'from the vein to the sack', as Jukumanis say, without any concentration. Mixed with stibnite, one finds objectionable impurities such as lead and arsenic, which decrease the value of the ore, but also alloying metals, such as auriferous quartz and silver, which are exploited as by-products, and sulphur, which is regarded as an asset because it eases combustion in smelting.

V Marketing risks and choice of technology

The most striking feature of these peasant mines is their rusticity, and their rusticity consists in old, inveterate technology, some of it dating back to pre-hispanic days. The mining equipment and tool kit are uniform across mines regardless of differences in mineralised reserves. There are thus no techno-logical notches separating, in a qualitative sense, one enterprise from another. Except, perhaps, for a couple of mechanised mills and one or two pneumatic drills, the extractive and dressing tool kit is strikingly simple and inexpensive (about US $30 in 1980 dollars): inside the mine, crowbar, mallet, shovel, wheel-barrow, windlass, axe; on the surface, screens, buddles, jigs, Cornish kieves, rolling crushers, rock-riffled sluice boxes.

Yet more striking than the rusticity of such antiquated technology is the complacency with which it is regarded. Peasants and mine owners cringe at the suggestion of modernising, not out of inability to borrow from state institutions and structured private capital markets the US $4,000 that would be needed to automate their sessile operations, but out of satisfaction with things as they are. These Stone Age tools and machines, they claim, are as good as any modern, automated contraption. What is more, even if they had shaking tables and flotation cells they would still use buddles and jigs to recover metal from tailings. To these crude miners and their patrons there is, thus, no conflict between a wish to minimise costs by using labour intensive pro-duction processes and a desire to use plants of recent vintage to facilitate responses to shifts in demand (Wells 1975: 81).

But to say the level of technology reflects a conscious decision ignores the constraints that make such choice logical. First, high-quality ore makes expendable investments in automated benefication plants. Simple hand-cleaning and washing suffices to raise the metallic content of ore to 60 per cent or more. Recouping 50 per cent of the pay dirt treated, buddles, jigs and kieves win from refuse what few modern mills could achieve. Furthermore, the pell-mell location of oreshoots limits the utility of open pit, tackless mining, diamond bits or pneumatic drilling. Second, automation entails problems and risks. Given the shortage of water during the dry months, modern mills, which require a constant supply of water to function optimally, can be burdensome. Moreover, reagents used in the flotation cells of automated mills, while raising the metallic content of ore, also increase the level of lead and arsenic impurities, thereby lowering the pulp's overall quality. Third, prices for mining equip-ment in Bolivia are higher than in other nations, making investments in technology expensive and vulnerable to escalating outlays for repairs and replacements (World Bank 1976: 81).

Finally, the price volatility of such minerals as antimony increases the riskiness of investment in new technology. Table 8.6 shows the realised mean and variance of antimony quotations in both nominal and real levels for a decade. Changes in nominal prices measure shifts in the cash flow profile;

Table 8.6   Mean and variance of nominal and real dollar price of antimony*

(1)	(2)	(3)	(4)	(5)	(6)	(7)
Year**	Average nominal price (P)	Average real price (p)	Standard deviation of nominal price (O(P))	Standard deviation of real price (O(p))	Coefficient of variation of nominal prices (O(P)/P)	Coefficient of variation of real prices (O(p)/p)
1969	20.63	18.85	9.18	7.84	.44	.42
1970	56.88	48.82	27.82	23.96	.49	.49
1971	10.98	9.25	2.81	2.52	.26	.27
1972	7.83	6.34	.53	.40	.07	.06
1973	14.35	10.75	3.02	1.92	.21	.18
1974	29.42	16.98	5.44	1.84	.18	.11
1975	22.77	12.26	2.71	1.49	.12	.12
1976	24.96	12.73	1.96	.74	.08	.06
1977	22.75	10.91	3.53	1.85	.16	.17
1978	17.48	7.68	.93	.40	.05	.05
1979	21.89	8.43	1.71	.38	.08	.05

*   Monthly average prices for lump sulphide ore, 60% sb c.i.f. Europe dollars per metric ton unit have been used for the nominal price series. (*Metal Bulletin Handbooks*, 1969–79). The USA wholesale price index has been used as a deflator to calculate real prices.
** A year includes twelve months, from January until December inclusively.
Figures may not be exact because of rounding errors.

movements in real mineral prices reflect consumption variability over time. The amplitude of the fluctuations is high, as the coefficient of variation (columns 6 and 7) indicates. The standard deviation of both nominal and real prices, on the average, is about 20 per cent of the mean, reaching 50 per cent during 1969–70.

Mine owners and peasants know prices change spasmodically, but they do not understand why: perhaps a war or an 'overharvest' of mineral eventuates price movements. And, as elsewhere in the peasant world, Jukumanis use past experiences as a prologue for what the future may hold (Ortiz 1980: 183–5). Mineral prices having changed capriciously in the past, the future market must therefore be equally risky. Mine owners and peasants thus choose simple and inexpensive technology as a hedge against inconstant metal prices, and point to the war years when several of them, in response to growing demand for antimony, hurriedly automated their mills. The drastic collapse of the market for strategic minerals after the first world war caught them by surprise, as it did many other antimony producers in Bolivia, making them 'vow they would never be caught that way again' (Reagan 1941: 42).

VI Risk spreading devices of property holders

Having outlined on a broad canvas the major features of these peasant mining operations, let us now turn to the major question of this chapter: the mechanisms employed by property holders for sharing and shifting risks to peasants.

Portfolio diversification

During the latter part of the eighteenth century, a perceptive observer of the Bolivian mining industry, Pedro Vicente y Dominquez Canete, wrote,

Anyone who knows the large expenditures needed to work silver and gold mines, and the likelihood of losing one's entire fortune if the vein disappears, will honestly confess that it would make more sense to exploit tin, copper, and lead deposits both because they were larger and have lower costs. (Canete 1791: my translation)

When Canete advised his countrymen to embark upon multiple mining enterprises, he perceived, accurately, that holding a mining portfolio with different mineral assets conferred upon the proprietor a larger return. By withdrawing funds from some mines and spreading these savings into other mineral ventures, owners traded smaller average income for a higher certainty of yearly earnings. The variability of total return narrows because the fate of different minerals and mines do not move in identical lockstep fashion.

Following the spirit of Canete's advice, mine owners in the Jukumani territory have divested rather than plough back investments into antimony ventures (Table 8.2).[5] As far back as the last century, a lucky placer miner continued to pan, but used his new wealth to '... buy some culture and cattle and (became) a coca trader'. (Chayanta 1810: 371).[6] A century later, when the mining of tin and tungsten became more profitable than gold, silver, copper or antimony, entrepreneurs began to invest in tin, wolfram, or 'fine buildings in La Paz' (Wepper 1914: 1252; Hess 1921: 497).

An intelligent, resourceful and energetic woman currently living in Llallagua, Julia Medrano, illustrates how an entrepreneur diversifies a portfolio of mining interests. Having inherited from her parents several antimony mines in the Jukumani region, Julia left them unattended until antimony prices increased in the late 1960s and early 1970s. To take advantage of these conditions, she borrowed money from her sisters and mother and offered premium prices for pure ore, ore that needed no further treatment before being sold to smelters. Workers flocked to her mines and, owing to a couple of windfall finds, she was able to make handsome gains. She did not reinvest profits in these mines, but used the proceeds instead to purchase real estate property, trucks, lands and pharmacies in Llallagua and the city of Cochabamba. When antimony prices plummeted in the mid 1970s, she sold and rented some of the concessions to wealthy mine owners, became a tin ore trader (*rescatiri*) in Llallagua, worked her tin mines harder, and entered commercial agriculture, wholesaling potatoes in major markets across the nation and barley to beer factories. By paying property taxes on idle concessions, she has retained title to her more lucrative antimony mines. She sometimes visits her concessions to purchase ore that local peasants may have produced in their spare time, but at present her heart is in tin, barley and real estate, not in antimony.

Maximising exploration acreage and explorers: limits and possibilities

Mine owners, granting prospecting rights to a maximum number of peasants, expect to increase the probabilities of detecting a vein. This is why these mines are buzzing honeycombs. One peasant drives a level deep into the hillside at the footstep of the mountain; another digs an adit a few feet above the first; a third sinks a winze connecting the first two tunnels; a fourth peasant, dissatisfied with the whole thing, hoping to hit bedrock, sinks a shaft from the surface; a fifth drifts randomly following the pay streak; and so it goes. Like scavengers in search of a morsel, some of them are likely to find something – eventually. And then, depending on how much the mine owner helped the worker during the exploration phase, the miner, in the spirit of *ayni* or reciprocity, will offer the ore to the concessionary for sale first.

Unfortunately, exposing the concession to numerous prospectors increases the chances of successful outcomes, but also dilutes the expected returns, for the rights to mineral discoveries depend on the share of exploration risks borne by each party (Godoy 1984b). In maximising both exploration expanse and number of explorers, owners reduce their involvement in each venture. When mineral is found, labour, having shouldered the prospecting risks, is free to trade the ore, and the property holder, until then largely a spectator, wishing to become involved in a joint undertaking, will want to share in the future costs of development and production to obtain equity rights to the mineral.

But if discoveries occur, the bargaining leverage shifts to the peasant, courted by mine owners to carry on a mutual enterprise. A paradox then appears: a peasant, having struck an oreshoot in *A*'s property, exploits it with *B*'s capital and support because, by prior personal and financial ties, the discoverer and patron, linked to each other through multistranded ties, expect to share in the gains from production. Lest the property holder reproach the peasant for 'stealing' mineral in the name of *B*, the peasant can retort, as an 'insolent' Jukumani teenager once did, 'this is my mineral because I found it and extracted it with my own valour'. Bluntly, the ore belonged to the youngster because he bore the risk of exploration. Or, in the language of Bolivian peasants, having had the 'courage' to enter the perilous underground world to search for mineral, they must now be rewarded with complete freeholding rights to the ore.

Only by sharing prospecting costs and risks – handing peasants cash, coca leaves, alcohol, dynamite, carbite – only through these financial and seigneurial gifts can mine owners hope to share the fruits of new discoveries. The strategy of 'the more explorers the better' is feasible only when a mine owner becomes socially and financially involved with the peasant. Otherwise, the bargaining weight tilts to, and remains with the peasants.

To secure rights to the mineral, owners must offer more than gifts and cash; receiving physical inputs for prospecting ventures compels Jukumanis only in a material, not moral, sense to share findings. Indeed, during the first quarter of this century, when international prices for antimony escalated, numerous creditors in Oruro, Llallagua, and La Paz advanced loans in the hope of winning a fast fortune. Not surprisingly, Jukumanis intentionally defaulted; the loan portfolio in arrears skyrocketed, driving dunners into bankruptcy (Voge 1926: 678). In the view of a peasant, why should he share with a mine owner the proceeds from the ore? Sharing is predicated on non-economic mutualities. The real owner not only supplies his peasants with materials but also with goods and services in time of need. The most successful mine owners provide Jukumanis with free lodging when they visit the cities, meals, good coca and medical care. If agricultural products are subject to price controls, these mine owners buy at black market rates. For festivities, they furnish corn flour for making beer, wheat flour for baking bread, cash with which to pay priests, and sundry items such as alcohol, clothing, wool and hats, without which no festival is complete. And during *ch'allas* they display magnanimity by offering free food and drinks to their workers, and wine, confetti and arcane ritual paraphernalia to the mine spirit, *el tío* (affectionately dubbed *niño* Jorge) and to his mistress, Maria Waluka. In the broader context, the peasant worker sees himself as an intermediary between the mine owner and these underground deities who control the supply of minerals. Just as cash, gifts and good will are necessary to make the peasant share the ore with the lawful owner, so also must niño Jorge and Maria Waluka be tendered offerings of confetti (their food), wine (their water), and cigarettes (their incense) before the patrons of the underworld reveal to Jukumanis the location of the hidden lodes. The model mine owner further reinforces his ties to workers through bonds of spiritual kinship (compadrazgo).

Mine rental

Holding legal title to tracts, but limited enthusiasms and funds for exploration, mine owners often lease rights to the claim to other outsiders and thus shift prospecting risks to lessees. In exchange for the privilege to explore and exploit a concession, the lessee pays the property holder in rent 8 to 15 per cent of the gross receipts from mineral sales. There is neither an option to purchase the tract, nor to renew the contract, because owners wish to maximise their take in the aftermath of discoveries, if they occur.

Such rental agreements lead to intense outbursts of work. Tenants, cognizant they can neither buy the mine nor extend the lease, have no abiding interest in the property and hire Jukumanis as wage labourers. Tenants work in their own short-term interest, scouring the concession, skimming its surface for rich patches, and rifling the mine for immediate gains. If ore remains in pillars, cut-fills or dumps, they take them first rather than exploring for new reserves or developing existing ore bodies. It is safer and cheaper to work old and known deposits than to survey for new ones since previous tenants and workers probably high-graded the mine. If tunnels are decaying, if shafts are flooded, if galleries are damp, tenants do not invest in rehabilitating the property because the payback period may be longer than the remaining life of the lease.

If mineable ore is found, short leases permit titleholders to share in the discovery. Having tapped a deposit, the tenant may not necessarily deplete the reserve owing to poor transportation facilities, low metal prices, shortage of labour, lack of knowledge or time. Hence, there may be residual minerals, of perhaps lower tenor, for the mine owner after the lease expires.

Benefiting from other people's discoveries: concession hunters (cazadores de mines)

Mine owners are classic examples of risk-averse individuals who try to profit from the 'spillover' of other people's discoveries (Devarajan and Fisher 1981: 70–1). Paraphrasing Hotelling (1931: 144), some of the concessionary's and peasant's rents come from discoveries made in the subsoil after observing the results of their neighbour's diggings. The economic intuition here is: wait until your neighbour strikes a vein, then explore in an adjacent area or else buy the property. In so doing one shifts the prospecting risks to a neighbour, or, more appropriately, the neighbour, having successfully borne the risks of exploration, gives you a free ride.

In peasant mining the cascading (spillover) effect is more complicated and less detrimental than has been assumed by economists (Peterson 1975, Stiglitz 1975). Looking at the vein, one notes it comes in the shape of an unclasped necklace embellished with randomly located beads. Having discovered a mineralised pocket, there are no wild rushes of prospectors or fledgeling mining companies because the next ore deposit, if there is one, may lie a few feet, yards or miles away in any direction. Unlike the obscure independent oil firm in the midwestern United States watching where the large companies drill successful wildcat wells to claim neighbouring terrain, among Jukumanis there are no such parasitic tendencies because the hidden resource is uneven in quality and distribution.

The cascading effect manifests itself at the level of conflicting claims over newly discovered reserves and at the level of marketing. Once payable ore is found, wily speculators, 'more anxious to sell the ground than to work it themselves' (Preumont 1908: 249), by bribing shyster officials, municipal gendarmes, and the like, try to evict concessionaries, not peasants, to have first purchasing options to the ore and to rent or sell the tract. Low property taxes coupled with the absence of minimum investment and performance standards favour unscrupulous speculation (Gillis 1975). Generally, the properties of mine owners who fall behind paying land rents are targets for takeovers. As Table 8.7 demonstrates, the privilege to work mines has sometimes lapsed into new hands for failure to maintain *patentes*.

Once again one must examine Indian and outsider perceptions of legitimate usufruct rights to see, in fact, where claim jumping is sufficient to allow risk-averse outsiders from enjoying the fruits of other people's discoveries. An

Table 8.7   Lapsing mining concessions in the Jukumani territory, 1900–25

Year	Name	Hectare extension	Location	New claimant		
				Name	Residence	Occupation
1902	Grupo Nuevo	116	Chukiuta	Dalio Fernandez	Sucre	Lawyere
1906	Constancia	1000	Ankayo	Teofilo Lozada	Cochabamba	Lawyer
1906	Santa Rosa	30	Ankayo	Jose Pacheco	Sucre	Merchant
1911	Oriflama	40	Ankayo	Eduardo Delgadillo	Oruru	Lawyer
1922	Seguridad	100	Luluni	Carlos Lemoine	Oruro	Lawyer

Source: Anuario 1900–25; Boletín de Minas 1900–25.

illustration will clarify the point. Two generations ago a *mestizo* from Llallagua, Eustaquio Raminez, claimed a mine close to the village of Wataria, and by offering good prices for the mineral, deigning to become the godfather of children and the *compadre* of workers, established himself as a just-minded owner, a reputation he perfected when villagers, involved in an internecine land dispute with the neighbouring community, began to demand his services, resources, time and links to the outside world. Eustaquio paid for the expenses of lawyers, legal fees, truck fares to courts, and provided needy families with food and cash to carry them through hard times. After his death, Eustaquio's son, German, inherited the mine and lived up to the reputation of a benevolent lord. Like his father, he helped the people of Wataria in their litigations. During a market lull in the 1960s, German, unconcerned with his properties, fell behind in land rental payments, thereby forfeiting his property rights, and a speculator, Juana Veizaga, also from Llallagua, wishing to acquire the devolving property, claimed the lapsing tract. Surprisingly, villagers reacted harshly to Juana's move and halted her from finalising her wish. The mine was theirs, they told her, and German the rightful owner 'because he and his father had paid for the *ch'ajwa* (land dispute)'. No, Juana could not replace German, unless he wished it so. Juana and German agreed to divide the work sites.

Capturing the value added of your neighbour's discovery is not simple: since the mineralisation is irregular one cannot sit and wait until a neighbour strikes a vein; it is not sufficient to evict the owner because villagers must approve the legitimacy of the new claimant; at a lower level, you cannot expel the peasant who discovered the shoot, otherwise you would have a small revolt in your hands, as once almost happened during field work. How then does an owner transfer the risks of exploration to someone else while, at the same time, reaping the benefits of new discoveries? Through the black market of minerals or *rescate*.

Rescate

The poor man's medium for shifting (rarely sharing) prospecting and marketing risks is rescate. In its classic form in the Jukumani territory, as elsewhere in Bolivia, the *rescatiri* is an individual with limited assets: cash to buy ore, three scales (two defective ones for buying and selling, and a good one for the police), coca leaves, bread, sugar, beer and coffee to clinch bargains. Operating from their homes in the Jukumani territory, they have no overhead liabilities and purchase from peasants only the finest of ores. Looked at broadly, rescatiris are intermediaries between petty commodity producers and the smelters or state mineral agencies; viewed myopically, they are interstitial links wedged between peasants and owners, though rescatiris are themselves often proprietors as well.

Rescatiris simply buy ore. Though they enjoyed legal status prior to the 1952 revolution, they were outlawed after nationalisation under the supposition that they were purchasing mineral from the newly nationalised mines. From a legal perspective, rescate foments theft; from a peasant's viewpoint, rescatiris are simply traders purchasing mineral that would otherwise remain in the ground. Like a problem child, rescate, despite its roots in colonial history, is viewed as an embarrassment by the Bolivian government, acknowledged in special circumstances, but usually blocked from consciousness in its waking hours.

From a rescatiri's perspective, the institution is ideally suited to an impoverished country, since mineral accrues without having to buy machinery or tools, or incurring fixed payroll liabilities.[7] Not advancing risk capital for exploration for fear of delinquency, rescatiris shoulder none of the prospecting risks.[8] Their local purchasing price is pegged to the London metal exchange quotation, rising and falling with international movements. Rescatiris thus shift systematic marketing risks to peasants and maintain income levels constant.

VII Conclusions

Like most entrepreneurs in developing countries, mine owners in the Jukumani territory attempt to shift and share risks. Except for complete divestiture outside the land of the Jukumanis, all other forms of apportioning risk force proprietors to enter the moral universe of peasants. The security-minded mine owner cannot claim the tract of a successful neighbour unless villagers approve the move; the owner cannot claim a share of equity to the mineral produced in his/her own property unless they participate in the venture materially, financially and morally and prove themselves munificent patrons. Although mine owners try to increase their stock of wealth and minimise uncertainty through such mechanisms as rescate and speculation, these strategies are constrained and moulded by the character of local culture and the ideology of Jukumani society. It is as though these capitalist entrepreneurs were adapting to the crucible of local folklore, rather than the other way around.

Notes

1 I would like to thank Susan Eckstein, Tom Greaves, and Sutti Ortiz for having read and commented on earlier versions of this essay.

2 *Cf.* Cancian (1972, 1979), Ortiz (1970, 1980), Chibnik (1980), Barlett (1980, 1982), Johnston (1971).

3 Even in the United States more than half of the large scale mining ventures grew out of enterprises initiated by small-scale prospectors (SSTS 1982: 11).

4 These are complex issues which are treated at length elsewhere (Godoy 1984b). But let me introduce a clarification. There are three actors to a small-scale Bolivian mining operation: labour, proprietor and ore regrator (buyer). The degree to which property holders attempt to exercise exclusive rights over their mining concessions depends on the size and quality of the reserve and the political instability of the country. In the area under consideration, the Jukumani territory, during the period of fieldwork (1979–81), antimony prices were unusually low. Consequently, proprietors threw open the mines to any person willing to search and produce ore. It therefore followed in the moral understanding of the Jukumanis, that, if they bore the geological risks and the costs of production, they also had the derivative right to market the ore at their own discretion, irrespective of who the rightful proprietor was. As we shall see below, the ore produced by peasants was sold by them to antimony-ore regrators (rescatiris).

5 I am only suggesting that portfolio diversification may reduce risk. The causes of portfolio diversification, as of many risk-reducing mechanisms, are unknown, but one of the consequences of portfolio diversification may be to reduce the variance of overall income stream (Berry 1980, Ortiz 1980).

6 Portfolio diversification among Bolivian mining entrepreneurs dates back to the eighteenth century, if not earlier. The wealthier and more successful businessmen achieved considerable vertical integration by moving into smelting and marketing, and acquiring agricultural properties to supply food and other inputs to their mines. (Bakewell 1973).

7 During Colombia's colonial era, gold dust traders (*rescatadores*) likewise captured the auriferous production of independent itinerant panners known as *mazamorreros* (Twinman 1976). Brading (1971: 149) discusses the role of rescatadores during Bourbon Mexico.

8 Though prior to the 1952 revolution they were among the most efficient creditors in the mining sector. (Ibañez 1943: 51–2).

References   * *See annotated bibliography for full details of all asterisked titles*

Anuario
1900–25 Anuario de Leyes, Decretos y Resoluciones Supremas de 1900–1925. Sucre: Archivo Nacional de Bolivia.

Association of Geoscientists for International Development (AGID)
1982 'Small scale mining'. *AGID News*, 30.

Bakewell, P. J.
1973 *Antonio Lopez de Quiroga, Industrial Minero del Potosi Colonial*. Potosi: Universitaria.

Banco Minero de Bolivia (BAMIN)
1966–75 *Revista Minera BAMIN*. La Paz.
1966–75a *Memorial Anual*. La Paz.

Barlett, Peggy F.
1980 'Adaptive strategies in peasant agricultural production'. *Annual Review of Anthropology* 9: 545–73.
1982 *Agricultural Choice and Change, Decision Making in a Costa Rican Community*. New Brunswick, New Jersey: Rutgers University Press.

Baxter, Michael William Peary
1975 'Garimpeiros of Poxoreo: small scale diamond miners and their environment in Brazil'. PhD dissertation, University of California, Berkeley, California.

Berry, Sara
1980 'Decision making and policymaking in rural development'. In *Agricultural Decision Making*. Peggy Barlett, ed. New York: Academic Press. Pp. 321–36.

Boletín de Minas
1900–25 *Boletín de Minas*. Sucre: Archivo Nacional de Bolivia.

Bolivia
1977 *Resultados Provisionales Total del País*. La Paz: Ministerio de Planamiento y Coordinación.

Brading, D. A.
1971*

Cancian, Frank
1972 *Change and Uncertainty in a Peasant Economy*. Stanford: Stanford University Press.
1979 *The Innovators' Situation: Upper Middle Class Conservatism in Agricultural Communities*. Stanford: Stanford University Press.

Canete, Pedro Vicente
1791 *Guía Histórica, Geográfica, Física, Política, Civil y Legal del Gobierno e Intendencia de la Provincia de Potosi*. Potosi: Editorial Potosi.

Chayanta
1810 *Razón Individual Que Remite al Gobierno …* Minas 2173, Tomo 81, V. Sucre: Archivo Nacional de Bolivia.

Chibnik, Michael
1980 'Working out or working in: the choice between wage labor and cash cropping in rural Belize'. *American Ethnologist* 7(1): 86–105.

Deere, Carmen Diana, and Robert Wasserstrom
1981 'Ingreso familiar y trabajo no agricola entre los pequeños productores de America Latina. In *Agricultura de ladera en America Tropical*. Informe Tecnico no. 11. Novoa A. R. and J. L. Posner, eds. Turrialba, Costa Rica: Centro Agronómico Tropical de Investigación y Enseñanza. Pp. 151–67.

Devarajan, Shantayanan, and Anthrony C. Fisher
1981 'Hotelling's 'economics of exhaustible resources': fifty years later'. *Journal of Economic Literature* 19(1): 65–73.

Eicher, Carl K., and Doyle C. Baker
1982 'Research on agricultural development in sub-Saharan Africa: a critical survey'. International Development Paper 1. East Lansing: Michigan State University.

Firth, Raymond
1950 'The peasantry of South East Asia'. *International Affairs* 26(3): 503–12.

Foster, George
1967 'What is a Peasant?' In *Peasant Society, a Reader*. Jack M. Potter, May N. Díaz, and George M. Foster, eds. Boston: The Little, Brown Series in Anthropology. Pp. 2–14.

Gillis, Malcolm *et al.*
1975*

Godoy, Ricardo
1984a 'Peasant mining: small scale mining among the Jukumani Indians of Bolivia'. Manuscript.
1984b 'Risk and moral contract in Bolivian peasant mining'. In *Research in Economic Anthropology*. Barry Isaac, ed. JAI Press, forthcoming.

Hess, Frank L.
1921 'Some unique Bolivian tungsten deposits'. *The Engineering and Mining Journal* 112: 429–9.

Hotelling, Harold
1931 'The economics of exhaustible resources'. *Journal of Political Economy* 39(2): 137–75.

Ibanez, D.C.
1943 *Historia Mineral de Bolivia*. Antofagasta, Chile: Imprenta MacFarlane.

Johnston, Allen
1971 *Sharecroppers of the Sertão*. Stanford: Stanford University Press.

Johnston, Bruce, and William C. Clark
1982 *Redesigning Rural Development: A Strategic Perspective*. Baltimore: The Johns Hopkins University Press.

Metal Bulletin
1966–80 Handbooks. London: Metal Bulleting Limited.

Nash, June C.
1972*

Neilson, James M., ed.
1982 'Strategies for small-scale mining and mineral industries'. Report No. 8. Bangkok: Association of Geoscientists for International Development.

Ortiz, Sutti
1970 'The structure of decision making among Indians of Colombia'. In *Themes in Economic Anthropology*. Raymond Firth, ed. ASA Monograph 6. London: Tavistock Publication. Pp. 191–228.
1980 'Forecasts, decisions, and the farmer's response to uncertain environments'. In *Agricultural Decision Making*. Peggy Barlett, ed. New York: Academic Press. Pp. 177–202.

Orwell, George
1937 *The Road to Wigan Pier*. New York: Harcourt Brace Jovanovich.

Peterson, Frederick M.
1975 'Two externalities in petroleum exploration'. In *Studies in Energy Tax Policy*. Gerar M. Brannon, ed. Cambridge: Ballinger.

Preumont, G.
1980 'Northern tin fields of Bolivia'. *The Mining Journal* 83(160): 249, 314.

Reagan, P.H.
1941 'Expansion of mining in Bolivia awaits mechanization'. *Engineering and Mining Journal* 142(5): 41–3.

Redfield, Robert
1956 *Peasant Society and Culture*. Chicago: The University of Chicago Press.

Shanin, Teodor
1971 'Introduction'. In *Peasants and Peasant Societies*. Teodor Shanin, ed. Baltimore: Penguin Books. Pp. 11–19.

Silverman, Sydel
1979 'The Peasant concept in anthropology'. *The Journal of Peasant Studies* 7(1): 49–59.

Stiglitz, Joseph E.
1975 'The efficiency of market prices in long-run allocation in the oil industry'. In *Studies in Energy Tax Policy*. Gerar M. Brannon, ed. Cambridge, Ballinger.

Subcommittee on Science, Technology, and Space (SSTS)
1982 *An Assessment of Factors Affecting Small Mining and Custom Milling and Smelting Operations in the Western United States*. Subcommittee on Science, Technology, and Space of the Committee on Commerce, Science and Transportation, United States Senate, 97th Congress, 2nd Session. Washington, DC: US Government Printing Office.

Tenth World Mining Congress (TWMC)
1979 *Tenth World Mining Congress*. Istanbul, Turkey. 2.

Twinam, Ann
1976*

United Nations
1970 *Mineral Resources Development With Particular Reference to the Developing Countries*. Department of Economic and Social Affairs.
1972 *Small Scale Mining in the Developing Countries*.
1978 *Small Scale Mining of the World*. Conference, JURICA. Queretaro, Mexico. Institute for Training and Research. November 1978.

Voge, Law
1926 'Antimony in Bolivia'. *Engineering and Mining Journal* 121: 677–80.

Wells, Louis T.
1975 'Economic man and engineering man: choice of technology in a low-wage country'. In *The Choice of Technology in Developing Countries (Some Cautionary Tales)*. Peter Timmer, ed. Cambridge: Harvard University, Center for International Affairs. Pp. 71–98.

Wepper, G. W.
1914 'Tungsten in Bolivia'. *The Engineering and Mining Journal* 97(7): 1251–2.

Wolf, Eric R.
1955 'Types of Latin American peasantry: a preliminary discussion'. *American Anthropologists* 57(3): 452–71.
1969 *Peasant Wars of the Twentieth Century*. New York: Harper Torchbook.

World Bank
1976 *Present Position and Prospects of the Mining and Metallurgical Sector of Bolivia*. Washington, DC.

CHAPTER 9

Industrial stagnation and women's strategies for survival at the siglo XX and Uncía mines

GUILLERMO DELGADO P.

Women who are the spouses of miners are under-recognised for their contribution to the viability of miners, both politically and economically. Nowhere has this been more apparent than in the violence-prone, state-owned mines of Bolivia, where the women of the mine camps have militated for action from the unions, from the mine management and from the government. On occasion they have been killed, as was María Barzola, martyred at the great tin mine, Siglo XX-Catavi.

This paper presents mining life as a environment in which coping strategies are employed by mining men and women, focusing on the women. It argues that women harbour a different mix of these strategies from those used by men, and proposes that the women's strategies can be traced to peasant origins. This is especially true of women in the mining families in the most marginal economic conditions.

Guillermo Delgado P. (Department of Anthropology, University of Texas, Austin, TX 78712, USA) is completing a doctorate on the basis of his participant research in a major Bolivian mine, including employment as a *carrero* (car hauler), the lowest job in the mining job structure. He is author of *Comentario a la Bibliografía del Movimiento Obrero Boliviano* (forthcoming, La Paz) and editor of *Cien Años de Lucha Obrera en Bolivia, 1876–1976* by Trifonio Delgado (1984).
— The editors

This paper advances the argument that the chronic economic crises in Bolivia tend to push the mining population, especially the female sector, toward reliance on survival strategies of peasant origin. For example, at first glance the COMIBOL-owned[1] Siglo XX-Catavi mining complex gives the impression of operating in a modern, well-organised, business-like manner. However, its failure as a rationally organised capitalist enterprise is revealed by the continuing persistence and reactivation of many coping strategies among its workers and their families. Among these strategies are a widespread reliance on gardening and peasant mechanisms of exchange, *juqueo* (mineral theft), *rescate* (ore collection outside managerial control), and *pasanaku* (an informal game women play in order to obtain cheaper commodities by circulation a pooled sum of money, allowing only one to benefit at a time).

While it is difficult to demonstrate with certainty, my observations at Siglo XX-Catavi suggest that the coping strategies of mine women are more often derived from agricultural peasants than are those employed by men. In Bolivia, miners overwhelmingly come out of peasant backgrounds, though in the several largest mines where the size of the labour force has been in long-term decline the connection to peasant ancestry may be two or three generations back. It may be, however, that women in peasant households and women in mining households find the *same* coping strategies workable, and, so, there has been more persistence of the peasant heritage among women than among men. Indeed, I would argue that the more impoverished the household, the more pronounced the tendency to maintain peasant strategies. It is possible, however, that this may be changing: new pressures of the consumer culture are an additional threat to women's solidarity. Mine women appear to have begun a process of proletarianisation. Too, they have increased their participation in the market, augmenting their household responsibilities with those of running a small business, while, also, maintaining their critical militancy through the housewives' union. Only since 1961 has their union gained legitimacy as a political party and been recognised as an important part of miners' struggles, although plagued by widespread belittling of their political potential by the men.

Can it be said, then, that strategies of peasant origin are reactivated by the women in mining populations? If so, which women reactivate them? And why or under what kind of circumstance? In order to develop my answers I will divide the discussion into three parts: (1) peasant antecedents to strategies of miner adaptation; (2) waged miners and unwaged women; and (3) women's reactivation of peasant strategies to cope with mining life.

I Peasant antecedents in strategies of miner adaptation

The great tin centre of Siglo XX-Uncia-Catavi entered the circle of mining life at the turn of the century after the drop in silver on the international market. Still earlier, two silver mines gave employment to the area's population; these were the Colquechaca and Aullagas (1865 and 1871). In the mid-nineteenth century, the mining elite of these mines numbered two hundred and twenty eight individuals who hired around 9,000 workers of peasant origin (Dalence 1851: 236–7). There has been a bitter tradition of capitalist presence in the area since the penetration of national capital, accelerated by the construction of railroads from Antofagasta (Chile) in 1889. In 1892 the railway reached Oruro, the mining centre where most production of the Northern Potosi mines was concentrated. According to Antonio Mitre, by 1870, 43 per cent of the work force in Pulacayo consisted of women, who also received the lowest wages (Mitre 1981: 146). Through this period, women were slowly being replaced by machinery in the mines,[2] and, as a response, they were building business in the trade of commodities available regularly through the *pulpería* (company store), but in scarce supply in the peasant villages. Women of miner families were sometimes able to accumulate reliable supplies of these commodities and barter with peasant clients. Thus the income earners in the mine camp at century's end included not only miners, but also these women traders, attached as spouses to the mining households. This practice still goes on at the Siglo XX mine. The presence of Quechua, Aymara and Spanish speakers in these early mining camps was thus due not only to workers, but also to businesswomen.

In current times, a practice of borrowing and sharing has developed that derived from the peasant *manay* practice. At present women in the mining camps use the term and practice *manarimuway*. One can easily circulate a 'borrowed bone' for soup from household to household. Often children are sent by their parents to the neighbour's door to borrow odds and ends, from blankets to money to perishable food. The items are returned after use, and in the case of food, replaced as soon as the family gets items from the *pulperia*, the peasant market or the city. The sharing of food is especially important at the mine. Whether in the celebration of fiesta days or at wakes it is closely linked to ritual peasant practices in the fields.

Not unrelated to these ritual peasant practices, gardening at the mining enclave is tied to the idea of sharing an abundance of tubers or grains.[3] Potato, *oca*, and *quinua* planting has become important for some generations of women at the camp. Some women in Uncia, Siglo XX, Cancaniri and Miraflores (Empresa Minera Catavi) devote some time to gardening in limited plots which belong to the company. A few share gardens outside the mining area with *comuneros* (recognised members of peasant communities).

Another interesting interaction between peasants and miners of the lower strata is evident at the *Tinku* celebration in the peasant villages of Chayanta, Aymaya, San Pedro and Amayapampa. Though no longer peasants themselves, the lowest strata in the mining camp show the greatest solidarity with the peasant class. Many of the women who garden have informal con- nections to land, and many wives of miners in the lowest strata of the mining camp still maintain ritual kinship ties with the peasantry, which – some- times – become exploitative of peasants (I observed five such cases among some *locatario* miners). A long period of stability at the mine tends to recreate linkages with the peasantry. With political unrest peasants tend to withdraw from the scene and relations become tense between the groups. Because political upheavals make a difference in the area, adding tension to relations between peasants and miners, women are caught in a system where they do not have definite political power. The independent *locatarios*, *venero* workers, *lameros* or *canaleta* workers, who do not necessarily belong to the company, mediate between the full-time miners and peasants. Wives of these 'surface' miners often work alongside their husbands, and because of poorer conditions they are more inclined to be accepted by peasants, who can even make fun of them. The degree of peasant–miner interface changes radically with fluctuations in the local levels of political participation. The distance between the 'señoras' at the mining camp and Indian women is great and reduced only through mediation by women of intermediate rank and under stable conditions. Class and ethnicity are at the base of the conflict.

Underground mining activities are destructive and devastating and males as a rule are openly exposed to such experiences (called by miners 'chasing death on a daily basis'), while females are in charge of and occupy the reproductive role. They constitute the unwaged workers at home (Cleaver 1979: 160). They are also in charge of maintaining the male labour force through their nurturing role. It is due to their unwaged worker status at home that women are able to avoid the fatal confrontations of the mine, the underground, lung diseases, *viento, susto, jap'eqa*, TBC, silicosis, colds, paranoia, depression, anguish and alcoholism. The miners have become proletarianised, but the miners' wives have entered a stagnated phase in the process of semi-proletarianisation. This allows them, according to their degree of pauperisation, the flexible use of peasant strategies as ways of coping with mining life.

II Waged miners and unwaged women

A miner's wife in 1959 cooked with *yareta* (*Bolax* and *Azorella*) firewood. She took almost the whole day to keep the fire alive so that she could have a warm dish of food for her family. Despite kerosene's wide distribution, poor households continued to use *yareta* until 1965. Women were able to cope with kitchen duties by sharing those responsibilities with extended kin or other neighbours. Time was invested also in the exchange of *yareta* with peasants, and to fulfil other needs such as washing, child rearing and carrying water. In 1980 women used gas stoves, but their other household duties remained as hard as before to fulfil. Women in depressed areas of mining camps (the poorest of the poor) consider gardening a resource for survival. These poverty sections can be seen at Cancaniri, Uncia, Morococala, Chocaya-Siete Suyus, Colquiri-Huayrapata, Pacuni-Caracoles and among their residents one observes gardening and even herding (such as in Atocha, Telamayu).

Most of these women belong to impoverished households, and most of them have picked up their agricultural knowledge from contact with the land or the peasantry in years past. While none of them own land, they are able to use land available from the company. In Santa Fe-Morococala occasionally they enter into '*al partir*' (sharecropping) relations with peasants, especially when there is a lack of cash or food at the camp. The *al partir* system means that peasants allow women to help with the duties of tilling, planting, harvesting, choosing seeds, *ch'unu* making and land rotation. Most older women in the Siglo XX camp, as well as Santa Fe, Morococala, Siete Suyus, Telamayu, Huayrapata (Colquiri), and Pacuni (Caracoles), know how to make *ch'unu*. Sometimes they are called to advise or teach the process to younger women. In Siglo XX itself there are few plots. Women engage in these land activities only when they will be planning for an important celebration (such as a wedding, baptism, or a widow's ritual emergence from the mourning period) for family and friends. It is a way of getting a large amount of tubers or grains to feed friends and relatives.

Couples who are recently wed and belong to the lower strata of the mining camp have closer ties to the peasantry. Often these relations fall under *compadrazgo* (ritual kinship based on godparenting), which can be used by peasants to get into independent mining activities at the camp (*rescate, juqueo*, grinding ore for monthly *entregas* (product shares)), or may be applied in reverse to enable a miner's wife to find work or easy barter of products in peasant areas. The level of poverty of unwaged women forces them to make concessions, so that some women look to agriculture before thinking about moving to the city. This is so particularly of widows and married women with marital problems.

Elba María, from Cancaniri, is one such woman. Remarried and lacking a formal education she looks to her brother's land in the Cochabamba Valley as a way of finding security every time relations with her husband Tomás sour. Tomás is a *jefe de punta* (shift boss), a position of high responsibility and prestige. He has to support his mother, Benigna, who, because she lost her husband ten years before, also lost rights to a household. The sudden death of her husband resulted in depression, social maladjustment, and the failure of her business at the Llallagua market. According to Elba María, who knew Benigna before she married Tomás, Benigna was unable to recuperate economically. Benigna now lives on charity from her three sons, does not have a place of her own, and feels rejected. Being the spouse of a mechanic, and a business woman, her ties with the peasantry were extremely weak, and she now has completely cut all ties with the peasantry, thereby removing any possibility of returning to the land. Interestingly enough, however, she will not hesitate to borrow money from her acquaintances at the mine to be able to pay for a 'luxurious' wedding for her last male child.

Tomás, the *jefe de punta*, is not well accepted by his extended family because Elba María is more 'campesina' than they. Elba María, because of her ties to the land in Cochabamba, is better able to cope with life than other females who sometimes face the same problems at the camp. The discrimination Elba María suffers has reinforced her relationship with the land. Her brother welcomes her every time she escapes from the mining camp to the valley. Elba María's brother was a *perforista* (driller) at the Siglo XX mine, but he chose to go back to the land as soon as his father died.

Domestic work such as child-bearing, nurture and the reproduction of labour power are female duties. To fulfil these duties at the mining camp some women take advantage of certain natural resources. Resources from the herbalist tradition are in common use. For certain diseases of a major clinical nature, such as those requiring long-term treatment, women and miners rely on the recommendation of western-type doctors and hospitals. But for mild diseases of a more psychosomatic nature, they still listen to native healers' and herbalists' recommendations. Women with knowledge of native medicine in Siglo XX or Santa Fe are not uncommon.

III Women reactivate peasant strategies

Nicolasa is an unusual illustration of Andean women at the mining camp making use of peasant-based adaptions. Her husband, an old-time miner with silicosis, has got a position as a door keeper at the mine's entrance at Socavon Patino (Uncia). He leaves home for his job at 6.15 am to work the *primera punta*, the first in the three-shift, twenty-four-hour cycle. Nicolasa prepares a food dish, *quni quqawi*, which she places in the *vianda* wrapped in woven *unkuna* of llama wool for warmth. In 1940, Nicolasa inherited her mother's home at the Uncia mine, where she lives with her husband and a *Charka* peasant boy who helps her with small jobs. Missing the land, Nicolasa was able to locate her mother's *chaqra*; it is a five-hour walk from the mining town at the foot of the Qala Cruz peak. While she does not have any *de jure* possession of the land because it legally belongs to the *comuneros* from Sunujuyu, the space has been used *de facto* since her mother entered exchange agreements with the *comuneros*. According to Nicolasa, her mother practiced the same strategy of shifting 'islands' of cultivation each year in this ecozone.

Her sporadic relationship with the land is not limited to the five-hour hike to the *chaqra* in the *puna* (ecological niche above 4,000 m). Nicolasa has inherited not only the plot rotation technique but also the knowledge of the ecosystem's output. Her knowledge of the ecosystem may be considered as a subsistence resource, a version of ecological control. From the political centre, Uncia, she is able to establish ties with the surrounding peasant communities and exchange several products for salt, fire wood, *quinua, tarwi, ch'unu, ch'arqui* and *quwikuna*. For instance, she trades dehydrated potatoes (*ch'unu*) which are cooked ceremonially during certain times of the year.

In addition to using her knowledge of the Andean high-altitude ecosystem as a subsistence resource, Nicolasa uses it to reinforce ritual ties of mutual aid. An example of this occurred with the up-coming wedding of one of her sons, a *contratista* miner at the *canaletas*, close to Socavon Patino, who had just passed his *tantasqa* (*sirwinaku*) state. Nicolasa put together products originating from different peasant areas, telling me she was interested in gathering all plants grown in the area in hopes that the assemblage might produce a cosmic equilibrium '*waqachinanpaq, qallarinanpaq*' providing the couple with products for a whole, tranquil life. With this in mind, during the year prior to the wedding, Nicolasa accumulated small quantities of cere-monially valued products such as: *ch'arqui, tarwi, ch'ichi mut'i, millma, t'unta, k'awi, haba, quinua*, and wheat (for *llusp'achi*). She was familiar with the places one could collect *llullucha* and *sik'i* as well as products not easily accessible due to distance. She knew that *coca* coming from Chapare was *k'allku* (bitter) and less appreciated by coca-chewers. Through second-hand references and her own experience she acquainted herself with the importance that shape, content, colour, flavour and thickness make in selecting the best coca leaves. Those turn out to be the ones from Yungas, 'the same ones my husband takes to the *Tio*, the sweet ones he likes'.

Unwaged women coping with mining life remain doubtful of becoming completely integrated into the proletarian process. They have learned to unearth and use resources that appear to be lacking from the mining life style. Child-rearing limits many women's capabilities and mobility; it appears that some of the older generations looked at the land as a real way of spending happy days after finishing their family raising years. With the resettlement of some mining families of *rentistas* (retired miners) in the Chimpa area of the city of Cochabamba, it was observed that gardening is an activity of many of those households of retired miners. A couple of them have even agreed to run a small business raising chickens. However, the closest ties to peasant culture for men remain limited to short trips to the *Tinku fiesta* at Chayanta, Aymaya, or San Pedro where some miners arrive in peasant *bayeta de la tierra* garments. Despite the years, the *Tinku* continues at Chayanta and, from time to time, acts as a sort of ethnic identity ritual.

IV Conclusion

In general, it can be said that most miners have become more interested in the cities or other mining centres than in returning to the fields. Nevertheless, due to long existing problems in the Bolivian economy, the lower strata of the working class at the mines seems pushed to scratch a meagre living from resources of the resilient peasant culture and economy surrounding the Siglo XX-Catavi mining complex. At the same time, the solidarity among women created by the use of these peasant coping strategies is undermined by pressures of the capitalist system such as new technology in the mines, competition for wages, consumption of sumptious goods, educational opportunities, and even for the means of subsistence. Scarcity, especially, tends to challenge community ties and may weaken kinship relations. Nevertheless, women are able to establish woman-to-woman communications which enable them to assist each other in times of stress or when coping with scarcity. Peasants at the same time have entered another level of political maturation and are able to engage in dialogue with miners. These movements occur largely due to the economic role of women in the mines. The high degree of pauperisation pushes the women to look for the best survival strategies. Coping through peasant strategies results in a positive gain; *juqueo*, *rescate*, reciprocal exchanges and gardening (wherever it is viable), will keep the mining population alive, but tin will still be Bolivia's wage.

Notes

1 COMIBOL (Corporacion Minera de Bolivia) is the state-owned mining corporation administering Bolivia's largest mines.

2 'Cornish' water pumps, and 'Whorthington' motors arrived in the Bolivian mines in 1898. Other brands came from England: 'Homman Brothers', 'Deissel', 'Sulzer SA Winterur'; and German 'Imperator' motors. In 1900 tin production was 9,739 tons; by 1918 Bolivia exported 29,280 tons. See Albarracin Millan 1972.

3 I use the term 'gardening' to stress the difference between *farming* and a more restricted agricultural activity current in the mines. This gardening is limited in the nationalised mining, more noticeable among medium miners, and extensive among *mineros chicos* (small mining).

Glossary

Bayeta de tierra Loose woven material used for underwear. At *Tinku* men's pants are rolled up to show the bayeta leggings.

Ch'arki Quechua = dehydrated meat.

Ch'ichi mutiy Quechua. Boiled lupinus.

Ch'unu Dehydrated potatoes. Can be stored for several years.

Huelgas Strikes. There are several kinds: huelga de brazos caídos, de hambre, escalonadas, de protesta, de horas, días, semanas, de rechazo, indefinida, de advertencia, parcial, general, repudio, etc.

Juqueo From the Quechua, juku = owl. It is an old practice of direct appropriation of ores. The juku, usually a waged worker, remains underground after his eight hour shift. He is capable of working for himself some eight more hours. Risks and danger are very high.

K'awi Quechua. Dehydrated oca.

Lameros or canaletas Independent workers. They collect ores from the small river's water waste flowing from the ore concentration plant.

Locatarios Spanish term. Individual and independent miners who rent a space from the mine in order to look for ores. They independently hire seasonal work force from the peasantry.

Llullucha, sik'i Quechua. A kind of water 'cress'. *Porphyra columbina*.

Llusp'achi Quechua. Boiled wheat soup. Mourning dish.

Millma Quechua. Wool.

Manay, manarimuway Quechua, to borrow. Attached to this term and practice is a meaning of reciprocity and retribution.

Oca Andean tuber. *Oxalis tuberosa*. Once removed from soil, it is necessary to expose them to the sun light and heat in order to sweeten. This belief is extended throughout the Andes.

Pasanaku Women's groups of ten get together for the purpose of diminishing expenses by putting together money to buy commodities at cheaper prices. It is seen as a type of game belonging to women.

Qallarinanpaq Quechua. To start.

Quinua Andean grain, *Chenopodium*.

Quwikuna Guinea pigs (an important festival meat delicacy).

Quini quqawi Warmth food.

Rentistas Spanish. Retired miners.

Rescate Lit. to rescue. Private ore collection; an arrangement among independent miners and wealthy ore buyers. Sometimes rescatiris (buyers) get the ore from jukus.

Tantasqa, Sirwinaku Quechua. Trial marriage. Lit. = together.

Tarwi Quechua. *Lupinus mutabilis.*

Tinku Andean blood ritual. Quechua term for 'encounter', it is also associated with the idea of the centre of a cross.

T'unta Quechua. A variety of dehydrated potato.

Unkuna Square woven cloth, usually of llama wool, used to wrap the vianda food containers to preserve food's warmth.

Veneros Independent workers. The ore they collect is sold to the Bolivian Ore Bank (BAMIN).

Vianda A fitted, covered tier of small metal pots used to pack a meal for eating away from home (typically at work).

Viento, susto, jap'eqa Psychosomatic diseases.

Waqachinanpaq Quechua. To keep.

Yareta Firewood, *Bolax* and *Azorella.*

References   * *See annotated bibliography for full details of all asterisked titles*

Albarracín Millán, Juan
1972 *El Poder Minero*. La Paz: Urquizo.

Cleaver, Harry
1979 *Reading Capital Politically*. Austin: University of Texas Press.

Dalence, Jose Maria
1851 *Bosquejo Estadístico de Bolivia*. Sucre: Imprenta de Sucre.

Mitre, Antonio
1981 *Los Patriarcas de la Plata. Estructura Socio-Económica de la Mineria Boliviana del Siglo XIX*. Lima: Instituto de Estudios Peruanos.

CHAPTER 10

Becoming a tin miner

THOMAS C. GREAVES, XAVIER ALBO
and GODOFREDO SANDOVAL S.[1]

Scholarly effort has been invested for more than a century in denoting
and understanding the relationship between peasants and proletarians.
The fate of guerrilla movements and political theorists has likewise
hinged on the propensities of each for political action. Bolivia supplies
an unrivalled opportunity to research the fundamental commonality or
distance between the peasant and miner, given that miners such as
those at 'Bocasa' in this study are substantially proletarianised, yet
many are recruited from isolated agricultural hamlets.

The authors' data illumine the transition from peasant to proletarian miner. The data indicate that the transition is quick, and
essentially trauma free. Little difference in attitude and group identity
is observed, whether a miner has been on the job for one year or
forty. The implications for our understanding of peasants and proletarians are explored.

Thomas Greaves, Xavier Albó and Godofredo Sandóval have
collaborated on various publications stemming from their joint research
at Bocasa and other Bolivian localities. Greaves (Trinity University,
San Antonio, TX USA) focuses on the anthropology of work and is
editor-in-chief of the *American Anthropologist*. Albó is a Bolivian
applied anthropologist and a director of the Centro de Investigación y
Promoción de Campesinado in La Paz (Casilla 5854, La Paz).
Sandóval (address c/o Albó) is a Bolivian applied sociologist
completing his degree at Louvain, Belgium.

– *The editors*

Miners the world over have been noted for their solidarity, their readiness to
assert and defend their collective interests, and for the frequency with which
their labour relations become political actions. The continuing pro-Solidarity
struggles of the Silesian coal miners are a current example, as are those of black
South African gold miners and collierymen of Yorkshire. One can, with
reasonable security, suggest that miners everywhere are prone to a strident
and stormy character. In our literature such aggressive labour unity has often
been attributed to the work environment, which isolates teams of men who
must depend on each other for their very survival, while counterposing them
to a management that appears (and often is) unmoved by the brutalising nature
of the toil and omnipresent mortal danger.[2]

Whether it is these or other factors which give rise to the solidarity and
proletarian characteristics of industrial miners, one would be hard pressed

to name another kind of worker with stronger claims to class consciousness, and with greater readiness to translate class consciousness into class action. Miners, we think, present a type case for examing the nexus between pro-letarianisation as an analytic concept and the ideological and behavioural realities of those who wrest minerals from the ground.[3] If we are to observe strident proletarians in a real world setting, miners are a likely choice.

Miners are fascinating for many reasons. For example, one cannot avoid noticing that in many countries the miners' rather singular militancy is about the only force sufficient to challenge the power of national and international elites to determine national form and the distribution of economic justice. From where does this militancy derive? When someone becomes a miner, how quickly does one acquire the individual loyalties and identities which undergird the unity of miners as a group?

Although miners come from many backgrounds, our interest has been with miners who come out of *peasant* backgrounds. In the usual case, this transition involves the replacement of the peasant's locality-based self-identity, un-conducive to long-term collective confrontation, with a vigorous class-based identity laced with militancy, a willingness to incur substantial personal sacrifice in the name of class-based objectives, and violence. The high contrast between peasant comportment and miner comportment makes this transition particularly instructive as one version of becoming a proletarian.

Bolivia is a country where this transition is especially vivid. Unionised Bolivian miners are highly proletarianised. Bolivia is perhaps the extreme case of a country where the possibility of the miners' political intervention must figure in every significant political stratagem of the government.[4] Bolivia is exceptional in that the ore produced by miners supplies most of the nation's foreign currency in a country dependent on such currencies for the import of virtually all manufactured goods (see Chapter 7 for further elaboration). Bolivia is a country with vigorous and insistent labour leaders usually facing military governments which most often regard these leaders as dangerous adversaries. And it is a country where mining is as harsh, as dangerous, as abrasive a livelihood one can find in mining anywhere.

As we have said, our interest goes beyond describing the features of proletarian behaviours to the process by which these behaviours are acquired. Here, too, Bolivia offers fruitful possibilities. A number of rather sizable Bolivian mines draw much of their labour force from isolated peasant areas. While the peasantry of Bolivia also has been politically active, the situation does allow one to observe quite remarkable changes of comportment as a peasant enters the mining life and becomes a miner.[5] It is this particular transition, from peasant to miner, which captures our attention in this paper.

The questions we ask of the peasant-to-miner transition are these: (1) What behaviours change as peasants adopt the mining way of life? (2) What attitudes and ideas change as peasants adopt the mining life? (3) What do these ideological and behavioural changes indicate about the nature of proletarianisation, and what do they indicate about the nature of miners and peasants?

The field work upon which this study is based was conducted in the months of May, June and July of 1976 in a Bolivian mine.[6] As is the case with most field data gathered in a politically active setting, our data are not as complete, as extensive or as systematic as one would wish. Nonetheless, we find them provocative and perhaps instructive regarding the three questions posed above. In the paragraphs that follow, we describe the research setting from which these data were derived.

I Bocasa, a COMIBOL mine

The pseudonym for the mine where we studied is Bocasa. Bocasa is a state-owned tin and tungsten mine belonging to COMIBOL (*Corporación Minera de Bolivia*), a wholly owned administrative appendage of the Bolivian government.[7] Compared with other mines in the COMIBOL network, Bocasa is among neither the largest nor the smallest. Bocasa can be reached by truck or tri-weekly bus from La Paz. The trip requires some twelve hours across unforgiving mountain roads frequently made more perilous with blowing snow. The vistas are spectacular. Several smaller, privately owned mine encampments are passed as the bus or truck pushes further toward the mountain precincts of Bocasa.

The Bocasa workings lie between 14,000 feet and 17,000 feet in altitude. The main population centre is home to some 725 miners, above-ground workers and salaried employees. Add the families of these people plus additional 'floating' residents (e.g. job seekers, itinerant merchants, etc.) and the total population of Bocasa exceeds 4,000.[8]

In addition to mine housing, the centre contains the administrative offices, equipment maintenance and storage yards, a sizeable ore concentration plant,[9] a catholic church, two evangelical protestant churches, a market area, elementary and secondary schools, a medical clinic, a movie theatre, the inevitable soccer field, and sundry other buildings. All land in the area is the property of COMIBOL. The settlement is a company town; the road in and out of the zone requires one to pass a garrison check point manned by armed police.

The entire built-up area is located on the floor and lower slopes of a glacial valley surrounded by snow-bound mountain crags. Snows and freezing temperatures are frequent in winter and the air of the valley is generally laden with haze, dust and noise emitted from the concentration plant. The mine shafts themselves (all ore extraction is sub-surface) are located at various points on the mountain slopes above the population centre. The air is thin at Bocasa, placing heavy physical demands on the body. The mine interiors are cold; ice is a problem in the tunnels.

Bocasa, like other mines in the COMIBOL chain, employs its labour force in several forms.[10] Most of the miners are *regulares*, payroll miners entitled to a variety of fringe benefits, company housing when available, and privileges at the company store (*pulpería*). They work directly under the supervision of mining engineers through foremen. Regulares are generally supported by capital equipment (e.g. compressed air drills, electrified ore trains, battery-powered helmet lamps, etc.) and specialised support personnel such as those who erect roof timbering and ladders, and who maintain and operate the trains. The term, *regular*, embraces not only the below ground worker but also those above ground in the concentration plant. A few work in office clerical jobs and as vehicle drivers. More regulares work above ground than below, and all are regarded as miners (*mineros*).

A second large category of workers are the *pirquiñeros*, contract miners who scavenge the older workings using hand tools and discarded company equipment. They must sell the ore they extract to the company. Pirquiñeros usually work below ground. They are divided into the *contratistas* who hold permissions from the mine administration to work designated mine faces, and the *peones* who work for the contratistas in small teams.

The third major labour category are the *maquipuras* or day labourers. Maquipuras work wherever they are assigned, enjoy no company benefits beyond a daily wage, and generally work above ground on cleaning, maintenance and construction jobs. Maquipuras are entry-level jobs. Some maquipuras have worked as long as twenty years waiting for the opportunity to become either a pirquiñero or a regular; other maquipuras make this transition within a year or less.

How long Bocasa has been in operation we do not know.[11] Local knowledge is limited because we could find no one who had lived at the mine before 1947. A shortage of field time and politically unsettled conditions prevented us from doing archival research elsewhere, but residents of the surrounding peasant areas suggested it may have been begun some years before 1900. The mine evidently passed through various hands and later became a part of the Patiñc mine holdings. The mine was nationalised along with the other major mining centres after the revolution of 1952, and has been administered through COMIBOL since October of that year.

The history of the mine is more strikingly portrayed in the visage which greets the eyes in the mine area. The mountain slopes have been reshaped by the rubble and tailing heaps. Ore extraction has been so extensive that one nearby peak, San Jorge, collapsed in on itself one Sunday afternoon some years ago. Spent machinery, ore cars, rusting trackage, crumbling tramway pylons, twisted cable and gaping tunnels, long silent but for dripping water and cold drafts, mark the decades of human toil this mine has consumed. Another testimonial to the length and nature of Bocasa's history is the huge and over-crowded cemetery perched on a steeply bounded shelf above the mining centre. Even though the cemetery appears disproportionately large for a settlement of Bocasa's size, burial space has been exhausted, requiring that new burials be interred where older graves already exist. Life is harsh at Bocasa; there is no indication that it has been otherwise in the past.

II The field work setting

Bocasa was selected as a field work site mainly because it met two essential criteria. First, Bocasa was characterised by a relatively high proportion of its labour force derived from peasant populations, especially from the surrounding rural area. This is unlike the pattern in most other COMIBOL mines of Bocasa's size or larger. Elsewhere, the ores and the labour force have been in general decline. Jobs are at a premium, and a son of a miner has preference over a newcomer. Because Bocasa's mineral yield has not suffered such declines, there have continued to be employment opportunities for peasants seeking to enter the mining life. A second reason was the supportive attitude of local labour leaders and the acquiescence to our work from local mine management. Both of these conditions were available at Bocasa.

The fieldwork itself began in May 1976 and continued into July. Shortly after we had settled in, the field situation was overtaken by a national miners' strike. There were deportations of the national mine worker leadership, troop occupations of major mines (including Bocasa), and a bitter confrontation between miners and government which eventually left the government very much strengthened and the miners' leadership very much weakened (Greaves and Albó 1979). Our last information indicates that the troops remain stationed at the mine.[12]

The confrontation and strike offered many more advantages to our research than hindrances. The Bocasa miner leaders allowed us into their strategy sessions and offered us their trust. Meanwhile, the company managers tolerated our presence and, in any case, probably found themselves too busy containing the strike action to pursue the revocation of our authorisation to stay at Bocasa. The strike allowed us to match what miners said about their attitudes and loyalties with the overt and often courageous actions they took during the strike. In general, intense and dramatic conflict throws social structure into sharper relief than it exhibits in more settled times. For us, the confrontation at Bocasa brought this structure out vividly.

To be sure, the confrontation was also a hindrance. It tended to focus the attention of the miners on the confrontation issue and reduce their patience for the broader interests of an anthropologist, indeed, the strike had the same effect on us. Security of field notes was a paramount concern. The use of a camera was carefully limited. We used official company records less often and less extensively than we would have preferred so as not to jeopardise the trust of the miners by appearing too often in the company of mine managers. Overall, we are aware that our data always have to be evaluated for the skewing caused by a tense and dramatic event, but on balance we believe the strike had a profoundly positive impact on our understanding of the realities of a Bolivian mine.

Fieldwork proceeded in a predictable pattern. The first several days were spent in introductions, network building, and becoming acquainted with the physical and practical realities of mines and the mining life. As the confrontation began, we conducted interviews, attended meetings and observed, and visited the underground work sites. During the several weeks of the strike we watched the deepening confrontation, increased our interviewing and discussions with miners and listened and learned much at their formal and informal meetings. We also visited the peasant areas from which many miners had come and conducted interviews and a survey there. After the strike had largely ended we applied a questionnaire to about one-fourth of the miners and examined various company employment records.

We are hopeful that as a corpus our data reflect with little distortion the realities of the Bocasa miners. There is good consistency between the questionnaire data and our less formalised data, although neither meets the requirements of a statistical sample. On the basis of these various data sources, we now present descriptive information on Bocasa mines. We describe the characteristics of the miners as a group, and then examine what differences appear when the miners are compared on the basis of the length of time they have been miners. If there is a transition from peasant to miner, its nature and velocity should be reflected in differences between responses of the new arrivals and those longer in mining.

III Characteristics of Bocasa miners

The characteristics of the Bocasa miners are derived from a combination of qualitative and quantitative data sources. In order that the reader adequately assess our characterisations, we present a brief discussion of the types and limitations of our data sources.

Data quality

Gathering systematic data in a Bolivian mining area is an enterprise fraught with compromises, even when the labour situation is tranquil. Official Bocasa records are not easily accessed, not systematically organised, and generally are single-purpose lists of limited information. One list generally does not correspond in sequence, length or scope to another, and apparent and unexplained contradictions are legion. Our use of the very poor administrative records was compounded by the tension raised by the strike, leaving us with little opportunity to extract a comprehensive and reliable picture from the records we examined. The information we have, thus, is helpful and fairly extensive, but short of the set of fixed parameters we would prefer.

Conducting a questionnaire among miners raises another set of problems. For one thing, there is never a place or time when all miners are in the same place. Miners live not only in the main centre, but at three outlying settlements and in numerous isolated house clusters. There are no maps indicating where they all live. House-to-house surveys are difficult because the miners work on different shifts and are amenable to interviews at times which we found difficult to predict. The strike compounded the matter because many workers used their idleness to begin temporary jobs in places other than Bocasa, or to perform various union-assigned strike duties at irregular hours. In light of these factors, we resorted to intensive participant-observation and focused interviewing, complemented by a questionnaire applied to those *of peasant backgrounds* who would consent to answer questions as they came off shift changes.

The questionnaire resulted in 128 interviews. We found that the respondents of the questionnaires included regulares, pirquiñeros and maquipuras in numbers not particularly deviant from the Bocasa population at large, and also have a reasonable representation of above- and below-ground miners, men and women, new hires and old timers. The distribution is provided in Table 10.1 below. The data are far from ideal, but they are the most systematic presently available on ex-peasants from Bocasa or any other mine population of which we are aware.[13]

Table 10.1   Worker types among questionnaire respondents

Regulares	
underground	22
above ground	60
Pirquiñeros	
contratistas	14
peones	15
Maquipuras	8
Others*	9

Total	128

* Bocasa has a number of *'particulares'*, who, in loose affiliation with the company, are allowed to scavenge ore from, for example, the plant effluent stream, to be sold back to the company.

For present purposes we will aggregate all 128 questionnaire respondents into a single population. We have divided the workers into the subcategories laid out above and found that on most dimensions they show little or no significant deviations from the group as a whole. The relatively few points on which there are significant differences among the worker types will be discussed later in 'Characteristics by job type' below.

Mine labour force characteristics

The average age of our 128 respondents is about thirty-five years. They range in age from sixteen to sixty-six with most distributed in approximately equal numbers in the ages of the twenties, thirties, and fourties. In our group there are twelve women (9.4 per cent) known as *palliris*, whose normal assignment is to pick ore-bearing rock from waste rock at the entrance to some of the shafts of little activity.[14] Of the 128 only five were born in towns, all the rest were born in small agrarian communities. None was born in Bocasa itself, although about a third come from rural communities in the immediate vicinity. More than three-quarters are married (either formally or in common law), a few are widowed, and less than twenty per cent are single. About three-quarters have children. By selection all were born in peasant (campesino) households and migrated as adolescents or adults to Bocasa.[15]

About 77 per cent claim to be literate, the same number who report having attended at least one year of school. Of those who have attended school, more than half have at least four years of primary schooling. All speak Spanish with reasonable fluency and all but three also speak Aymara, the preponderant indigenous language in this region. The remaining three speak Quechua in addition to Spanish. The degree of bilingualism and education is also revealed in language use patterns: 85 per cent of the group report that both Spanish and Aymara are spoken in the home. Ninety-three per cent identify themselves as Catholic.

About half of our 128 had come to Bocasa from their natal community. The remainder had been elsewhere first. Twelve had resided in a major city, ten had worked in another mine, and most of the rest had been in another rural area. Interestingly, some 60 per cent continued to have some claim to land in their natal communities, although in the main this consists of shared or token claims inadequate to provide a livelihood. In the agricultural areas adjacent to Bocasa the shortage of land is acute. One offspring is designated the heir to work the land and the remaining sons generally migrate. The migrants retain some rights to the land but the yield mainly goes to the support of the sibling designated to work the property. Ninety per cent of the 128 return to their natal communities from time to time. More than three-quarters exchange food and manufactured goods purchased at a discount in the pulperia for agricultural produce from their kinsmen or community members, or with peasant traders who come to Bocasa for the Sunday market. Company rules prohibit reselling items from the pulperia, but the practice, though economically small scale, is very wide spread. (There is a nearby market centre which exists *principally* as an *entrepôt* for the exchange of local agricultural products for goods originating in the pulperias of three adjacent mines.) One concludes that Bocasa miners typically maintain ties of membership with natal agrarian communities and engage in a small way in the exchange of products with them. We will return to this point later.

On the whole our questionnaire turned up rather high regard for both the peasant lifestyle and the miners'. In response to the question, *'with whom do you feel more united, the peasant or the miner?'* some 85 per cent declared that they felt united with both. (Of the remainder more aligned themselves with miners than with peasants.) Nearly 80 per cent felt the life of the peasant was either unchanged or had improved, and about 90 per cent felt life at Bocasa had also remained unchanged or improved.

Taken as a whole, the questionnaire data describe a population with a significant average level of education. They are mainly young-to-middle-aged adults, primarily married, usually with families, who continue to maintain links with natal villages and kinsmen. In general, they have a positive view toward both the peasant's and the miner's lifeways.

Characteristics by job type

The reader will recall that our group of 128 contains workers holding several different types of occupational categories and status levels at Bocasa. It is essential that the occupational types be examined individually lest sharply different groups are offsetting each other in the aggregated analysis. What we find is that the main contrasts are age-related and predictable. Attitudes, on the other hand, are very similar among all of the groups.

The maquipuras and the pirquiñero peones are younger, not surprising since these are entry-level positions. Few regulares continue in the strenuous and dangerous underground jobs after their forties. Where there is a concentration of younger individuals there is, predictably, a higher level of education, reflecting the large expansion of rural schooling in Bolivia that has occurred since 1952.[16]

By contrast, on attitudinal questions the various occupational groups showed similar distributions. For example, in response to the question, *'What advice would you give to peasants thinking about coming to the mines?'* the distribution was as shown in Table 10.2.

Table 10.2   'What advice would you give to peasants thinking about coming to the mines?'

	Regulares	Pirquiñeros	Maquipuras	Others	Total
Discouraging plan to come	26	7	3	1	37
Ambivalent on advice	17	9	2	2	30
Encouraging plan to come	31	10	4	5	50
No answer	8	2	–	1	11
Total	82	28	9	9	128

Distributions showing a similar lack of dissimilarity between the job types appeared in perceptions about the better life (peasant's *vs*. miner's), sense of unity (with peasants *vs*. with miners), attitudes toward the union, the company, whether miners are united or divided, and others. The conclusion we draw is that, other than predictable demographic variables based on age differences, these occupational subgroups are otherwise quite similar.

Tabular provisos

Before launching into the tabular compilations that follow, a few comments on data treatment procedures are in order. First, statistics have not been applied because (1) the 128 ex-peasant miners we interviewed cannot be demonstrated to be a random sample, (2) even with considerable aggregation, the numbers in some cells are very small, and, (3) the variables used in this paper (other than years in mining) are a mix of nominal and ordinal lists, but no ratio scales. Similarly, although we have provided the marginal totals for each table, we have not entered percentages because the number of cases in the columns are often small, producing meaningless leaps in percentages with the shift of one or two cases. It seemed to us most sensible to look at the distribution of responses given by these 128 individuals without reducing them to shorthand relationships.

The number of years working at Bocasa (the x axis in the following tables) is by self report. Because the questionnaire was answered anonymously, we did not cross-check their reponses with other records. However, given our rather solid acceptance in camp by the time we took the data, and the endorsement of our research by union officials, we do not suspect much misinformation. We did ask the question in two separated ways (year of arrival and number of years worked) and the two reports correspond closely. Indeed, the openness and thoughtfulness the miners displayed during the interviews increase our confidence in the data. The questionnaire was administered in Spanish on several successive days to consenting miners as they came off shift. When the respondent elaborated on the reasons for an answer (which was typical) we wrote it down.

The grouping of the responses into the small number of categories on the y axis of these tables, likewise, has not been subjected to methodological rigour. Only one coder was used, the first author of this chapter. Grouping did not, for the most part, entail borderline choices and, where uncertainty or ambiguity occurred, the case was set aside.

IV Becoming a tin miner

Having described the basic characteristics of our 128 miners, and the nature of this data, we now turn to the process by which these characteristics are acquired. The focus is on the changes that occur when an incoming peasant arrives at the mine and starts work as a miner. If proletarianisation occurs, much of the process should transpire across some period of time after mine employment begins.

Our strategy is to sort the 128 respondents into categories ranging from those who have worked a very short time at Bocasa to those who have worked a very long time. The range is from less than one year to nearly forty years. The actual distribution is as shown in Table 10.3

Table 10.3

Years as miner, (interval width)		Number
0–1	(2)	13
2–4	(3)	11
5–9	(5)	26
10–14	(5)	27
15–19	(5)	12
20–4	(5)	21
25–9	(5)	8
30–4	(5)	7
35–40	(6)	3

Our questionnaire was designed to tap a number of factors we would expect to be associated with the transition from peasant to proletarian and the emergence of class consciousness, but the questionnaire also included a variety of demographic and personal history items. Many of the latter understandably correlated with length of service at Bocasa. Among them were:

Age. Not surprisingly, the longer one had been at Bocasa, the older one tended to be.

Marital status. Predictably, being single was associated with having worked at Bocasa for only a few years.

Persons in household. The longer one had worked at Bocasa the more persons lived in the worker's household.

Education. The newest workers usually had completed 3 to 6 years of education; those working longer, 0 to 4 years.

Language used at home. Those who spoke only Aymara at home tended to be the workers of long tenure at Bocasa.

Literacy. Workers reporting themselves to be illiterate tended to be those who had been at Bocasa a long time.

Exchanging products with peasants. Those who had worked at Bocasa for fewer years tended not to participate as often as those who had worked a longer number of years.

Birth place. Those workers who had worked only a few years tended to come exclusively from the nearby peasant areas; workers of longer standing come from more distant areas too.

Present Bocasa job type. The most recent workers were predominantly maquipuras and the peones of pirquiñeros.

We report the preceding correlations mainly to show that the questionnaire is providing plausible data, and to expand on the general features of the population. We now turn to the questions dealing with the transition from peasant to proletarian and the development of those attitudinal correlates of tin miner behaviour. Our reasoning is that any transition from peasant to proletarian might be reflected in personal stresses that would seem to flow from changing from one status to the next, with the readiness to beckon or dissuade peasants from becoming miners, evaluation of the comparative merits of the two categories, with degree of disenchantment with the mine administration, the attitudes of solidarity with a class rather than a natal community, and degree of solidarity with the labour union.

Against each of these factors we distributed the 128 miners according to their length of service as a miner. The groupings presented here cut finely in the first years of service, and grossly for old-time miners. The greater discrimination on the early years is so that a transition could be detected even if it were completed in a short period of years. As will now become evident, the answers are about the same regardless of whether the miner has been on the job for one year or thirty. After a review of the data, we will proceed to the implications these findings may have for the nature of the proletarianising process.

We thought answers to, *'What is the most serious problem you encountered when you came to work as a miner?'* could be indicative of the socio-cultural gulf between peasant and proletarian, and the psychological stress that would logically accompany such a life change. The marxist perspective would lead one to assume that the gulf is wide, necessitating substantial re-education and political 'conscientisation' of peasant peoples to effect their transfer to the ranks of the proletariat. The response to our question, however, shows an almost total absense of reported difficulty with the transition or symptoms of psychological stress such as loneliness, fear or unhappiness. Nearly everyone firmly reported that there was no problem. This is supported by our ethnographic interview data, field observations of new workers, and prior research in other settings (Greaves 1972 a: 39 ff). What few problems were mentioned centred around adapting to the cold climate or high altitude. The overwhelming propensity is for both new hires and long-employed miners to report no problems.

Why is the transition so apparently routine? That becomes easier to explain when the reponses to the remaining questions are considered. In

Table 10.4   'What is the most serious problem you encountered when you went to work as a miner?'

| | \multicolumn{4}{c}{Years Working at Bocasa} | | | |
	0–1	2–4	5–9	10–40	Total
No problems	12	11	22	70	115
Learn job	–	–	–	3	3
Illness	–	–	–	2	2
New ambience	1	–	1	3	5
Total	13	11	23	78	125

Unclassified answers = 3

Table 10.5, the distribution is given for the question, *'What advice would you give those peasants who are thinking of coming to work as miners?'* The wide variety of answers was classified into those who, for various reasons counselled against coming, those who offered pros and cons (ambivalence), and those who positively recommended the mining life.

Table 10.5   'What advice would you give those peasants who are thinking of coming to work as miners?'

| | Years working at Bocasa | | | | |
	0–1	2–4	5–9	10–14	Total
Don't come	2	2	11	22	37
Ambivalent	4	1	3	21	29
Yes, come	6	4	7	32	49
Total	12	7	21	75	115

Unclassified answers = 13

There is a slight tendency for newcomers to be more ambivalent or more positive – but not more negative – about mining at Bocasa than those workers of longer standing (though not at a statistically significant level). This seems to corroborate our responses to the previous question: the transition to mining is not regarded as difficult, stressful, or ill-advised.

The responses to, *Have you noticed differences between peasants and miners? [if yes] what are they?'* were classified into differences which carried positive evaluation (e.g. miners make more money, peasants have a more tranquil life), those with a negative connotation (e.g. peasants' land is not secure, miners lose their health). Twenty-three were not classifiable as positive or negative, and are excluded from our tabulation (Table 10.6). Interestingly, sixty-five of the usable replies declared that there was no difference between peasant and miner. We take this to add support to the pattern of the previous questions, that a peasant becoming a miner is not undergoing a soul-wrenching transition. Finally, while the older miner may have a slight tendency to give favourable nod to the peasant life, the linkage is meagre.

Table 10.6   'Have you noticed differences between peasants and miners?' [If yes] 'What are they?'

| | Years working at Bocasa | | | | |
	0–1	2–4	5–9	10–40	Total
Miner best	2	2	2	6	12
No difference	8	5	9	43	65
Peasant best	1	1	6	18	26
Total	11	8	17	67	103

Unclassified answers = 25

We then asked more simply: *'Which is better: the life of the miner? the life of the peasant?'* Here we eliminated the option of fence-sitting and found (Table 10.7) that they swung mainly to the pro-peasant side. When the responses are sorted by years in mining, we find that there is a weak tendency for those who favour the mining life to be newcomers (the reverse of what a difficult transition might predict), but more of both new and old miners are pro-peasant.

Table 10.7   'Which is better: the life of the miner? the life of the peasant?'

| | Years working at Bocasa | | | | |
	0–1	2–4	5–9	10–40	Total
Miner's life	5	3	2	15	25
Peasant's life	8	8	22	63	101
Total	13	11	24	78	126

Unclassified answers = 2

Perhaps this is nostalgia, but remembering that 90 per cent visit villages of birth at least yearly, it is probably more than myth-making. Peasant life is not regarded at Bocasa as a rustic backwater to proletarian life in the fast lane.

The distribution for the next question, *'Do you think the company treats its workers well?'* shows a strong tendency for newcomers to rate the company highly. At the same time, two-thirds of the long-time miners also were pro-company.[17]

Table 10.8 'Do you think the company treats its workers well?'

| | Years working at Bocasa | | | | |
	0−1	2−4	5−9	10−40	Total
Well	13	9	12	48	82
Badly	−	2	12	26	40
Total	13	11	24	74	122

Unclassififed answers = 6

What may be indicated is that the mining life is, like peasantry, a tolerable option. Taken together these two questions seem to suggest that the move from peasant to proletarian is not 'up', but 'over'.

Another question, *'With whom do you feel most united: with the peasants (campesinos) of your natal community? with your workmates at Bocasa? with both? other?'* (Table 10.9) yielded a strong preference for 'both', and no tendency for newcomers to answer differently from miners of long standing.

Table 10.9 'With whom do you feel most united: with the peasants of your natal community? with your workmates at Bocasa? with both? other?'

| | Years working at Bocasa | | | | |
	0−1	2−4	5−9	10−40	Total
With peasants	2	−	1	2	5
With miners	3	1	2	6	12
With both	8	10	21	69	108
Total	13	11	24	77	125

Unclassified answers = 3

The pattern of the earlier two questions seems reasserted. We are not seeing evidence in these data of a dramatic or traumatic peasant-proletarian transition.

The questionnaire, containing forty-eight items and item areas, offers many further analyses which remain in preparation. With respect to the small number of items selected for discussion here, the pattern correlating length of service with questionnaire items is very consistent. Items dealing with demographic, educational, work type, birth place, marital status and similar matters provide linkages which are clear and expectable. Meanwhile, the linkages to attitudinal questions intended to tap the process of proletarianisation fail to reflect a transition that is difficult, traumatic or protracted.

V Discussion

Perhaps the first point to make is that these questionnaire items, the data from other sections of the questionnaire, and the broader observations and interviews of our fieldwork all argue that the social and, perhaps, psychological distance between Bocasa miners and their peasant neighbours is not great. The continuing recruitment from peasant to miner, the ongoing mutualist exchange between them, and the intent of a majority of these miners to retain some claim to land rights in their natal villages all argue that the interlock between Bocasa miner and peasant cultivator is collegial and miscible.

How is this to be understood? One's first explanation might be that Bocasa miners are really not very proletarianised — that we are dealing with a group of miners who are so closely linked with their peasant cousins that they could only be spectators to the indisputably proletarianised miners of the largest COMIBOL mines (e.g. Siglo XX-Catavi, Uncia, Llallagua, Huanuni etc).[18] This characterisation, however, is not borne out by events both during our fieldwork and in the years since. Elsewhere, we have described a protracted strike at Bocasa (Greaves and Albó, 1979). Great sacrifices were made; strident efforts were made to unite in solidarity and brotherhood with their *compañeros* in other mines. Bocasa miners were outraged when national radio listed all the mines on strike and then announced that Bocasa miners were still at work. Messengers were immediately sent across remote mountain trails carrying official union documents affirming Bocasa's participation in the COMIBOL strike. Class solidarity and sacrifices to back it up were certainly present among Bocasa miners. Less than four years later, during the violent coup of Garcia Mesa, Bocasa miners put up concerted resistance, sustaining deaths and many taken prisoner by the attacking troops. Whitehead recently wrote, 'The worker miners of Bolivia possess a reputation for greater labor militancy and radicalism than any other class of worker in Latin America' (1980: 1465, our trans.). While it is true that some other, larger mines have histories of greater militancy, Bocasa miners are members of that proletarian tradition.

One might then ask, how traditional are the peasant regions from which so many Bocasa miners have come? One can say with assurance that no part of the highlands remains unaffected by the 1952 Bolivian national revolution of 1952, itself a product of changes underway since Ucureña in the 1930s (Dandler 1969). Without doubt many parts of contemporary rural Bolivia are inhabited by 'rural proletarians' (Greaves 1972 b). Yet the rural change process has not been uniform. Some areas are strongly mobilised, others are much more on the periphery (Malloy 1970: 207–15). The hinterland area from which Bocasa draws the majority of its miners is isolated and, although we lack intensive study of the zone, it is no wellspring of national involvement. The zone consists of a mixture of former haciendas and freehold peasant communities. Their rural syndicates appear localised, quiescent, frequently concerned with boundaries and land titles.

Our best assessment is that in comportment styles those who live in Bocasa's adjacent agrarian villages and hamlets are reasonably well characterised as living peasant lifeways. Meanwhile, the Bocasa miners can be reasonably described as proletarians. How, then, can the transition be so rapid, so undisruptive?

A possibility that need not detain us long is that the Bocasa miners may come from peasant locales, but were atypical. One could posit that those who leave the peasantry are those who are estranged from their home communities, and from the lifeway which is found there. To some degree this is bound to be true; the literature tells us that in nearly every instance of migration, those who do the migrating are atypical compared with those who stay behind. One suspects that those who migrate are usually those who have few satisfactory options to stay behind, who are open to a change of lifestyle, and who may have a higher level of education than those who were left behind.

Bocasa migrants do have an unusually high degree of education. In addition, when we asked what reasons caused them to leave their communities of origin, 73 per cent gave economic reasons. Yet we also found that miners retained social and economic links to their home villages and the large majority hailed from very traditional areas. We visited the villages of the province nearest the mine, an area containing the home communities of about one third of our 128 miners. We administered a questionnaire to a small number of residents of each village in which we probed local perceptions of villagers who had become miners. None of the responses indicated that those who had left for the mine were perceived by villagers to be unusual in ways other than, typically, that they would have faced serious economic constraints had they remained at home. That our Bocasa miners are different is readily apparent, but in our view there is no indication that before they migrated they acted in ways particularly different from those left behind.

We are therefore disinclined to believe that (1) Bocasa miners are simply peasants with hard hats, (2) the villages from which most come are not reasonably well described as peasant in lifestyle, or (3) Bocasa migrants were behaving fundamentally like miners while still in their home villages. Rather, we incline toward another explanation. We would suggest that our literature has tended to assume that peasants and proletarians are more profoundly distinct than they are. The Bocasa data suggest the existence of an underlying commonality between peasants and miners. Despite fairly sharp differences of behaviour between peasants and miners, there is no evidence of a post-arrival transition in the attitudes we measured. It seems difficult to avoid concluding that the transition from peasant to miner, at least in this setting and with these people, is rapidly accomplished, and fundamentally not a very large transition. Underneath quite different comportment styles, peasants and proletarians may be fundamentally quite similar.[19] One need not assume that only one set of behaviours, specific attitudes and information inventory can be supported by that set. Why could it not be the case that becoming proletarianised from a peasant background is more like code switching than what we understand as dialetical evolution?

The frequency and importance attached by miners to the maintenance of participatory ties and rights in their natal villages, the visiting patterns, and the equal valuation made by miners of both lifestyles are also easier to explain if peasants and proletarian miners are understood from this perspective. It would then follow that the greater incidence of miner–peasant interaction at Bocasa is not because they are only modestly proletarianised, but because a continuing availability of entry-level jobs, the continuing flow of ex-peasants to fill them, and the proximity of peasant homelands, setting the conditions for Bocasa miners and peasants to stay in active contact and to muster resources from either zone in order to cope with the harsh realities of the two lifestyles.

Notes

1 The research upon which this paper is based was generously funded by the Social Science Research Council, whose assistance is gratefully acknowledged. The council takes no position on the contents or conclusions of this research.

2 A good recent discussion of such issues is given in Mann 1973 and also in Nash 1979, Chapter 9. Also see Wallman 1979, 16–20.

3 We are the beneficiaries of many whose work preceded us. We mention, for example, the work of Gregorio Iriarte (1972), June Nash (1979), Heraclio Bonilla (1974) and Sergio Almaraz Paz (1969). We also find the insights offered by novels of Bolivian mining very illuminating. The reader is directed to such novels as *Koya Loco*, and *La Khola* (Poppe 1973 and 1975), *Metal del Diablo* (Cespedes 1969), *Mina* (Guillen Pinto 1953), *Minero (y Otros Relatos)* (Millan Mauri 1971), and for a larger sense of the Bolivian spirit, the classic *Raza de Bronce* (Arguedas 1909).

4 The political activity of Bolivian miners is amply documented. Among the many good sources are Alexander (1958), Malloy (1970), Kelley and Klein (1981), Mansilla T. (1978), and, especially, Lora (1977).

5 The political integration of Bolivian agrarian areas is described in Harris and Albó (1975), Malloy (1970), McEwen *et al.* (1969), and Patch (1963). See also Preston (1978).

6 There are various sources based on ethnographic fieldwork in Bolivian mining centres in this volume. The reader might begin by consulting Nash (1979), Vasquez (1965) and Delgado's and Godoy's chapters in this volume.

7 Among the available studies of COMIBOL are Canelas O. (1966), Ruiz González (1965) Gómez d'Angelo (1978), Ford, Bacon and Davis (1956) and, in part, Alurralde Anaya (1973).

8 This estimate is supported by the 1976 national census which lists 'Bocasa' above 4,000, with probable under-enumeration of the more mobile elements of the population (*cf.* Instituto Nacional de Estadística, Resultados provisionales, Departamento de La Paz, 1977).

9 No Bolivian mine includes an ore smelter. At the larger mines the ore is processed through various mechanical and chemical steps into concentrates. Until the past decade most Bolivian concentrate was shipped to smelters in Europe for smelting. The expansion of Bolivia's only tin smelter at Vinto has just recently enabled the country to smelt 100 per cent of its own tin ores. See the paper by Fox in this volume for further information.

10 The complex typology of mine jobs is described in Harris and Albó (1975: 14–22). The precise meaning of terms shows some variation from one mining region to another.

11 Historical data on COMIBOL mines can be found in a number of the sources cited (e.g. Nash 1979 for recent history, Almaraz Paz 1969 for the earlier phases, and, as a period document, *Estaño en Bolivia* 1953.

12 This did not prevent, however, the evident regrouping of miner solidarity at Bocasa. During the bloody coup of 1980, Bocasa miners confronted armed troops and fought with great valour (see Asamblea 1980).

13 Questionnaire data from Bolivian miners are available in some other studies. See, for example, Vasquez (1969).

14 The descriptor, miner, *minero(a)*, includes both men and women. The most common job for women miners is the *palliri*, a woman who is usually assigned to the entrance to a small mine shaft, picking the ore-bearing rock from the waste dumped by male miners working inside the shaft. In a political sense, however, the female spouses of miners need to be included in any analysis. Bolivian labour conflicts at the mines usually include the aggressive and often organised participation of women. Many courageous acts by women of the mines are on record. See Viezzer/Barrios de Chungara (1977), Nash and Maria Rocca (1976) and Mansilla T. (1978) for examples.

15 In this paper we make the assumption – as do our colleagues – that every migrant born in a peasant household is an ex-peasant. This is a frail assumption, first because peasant households are far from socially homogenous (often including 'peasant elites', civil functionaries etc.), and second, because migrants may have left at an early age. The latter raises a question about the point at which a person may be said to become peasant in the life cycle.

16 See McEwen (1969), Buechler and Buechler (1971), and Preston (1978) for descriptions of broad social changes in education and other aspects of rural Bolivia.

17 This seems curious given that these miners had just endured an extended strike. The explanation may lie in the success with which the local managers were able to shift blame to the La Paz officials and the government for opposition to union demands.

18 The centrality in labour politics of the miners of Catavi, Siglo XX, Huanuni, Llallagua and others has been recognised elsewhere. Written sources commenting on this pattern include Almaraz Paz (1969), Nash and Maria Rocca (1976), Iriarte (1972), Lora (1977) etc.

19 The distinction between comportment styles and underlying commonalities has been discussed further by Greaves (1972). Dual comportment styles for rural agriculturalists in the Andes are described by Doughty (1965: 14) and Maynard (1966: 1–13) among other sources.

References   * *See annotated bibliography for full details of all asterisked titles*

Alexander, Robert J.
1958 *The Bolivian National Revolution.* New Brunswick (NJ): Rutgers, the State University.

Almaraz Paz, Sergio
1969*

Alurralde Anaya, Antonio
1973*

Arguedas, Alcides
1909 *Raza de Bronce*. La Paz: Ediciones Puerta del Sol. (1970).

Asamblea Permanente de los Derechos Humanos de Bolivia
1980*

Bonilla, Heraclio
1974*

Buechler, Hans Ch. and Judith-Maria Buechler
1971 *The Bolivian Aymara*. New York: Holt, Rinehart and Winston.

Canelas, O. Amado
1966 *Mito y Realidad de la Corporación Minera de Bolivia* (COMIBOL). La Paz:
   Editorial Los Amigos del Libro.

Céspedes, Augusto
1969 *El Metal del Diablo*. La Paz: Ediciones Puerta del Sol.

Dandler, Jorge
1969 *El Sindicalismo Campesino en Bolivia: Los Cambios Estructurales en Ucureña*.
   Mexico: Instituto Indigenista Interamericano.

Doughty, Paul L.
1965 'The interrelationship of power, respect, affection and rectitude in Vicos'.
   American Behavioral Scientist 8(7): 13–17.

Estaño en Bolivia, El
1935 *El Estaño en Bolivia*. La Paz: Imprenta Renacimiento.

Ford, Bacon and Davis, Inc.
1956 *Mining Industry in Bolivia*. Report no. 36631, 9 vols. Prepared for the Ministerio
   de Minas y Petróleo de Bolivia.

Gómez D'Angelo, Walter
1978*

Greaves, Thomas C.
1972a 'Pursuing cultural pluralism in the Andes'. *Plural Societies (summer): 33–49*.
1972b 'The Andean rural proletarians'. *Anthropological Quarterly* 45(2): 65–83.

Greaves, Thomas C. and Xavier Albó
1979

Guillén Pinto, Alfredo
1953 *Mina (Novela Postuma)*. La Paz: Talleres Gráficos Bolivianos.

Harris, Olivia and Xavier Albó
1975*

Iriarte, Gregorio
1972*

Kelley, Jonathan and Herbert S. Klein
1981 *Revolution and the Rebirth of Inequality; a Theory Applied to the National
   Revolution*. Berkeley: University of California Press.

Lora, Guillermo
1977*

Malloy, James M.
1970 *Bolivia: The Uncomplete Revolution*. Pittsburgh: University of Pittsburgh Press.

Mann, Michael
1973 *Consciousness and Action among the Western Working Class*. Studies in
   Sociology. London: Macmillan.

Mansilla T., Jorge
1978*

Maynard, Eileen
1966 'Indian–mestizo relations'. In *The Indians of Colta; Essays on the Colta Lake Zone, Chimborazo (Ecuador)*. E. Maynard, ed. Ithaca, New York: Department of Anthropology, Cornell University. Pp. 1–36.

McEwen, William J. *et al.*
1969 *Changing Rural Bolivia; Final Anthropological Report for the Peace Corps*. New York: Research Institute for the Study of Man.

Millán Mauri, José
1971 *Minero (y Otros Relatos)*. (2nd edition.) La Paz: Empresa Editora Universo.

Nash, June C.
1979 *We Eat the Mines and the Mines Eat Us; Dependency and Exploitation in Bolivian Tin Mines*. New York: Columbia University Press.

Nash, June C. and Manuel Maria Rocca
1976*

Patch, Richard W.
1963 'Peasantry and national revolution: Bolivia'. In *Expectant Peoples*. K. H. Silvert, ed. New York: Vintage Books. Pp. 95–126.

Poppe, Rene
1973*
1975*

Preston, David A.
1978 *Farmers and Towns; Rural–Urban Relations in Highland Bolivia*. Norwich (England): Geo Abstracts, Ltd.

Ruiz González, René
1965 *La Administración Empírica de las Minas Nacionalizadas*. La Paz: Empresa Burillo.

Vasquez Varela, Mario C.
1969 *The Kami Mine: A Bolivian Experiment in Industrial Management*. Ithaca, New York: Department of Anthropology, Cornell University.

Viezzer, Moema (compiler); Domitila Barrios de Chungara
1977*

Wallman, Sandra, ed.
1979 *Social Anthropology of Work*. ASA Monograph 19. London: Academic Press.

Whitehead, Lawrence
1980*

Widerkehr, Doris E.
1975*

CHAPTER 11

Together we work, together we grow old; life and work in a coal mining town[1]

PATRICIA SACHS

We are used to characterising miner consciousness, proletarian or otherwise, in terms of antagonism toward capitalist employers, mal-distributions of power, attitudes about the work place and the work done there, and active identity with fellow workers. The author of this chapter asks us to expand the compass of miner consciousness to include not only company-directed, remunerative labour, but also the work at home − the unpaid and volitional work − and the cognitive links between paid and unpaid work.

Patricia Sachs explores the texture of the paid−unpaid work connection, first, in the context of retirement, when paid work has ceased and, second, retrospectively when these West Virginia coal miners were still on the payroll.

Patricia Sachs is a doctoral candidate at the City University of New York Graduate School (Department of Anthropology, 33rd W. 42nd St, New York, NY 10036, USA). She has worked with retired coal mining families in West Virginia for two years and pursues studies of unpaid labour, the relation of industry to community, and the process of growing old. A related article, 'Is volitional work voluntary?' appears in *Anthropology of Work Review* 5 (2): 19−24 (1984). − *The editors*

Research on miners tends to focus upon the labour of men in mines, the choices they have or do not have about being miners, the consciousness they have about being exploited, and the ways in which they cope with the danger of mining.[2] This view of a mining life centres on the productive, formalised labour of men during a particular phase in their lives. Such a focus makes sense because the labour of men in the mines is extremely absorbing. By looking at mining only through waged labour, however, the importance of more prosaic forms of work, such as the unpaid labour of both men and women, is glossed over. Unpaid, usually domestic labour is integral to the experience of mining work and to life in a mining town. Further, the focus of research on the middle of the life-cycle (when men are employed in remunerative labour) diminishes the importance of those parts of the life cycle during which one is not directly involved in the wage economy.

This article examines the lives of mining families during retirement. Looking at retirement, I suggest, sheds light on social organisation at other points in the life cycle. For miners, this illuminates aspects of life other than mining itself. The employment of mining work and life in a company town can be seen as more clearly intertwined when viewed from the perspective of old age, when mining is no longer present, but the perceptions of work and the feelings about mining are. In the community under investigation, a former company town in northern West Virginia, retired life is defined as a life of work. Why is it work? Why isn't it leisure? Why does it matter that they were miners?

Retirement is a familiar but odd phenomenon in our society. It is broadly defined as exclusion from the labour force on the basis of age. A great transition from one life structure into another, it is a passage which occurs with minimal ritual (even the gold watch ceremony is waning). Retirement is largely socially undefined. What does one *do* when retired? The myths and ideologies of our culture offer little help: one does what one always wanted to do, but couldn't. Perhaps this means taking a vacation, pursuing a hobby, or fussing about the house. Such notions presume a leisurely sort of life. This seems appropriate, since retirement is the 'opposite' of work, just as leisure is the 'opposite' of work. As appropriate as it may seem, it isn't what happens in Thomas Creek, West Virginia. While the tasks which its residents carry out might be defined by others as leisurely, to the residents themselves they are work, and comprise the basic requirements for being a member in good standing within the community.

Work has been the keystone of social relations, values and identity for the residents of Thomas Creek for many years. When the men of the community worked as paid labourers the organisation of these values stemmed from the importance of the coal company to the community. When the men retired, these values were organised around unpaid, domestic work (see Sachs 1984 for further elaboration of this point).

Unpaid work exists in many forms and in many communities throughout the world, and the relationship between paid and unpaid work is a provocative one. This study focuses upon the way in which paid and unpaid labour have intertwined throughout peoples' lives in the formation of their community, and the way in which work is a source of dignity for the Thomas Creek residents, whether paid and highly exploited or unpaid and marginal.

I Retired life

Thomas Creek is a small town, housing seventy-five or eighty people in forty houses. The houses line half a mile of a coal route upon which coal trucks pass throughout the day and night. The town is tree-lined and shady. The houses are well-kept, many have new aluminium siding and shutters. The lawns and hedges are trimmed, the sidewalks swept and the gardens tended.

Half the population of Thomas Creek is elderly, between sixty and ninety years old. The elders occupy 70 per cent of the houses. Too old to continue working in the mines, or too ill with arthritis, black lung or silicosis to perform any wage work whatsoever, these people devote their time and emergies to keeping their town in shape and keeping in touch with each other. At first glance, the rhythm of this community seems easy-going and quiet. Upon closer inspection it becomes apparent that the vigour and intensity with which the people work is a subtle, but communally defined way of life.

The days start early in Thomas Creek. Up at 6.00 am, these older couples have a good breakfast – coffee-soup, perhaps some eggs, and cereal before starting the day's work. Men begin work in the garden, doing whichever tasks are 'suitable' for the day. The suitability of tasks depends upon the phases of the moon as well as the weather and season. For example, certain crops should be planted when the crescent moon points downward (below-ground crops such as potatoes, onions and garlic). Others, such as beans or peppers, should be planted when the moon points up. The full moon may indicate a preference for planting one crop over another, while the new moon indicates yet other crops. In addition, the days of the lunar cycle are represented by the signs of the moon such as the ram, bull, fish and so on. These indicate good days for weeding, harvesting or planting. To weed on a day when planting is indicated is somewhat foolish, for the weeds will grow back again too quickly. To plant on a day good for weeding will only result in poor growth. These lunar rules for planting are easily obtained in the almanac, although not all the Thomas Creek residents used it. Some simply 'read' the moon by looking at it. Not all residents believed equally in planting by the moon: some ignored it altogether. For many, however, gardening not only involves the labour of tilling, planting, weeding, pruning and harvesting, it also includes staying abreast of the lunar cycle, and planning when the great variety of tasks required to keep a good garden should be carried out over the course of a month. This sometimes meant planting in the rain, if it happened to rain the only 'good' day that garlic could be planted over the month, or preparing near-frozen soil for tilling, if one were to get the right crop in on time.

Men usually worked in the garden until 10.30 or 11.00 am, and then came in for a snack and a rest. They shared this with the women, who were taking a pause from their early morning's work. Women's work generally started with laundry, which was hung out on the line in the back yard by 8.00 am. This was usually followed by a round of doing dishes, cleaning walls and preparing garden vegetables for canning. After a snack, more of this same work would continue until lunchtime, around noon, when a couple would sit down for a meal of home-canned hot peppers, or fresh salad and sandwiches, coffee and homemade cookies. Lunch lasted an hour to an hour and a half, and was followed by a nap. In the afternoon, the couple would often go shopping or have a doctor's appointment, and return home for another two or three hours of gardening, mowing the lawn, trimming the hedges or repair around the house. Dinner was at 5.00 pm. Evenings were usually spent roaming up and down the 'alley', a long public space behind the row of houses defined only by one's knowledge of its existence rather than by any physical markings. During these walks, one might stop and chat with one neighbour or another, talk about how the gardens were doing, whether a neighbour was ill or doing better, ruminate about the mining days, or the upcoming visits of a grown child and grandchildren. Along these walks one might hear, 'Oh, her flowers are always blooming and her lettuce grows all year, I don't know how she does it', or 'he just lets those tomatoes rot there, why doesn't he tie them up?', or 'she keeps such a clean house, I'm afraid to let her come into mine'. Neighbours take a generous interest in each other's work, commenting on the gardens, houses, laundry and hedges. The carrying out of such tasks seems to indicate the identity of an individual within the community.

'A man just ain't a man when he can't keep his garden anymore', Marion told me one afternoon after a hard rain, when neighbours were huddling around puddles comparing notes on the storm's effect. The storm had flattened the corn and mud had slid off the hill behind the houses into the gardens. Cucumbers and peppers were lying in the stick mess. Marion was lamenting the condition of her husband who was unable to tend his garden. He suffered from black lung, and had to spend twenty-four hours a day hooked up to the oxygen machine. Marion made the rounds, however, talking with her old friends who had the biggest and best-kept garden in Thomas Creek, and checking out who was working at getting the mess cleaned up. The clean-up effort was important to the older members of the community. 'You're lucky you didn't get it too bad', I was told, 'but we know you'd clean it up. Alice over there, she didn't never clean up her garage last time it got flooded, and she let that mud harden and it's still sittin' there. Hard as a rock'.

The work of maintaining one's home is highly visible work, especially in such a tiny community. Because this work is private, self-directed and carried out separately by members of each household, the social value of such labour is made apparent in the product of the work itself. The work, which results in well-kept gardens, clean houses and trimmed lawns is shaped by the seasons and the weather rather than by ceremonial or ritual events. It is also shaped by one's neighbours. While the tasks are not prescribed – one doesn't report to work – the communally-shared values and morals about work behaviour constrain the extent to which the work is 'volitional'. Living up to communal standards provides a limit and goal to achieve *vis à vis* such labour.

During chats and visits, food or some other small token is frequently offered. This might be produce from the garden, a piece of cake or bread, or a handmade towel or pillow. These small gifts are part of a system of reciprocal exchanges which serve to circulate the products of the gardens, kitchens and handiwork. The exchanges reinforce the neighbourly, kin and ethnic ties which already exist as well as the pride in the work. These transactions weave the participants in a web of reciprocal relations that serve as a tacit guarantee for aid from others during times of crisis; thus the gift exchanges are the warp and woof of a voluntary support system. The exchanges enable the elders to depend upon one another in a graceful way helping them avoid a dependence upon the social services provided by outsiders.

To an outsider unfamiliar with the lives of these families, their work during retirement might seem ordinary. Their reliance upon each other in old age might be interpreted as predictably rural, perhaps quaint, and possibly nosey. They are not viewed as behaving as 'mining' families because they no longer mine. Indeed, newcomers to the community over the past few years find the interest which the older residents take in their lives, in how they care for their houses and lawns, or whether they want to partake in the exchange of produce and information about when to plant, both unwarranted and unwelcome. The work world the newcomers to Thomas Creek come from is one which fosters and rewards individual achievement, upward mobility and the severance of non-instrumental ties. To them, neighbours really don't have anything to do with going to work or surviving in today's world.

The nature of the interdependence of these families is very characteristic of mining, however. The community of Thomas Creek has changed from a company town to a retirement community, and is now changing into a dormitory community for Morgantown. The older residents have lived in the community throughout its various phases. For them, the experience of 'community life' is rooted in the understanding they have of their community as it was when it was a company town.

II The company town

During the first half of the century Thomas Creek was one of several mining communities or 'camps' in the area. The men worked in the nearby Thomas Creek mine. The women worked at home, managing the household and raising children. Like other camps of the time, Thomas Creek was company-owned and company-managed. It was built in 1906 by Bethlehem Steel, and remained a company town until 1948. It was then sold to a local entrepreneur who sold the houses to the people who lived in them.

Thomas Creek was similar to many other mining camps. People could live there only if they had a job in the Thomas Creek mine. Without the job, there was no house for the miner's family. The houses, such as they were, provided minimal shelter. One resident described them this way:

They repelled the rain most of the time. There was nothing to turn the cold except one layer of construction boards and many times these were not put together very tight, with the result that you waited until a warm day to take a bath. There was no running water in the houses. The water came from a pump that served at least four houses by means of a water pail. All the water used for cooking, bathing or washing clothes was carried into the house by the housewife, or if a family was lucky, by one of the boys ... Sometimes the kitchen had one electric light, which many times was supplied by the company and was hooked into the mine electric lines. So, when for any reason the mind did not work, there was no power to the houses ... most of the lighting was done by kerosene lamps. (Densmore 1977: 8–9)

In Thomas Creek, as in other company towns of the era, the miners were paid in scrip, i.e. rent for one's house and the purchases at the company store were debited against the miner's paycheck. Since the prices at the company store were two to three times the prices at the stores in town, the miners usually drew no cash on their paychecks. They were constantly indebted to the company. Densmore noted:

there was a record of purchases of each family that lived in the town. Most times when a man happened to draw some pay on payday he was called into the company office for an explanation of the fact and given warning to the effect that it better not happen again if he expected to keep his job.

Some single men were allowed to draw pay to meet boarding house expenses, 'but if the family man tried, there was trouble to pay'.

The difficult circumstances of company town life were not new to the residents of Thomas Creek when they first moved there in the late 1930s and early 1940s. They had grown up in company towns much like Thomas Creek. One woman described it like this:

The house that I was born in was at the west end of Tunnelton, and there was three of us – three children, and I lived next door to my grandmother. My mother's mother. When I was twelve years old we moved to Tunnelton proper. The west end of Tunnelton was a tunnel that cut through the hill, and railroad went under there, and we lived at the far end of that tunnel. There was very few houses at the west end, maybe a dozen or so. There was mines around. I think the closest one was what they called Austin, and Gorman, two old mining towns below us, on down the railroad track. We were just between. Then we moved to Tunnelton, and Dad couldn't find a house, I mean we rented for awhile. There was only one street right through the middle of town, we lived on it, and several years after that – 10 years – mother had three more kids, so Dad finally bought a house, the banker sold his house on the far side of the railroad, and that's where we grew up. I lived there 'til I was married and when I had two children of my own, and when the baby was six weeks old, we came down to Mongngehela County. J. and I lived in a small apartment over the Dobson's garage. J. was in the mines.[3]

Another man had lived in several towns: 'We came here in '36, '40 maybe. I'd worked the coal mine at Fairmont, then Canyon. Then I moved to Masontown and Bretts. I started here on Ground Hog's Day in '43. We took this house 'cause it was ready.'[4]

Most of the older residents were born between 1910 and 1915 of immigrant parents (Italian, Yugoslavian, Hungarian, Welsh). For children of miners, living in one or several company towns, life was rough and any labour the children could provide helped the family enormously. Parents sometimes asked the girls to stay home from school – and school authorities would come around looking for them. 'I remember one day my sister A. stayed home to help mama with the washin' (it was scrubbing on them boards, you know). Well, mama just said, I'm sorry, but she can go back tomorrow, I need her at home today.'

The boys would stay in school until about the eighth grade, when they were fourteen or so, and then go into the mines.

Loading coal, that's what I was doing at the age of fourteen. That's what I helped my step-daddy do is load coal. That was back in '24. I don't know what he got for a ton of coal back then. [Did your daddy get paid for the work for the work you did?] Oh yes, he got the money, that's why he taken me in the coal mine. At that time we wanted to go to Italy, and I went about six months in the coal mine so he could make extra money, load more coal cars, so we could have money to take to Italy with us.[5]

Although it was illegal for children to work in the mines, sympathetic mine bosses would call out a warning when an authority would come around to see if children were working, and the boys would hide.

By the time the boys reached manhood, and girls womanhood, they were experienced at working hard at the chores which would become their life's work. Social adulthood came early. Girls married and 'set up housekeeping' between ages fifteen and nineteen, and boys married at age twenty. Most never seriously questioned the pattern.

Although they were familiar with the way of life in a company town, moving into a new 'camp' was nonetheless initially difficult. 'I didn't like it when I first got here. I don't remember why. I just hoped we wouldn't have to stay too long. But, it's okay now, I guess.' Relationships became very important, helping them cope with the exigencies of camp life.

You know, so many things were scarce – coffee, sugar, yardage, muslin was ... you couldn't beg a piece of muslin hardly, back during the war; and he (the company store manager) used to get it in by the bagload up here, and the dirty devil would sell it out you know, and make a profit on it for himself, and the people here never got it. You just couldn't hardly buy anything. Well, he got kicked out because of it. I was there one day and I wanted some material, we used to wear little housedresses you know, and I wanted some to sew up things like that for everyday things, and I was told through the underground that there was some that come in, you know. I went up there and he was there and I asked about it, and he said no such a-thing, where did I get such an idea? and all that business, you know. And, uh, I don't remember, I said some pretty snotty things to him. [You didn't worry about keeping your mouth shut?] Well no, 'cause I know the butcher in the store!'[6]

Not only did knowing one another, depending upon one another as a kind of buffer against the company people, provide an important resource for the residents, but men depended on women and husbands on wives. Their work intertwined during the course of a day.

In the morning, at about 5.30, the men would head for the mine. Unlike mining today, these miners dug coal by hand and were paid by the ton. They would stay in the mine most of the day, blasting, shovelling and loading the coal into mule trains which pulled the ore from underground.

Although hand loading coal was difficult, exploited work, the miners took pride in it. They were relatively free from supervision and held a certain degree of control over the actual work process. One observer, writing in the 1920s noted:

The miner is first of all a workman in a unique relation to the working place which he is said, with little exaggeration, to 'own and inhabit' ... The mine is made up of 'rooms', which may be as wide as 24 feet, and which are held up by solid walls of coal. One or two men might work each 'room', so that isolation and independence remain very much the same, and the miner and his buddy work month after month, pressing forward slowly into the seam of coal by the light of their caps, left alone except for the track and the boss's brief visit. (Goodrich 1925: 20–2).

The men at the face were called loaders, or pick miners. These men, working alone, cut along the underface of the coal — with either a pick or a blast of dynamite — to loosen the coal which they then shovelled into the waiting cars. From this point on, the job was mainly shovelling,

mere shoveling, if you like, although in the process the loader must also clean the coal by throwing aside at least the larger pieces of slate and other impurities, shoveling itself may involve a considerable knack under underground conditions — particularly in the many cases where it is necessary ... to throw shovel after shovel of coal against the roof from which carooms into the car. This cycle of shooting, picking, cleaning and loading is repeated until the place is 'cleaned up' several days later and ready for the next cut. (Goodrich 1925: 26–7).

When hand-loading coal, the men were responsible for their tools. When they 'shot' the coal, they used dynamite they handrolled themselves, kept their tools sharpened at home after work, kept their lights in good shape and their clothes mended. Paid by the ton, the miner was a pieceworker and it was his own living, as well as his life, which was immediately affected by what he did or failed to do at the worksite. Although exploited, the miner controlled the process of work. Knowing how to mine and knowing the importance of coal to the national economy gave him a sense of great worth, and pride in his work.

The men would arrive back home about 3.30 in the afternoon, clean up and go out into the garden. Gardening has always been considered men's work in Thomas Creek, although women would occasionally help in planting or harvesting and occasionally plant and tend gardens themselves. One woman told me:

Well, the women around her putt around, but they don't, I don't think they actually *make* the garden, what you call make it. The men usually put it in, and, I know Jo just this summer would get out there and help Tony plant and things like that, she never did that before. And I never, I just don't have a knack for that sort of thing, raising flowers and things like that. It's not up my alley!'[7]

Women's work involved washing out the clothes the men brought out of the mine. This task involved carrying water in from the pump, heating it on the coal stove, washing garments by hand and hanging them out to dry. Women also cleaned house without the aid of labour-saving appliances, prepared and canned vegetables from the garden, cooked dinner, and cared for children.

The people in Thomas Creek regarded each other's work very highly. Women were conscious of the danger of mining and the consequences to themselves and their children should they become widows. Men were conscious of the difficulty of domestic life, and their dependence on the work which the women carried out. Not only did men and women utterly rely on one another, they relied on other families in the community as well. To survive and resist the tenuous conditions provided by the company, the people in the mining community developed a web of reciprocal relations involving the exchange of food, aid, comfort, joys and sorrows.

Perhaps one of the strongest forms this took was when men and women participated in fighting for the union, in order to protect the men at work and the families at home. The company fought back in ways which affected both the worker and his family:

It was the custom of the coal company officials, when they learned of men who joined the worker's union, to have their families moved out into the road. They cared not for children or sick folk or elderly people. To help alleviate this condition, the union built shacks for these people to move into and supplied tents for hundreds of families. I remember my uncle living in such a tent town and walking for miles to get groceries that were shipped in by boxcar on the railroad to the nearest town. (Densmore 1977: 8–9)

The domination by the company over the workers thus extended directly over family and household, even at the most basic point of the mere provision of shelter.

The relation of the company town to mining work and domestic life was important. The town provided the arena in which one's daily life was played out. For women this meant doing the hard work of maintaining a house without amenities, and living with the knowledge that they could become widows at any time; for men this meant working in dangerous circumstances; and for both it meant living at the whim of the company. Both men and women were equally affected by whether or not the mine operated. They were both affected by layoffs, and neither had a great deal of choice about living another sort of life.

These circumstances provided the setting for mining families to develop a great deal of solidarity. Even while experiencing exploitation they were proud of their work − both the men in the mines, and men and women in the town. The adverse circumstances in which they lived served to encourage the development of close social ties, sharing of resources, reliance upon one another. Being able to survive such circumstances, and feeling important within the national economy, underlay the sense of dignity they had about their work.

III Discussion

Over the last forty years Thomas Creek has undergone some profound changes. When the older people with whom I lived moved there, it was a company town. In 1950 the mine shut down and they bought their houses, and worked at other mines. During the mid-1960s, most of them retired.

Although the men were only about fifty-five years old at the time, they suffered from black lung and silicosis, and could no longer keep their jobs. They lived on their pensions, social security and, after 1969, federal black lung benefits.

From the mid-1960s until 1978, when I began my research, Thomas Creek was a company town grown old; a town in which neighbours had lived and worked together for thirty to forty years. They had been retired together for about ten years. The patterns of exchange, the importance of gardening, canning enough food to last them through a sparse season, became a set way of life. No longer living under the domination of the company, the solidarity which they had garnered during those black years remained intact. Rather than supporting each other's welfare in the face of an overly greedy company store manager who kept goods from them, their joint concerns are diminishing social services, and rising prices.

In the late 1970s, the community began to experience another change: as the nearby university began to grow, young newcomers to Morgantown started looking for housing. When a house became vacant in Thomas Creek as an elder died, or moved out to live with son or daughter when becoming widowed, a young couple would move in.

These newcomers were not only fifty years younger than the older residents, but they were also not interested in participating in the community. The elders found it hard to understand why or how anyone who lives next door might not wish to visit, to exchange food, or to simply chat. Their behaviour was indeed so odd that on one occasion I heard a young woman referred to as 'queer-like, she acts like there's really something wrong with her. I sometimes wonder if she ain't going mad.'

The values which the older residents of Thomas Creek share − working hard, taking care of one's house, keeping an eye on one another − would lead one to predict that derogatory comments would be directed toward those people who don't work hard, who don't help and who don't participate in the 'community way'; that is, the way as it is defined by the older residents. Work, to these older people, is a way of taking care of one another, a way of maintaining a relationship. Not to work is a sign of alienation from the community. The fact that these mining families communicate and sustain each other through work is an upside-down grammar when compared to the usual conceptions of work in our society. One normally links work to alienation, and escape from the workplace with sustenance, which is generally expressed through leisure.

In Thomas Creek the opposite occurs. Work is the means by which the residents communicate, take care of and invest in their community. The underlying 'rules' which provide the logic for their behaviour stem from the shared background of living in a dominated community, performing work which frequently results in injury, disability or death, and living in a family whose members are as affected as the worker himself by haphazard availability of work and the whims of the company, because their houses were dependent upon and sometimes attached to, the workplace. This shared experience of having grown up in company towns is a kind of 'grammatical knowledge' for the residents of Thomas Creek. The way they structure their relations – organising the home-as-workplace, allowing work to provide sustenance and be a measure of one's moral stock in the community – is a transformation of the pattern of having to sustain one another under the heel of a dominant company. Work has become a visible and nurturing symbol for stating one's investment in the community for the elders. By working and engaging in exchanges with one another, the residents have been able to plant and cultivate resources in each other which can be harvested in time of need.

The emergence of solidarity in dominated conditions is not unusual; Wolf described this nearly thirty years ago (Wolf, 1957). It is interesting that the social patterns developed during the earlier lives of the people continue to order their social relations toward the end of their lives as strongly as they do. It makes sense that these patterns continue, given the new constraints which the people have faced in recent years. Early retirement produced decreased incomes. Gardening and canning – aspects of domestic work which had been part of their lives for many years – logically became the focus around which people organised their labour. They provided an immediate cure for financial ills as well as a visible activity which was socially credible.

The influx of 'the newcomers' to the community was an affront to the elder's understanding of social solidarity since most of the newcomers have not been interested in participating in the way of life established by the older residents.

In 1984, the elderly retired mining families of Thomas Creek face living the remainder of their lives in the community in which they have grown old, but in which they experience a decreasing sense of 'belonging'. This sense of loss tends to augment the importance of what social ties they do have and 'the community' becomes more important to them as they grow older. While lamenting the present, they praise the past – the same past which they found so difficult to endure: 'You whould have seen Thomas Creek back then. It was boomin'. The mine was goin'. And everybody, oh just everybody, just took care of each other. It was really somethin'.'[8]

Notes

1 The research for this article was carried out in a community which I call 'Thomas Creek', West Virginia in 1978 and 1979. I am indebted to the residents of that community who shared their lives with me. I would like to thank Faye Ginsburg of CUNY Graduate Center for her critical comments on this manuscript.

2 The work of Dennis *et al.* (1956), Nash (1979) and Greaves *et al.* (1982), however, provide a full treatment of the mines and the community.

3–8 Taped interviews from author's field research.

References ** See annotated bibliography for full details of all asterisked titles*

Dennis, N., Henrique, F. & Slaughter C.
1956 *Coal is our Life*. London: Tavistock.

Densmore, Raymond E.
1977*

Goodrich, Carter
1925 *The Miner's Freedom*. Boston: Marshall Jones Company.

Greaves, Thomas, Xavier Albo and Godofredo Sandoval
1982 'Becoming a tin miner'. Paper presented to the 44th International Congress of Americanists, Manchester, England, September 5–10.

Nash, June
1979*

Sachs, Patricia
1984 'Is volitional work voluntary?: the case of retired coal miners'. *Anthropology of Work Review* (Stewartsville, New Jersey) 5(2): 19–24.

Wolf, Eric R.
1957 'Closed corporate peasant communities in MesoAmerica and Central Java'. *Southwestern Journal of Anthropology* 13: 1–18.

CHAPTER 12

'Troncos' among black miners in Colombia[1]

NINA S. de FRIEDEMANN

Mining research that ventures beyond the orbit of capital intensive, complex organisation to examine marginal, small-scale mining is uncommon in our literature. It is doubly rare when the setting involves indigenous blacks and highly distinctive cultural mores. Dr Friedemann's research on Colombia's south-west coastal frontier is, therefore, of special interest. Her paper describes the intricate adaptation that exists between mining activity and the cultural web of kinship, reciprocal labour customs, and domestic life for the blacks of the gold-bearing Güelmambí River.

One contribution of her research is to direct our attention to the important role kinship and marriage necessarily play in mining life. Our literature takes such a consuming interest in a miner's class, union and locality membership that the personal matrix of kinship resources and identity is given scant notice. Yet, these assets are a part of, and used by, miners everywhere. The Güelmambí blacks have lessons for all studies of miners.

Nina S. de Friedemann (Apartado Aéreo 100.375, Bogotá, Colombia) is a Colombian anthropologist whose work includes a major monograph (1974) on the Güelmambí blacks and continuing research and writing on Colombia's black population.

— *The editors*

We are trees rooted in our mines
along the gold rivers.
Each trunk is a brother.
They were the founders
of our families and mines.
We are branches, limbs and twigs.
We are the descendants.   — *A miner of the Güelmambí River*

I Introduction

The black miners of the gold bearing zone of the Pacific Littoral of Colombia have a social organisation based on cognatic groups known as *troncos*. Mining, natural environment and the socio-cultural context are critical factors for their family organisation. Associated with this system of relations is an exclusive dependence on human energy and a technology so rudimentary that even the wheel has not been incorporated. Their habitat, a tropical rain forest, is one of the most humid of the world.

Troncos are defined by the miners as groups of families with rights to live, mine and do agriculture in a specific territory inherited from ancestors who took possession of the land more than 150 years ago. Each tronco claims a particular founder to whom every individual traces his line of descent through a male or female link to exercise his latent or active rights.

The miners in the Güelmambí River of the extreme south-western corner of Colombia's Pacific littoral, are part of the South American black frontiersmen described by Whitten (1974). The tronco functions as a strategy for survival in the regional, boom or bust economy that pervades the whole society of the Pacific Littoral. The sporadic international demand for gold, tagua, wood or oil has influenced their adaptation, as in the following examples:

(a) When multinational dredges destroyed fertile land and subsistence crops in the river lowlands the miners were forced to abandon their settlements and re-establish elsewhere.[2]

(b) When the sharp increases in labour demand in oil camps or similar venture capital enterprises attract miners to leave their own mines, the mining force becomes too small, and some local mining operations have to be closed temporarily. The remaining members of the descent group have to work in other mines.

(c) When the rivers flood and the houses become destroyed, the miners have to find new housing sites.

(d) When the mining area becomes depleted, the miners have to move to other sites in search of gold.

To cope with these problems arising from the larger society, and from their own physical and social environment their tactic has been to use the resources offered by their troncos.

The Güelmambí River is located in the Barbacoas region of south-western Colombia. The region is a wide alluvial delta formed by the deposits of the Patía River and its tributaries as they flow out of the western cordillera of the Andes toward the Pacific Ocean. The Güelmambí flows into the Telembí River, a tributary of the Patía. The upper and middle portions of almost all the waterways of the western slope of the cordillera are gold bearing. The most important gold zones are the ancient gravels of the interfluvial areas between the modern streams. Most of these gravels were deposited at the end of the Pliocene or during the Pleistocene by streams cutting through the western cordillera gold zone. They extend along a belt that runs the length of the cordillera, from the Upper Atrato River to the Colombian–Ecuadorian border. Barbacoas is located in one portion of this gold belt (West 1952, 1957).

II The miners

With the arrival of the Spaniards in the sixteenth century, the Indians began to disappear rapidly, victims of epidemics, suicide and forced labour. Although they were being replaced by black slaves in the mining centres, during the first part of this period a number of Indians worked together with blacks in the mines. The aboriginal mining technology was thus transferred to the colonial technology of mining and agricultural system and thence to the material culture of the blacks.

After abolition in 1851 many blacks dispersed over the littoral forests and settled along the banks of the rivers, much as other blacks had done before abolition, when they were allowed to buy their freedom from their masters by working on Sundays and holidays. Soon the mining centre of Barbacoas became depopulated, and commercial mining was abandoned. In the forest, blacks recreated a mining economy on their own with aboriginal skills and traits, patterns from the colonial mining system, and their own cultural and social elaborations.

The various mining villages are strung out over a distance of thirty kilometres along the banks of the Güelmambì. They vary in size, the larger ones from twelve to fifty houses and the small ones from two to six. In 1980, approximately 1,200 miners lived in these settlements on the Güelmambí, i.e. 1.6 per cent of the total population of the municipality of Barbacoas. Through the innumerable labyrinths of streams, swamps and rivers that flow through the soggy jungle, the inhabitants paddle their dugout canoes, the only vehicle for transportation. The rain drips continually from the palm thatched roofs of their huts which are raised on wooden pillars two or more metres above the ground. Pigs, chickens and canoes are kept beneath the huts. Some households have a hand press for processing sugar. Others have an earthen oven for baking bread. The rocky debris left from the mining process is used to pave the areas around the house.

In the morning before adults and older children set off for the mining cuts, the women do the washing in the river. In the meantime the men and older boys work in their *chagras* (agricultural plots) and check their traps set up the night before to capture small animals such as agouti, rabbits or armadillos.

During the rainy season, mining begins at six in the morning, three days of the week in the mine of their kin group and the other days in the gold workings and fields assigned to each family. If the cut has enough water, work continues until six in the evening. When the rains slacken during the dry season, people stop going to the mines every day. When they go, they stop work between one and three in the afternoon. The division of labour tends to conform to the following pattern: men plant the crops and weed the chagras while women harvest the products and take them home; children gather fruit in the forest; and young women go to the river to catch small fish with large round nets.

At least once a month there is a *velorio*, a celebration with singing and drumming in honour of one or more of the catholic saints. The velorio starts Saturday night and lasts until dawn on Sunday and brings together people from different troncos, settlements and rivers (Friedemann 1966–9, 1974). This is an occasion to renew genealogical knowledge on troncos, ancestors and rights. Tuesday is market day in Barbacoas and some of the miners go to sell gold dust and to buy supplies such as candles, salt and oil.

Mining work is done with an iron bar, a digging stick, an iron spoon with a wooden handle, *cachos* (gourd scoops), a panning tray and an oblong wooden tray. Men break up the slopes or terraces with the iron bars. Stones are removed and passed from hand to hand along lines of young men and women and piled up to make terraces far from the original 'cut'. The finer materials are washed and the sand separated and deposited in a trench made for this purpose. Here men and women patiently stoop over the water, sand and debris, picking the remaining small stones from the mix, called *mazamorra*. From this mazamorra the women finally separate the gold dust, using slightly convex wooden panning trays that they work rhythmically to move the *jagua* (a mixture of gold dust and particles of iron oxide) to one side of the pan.

III Mine and tronco

Existing literature on placer mining in the Pacific Littoral talks only about the work of groups of families and friends on public lands. The geographer, West, realised that among the miners there existed a property system regulated by descent rules, but he made only a passing reference to the system (1957: 154).

Over the last twenty years the anthropological understanding of non-unilineal descent groups with a focal ancestor has been expanded (*cf.* Davenport 1959; Sahlins 1962; Fox 1967; Buchler and Selby 1968; Goodenough 1970; Keesing 1975), including further study of a type of non-unilineal descent group, the ramage, known especially from Polynesia.[3] The organisation that has been defined as ramages in Polynesia is similar to that of the troncos in Colombia, particularly with respect to the choice between patrilateral or matrilateral affiliation alternatives and the connection of group members with a common ancestor in a chain of parents and children that is basic to the definition of ramages in non-unilineal or cognatic systems (Buchler and Selby 1968: 90; Goodenough 1970: 42). Thus, troncos in the Güelmambí may be described as consanguineal kinship groups whose members trace their descent to a common ancestor through a line of males or females in a series of parent – child links and, therefore, may also be classified as ramages.

In this article, I will analyse the San Antuco Mine, which is a *mina grande*, containing twenty-four households divided among three troncos designated Cristino, Otulio and Leonco. These names stem from the first name of the ancestor of each descent group. Traditional narratives relate the origins of each tronco and its lands, and tell of the descendants of freed slaves who came to the area, took possession of former Indian territories and formed the families that have lived there ever since.

Kinship ties are important and often discussed. At the frequent velorios genealogical histories are recounted in detail. When a man or woman arrives at the tronco's mine workings, the greeting from those at work is accompanied by remarks about the kinship tie that underlies the right and expectations on which the arriving worker's participation is based.

Each tronco owns a territory where all the family units live and work. The three troncos constitute the San Antuco Mine (Fig. 1), known generically as a mina grande. A mina grande is composed of more than one tronco and includes each tronco's communally worked *compañía* mine, as well as the *comedero* mine alloted by each tronco to its member families, together with their agricultural plot (chagra) and house-garden-patio site (see Figure 12.1).

A document exists that legitimises the miners' claim to the San Antuco territory, dating from 1899, twenty-five years after the founders had settled in the area. However, this document, like many other property titles, is no longer regarded as valid by the national government since the miners did not keep up with tax payments and other bureaucratic requirements. Consequently, their official status is defined as squatters on public land.

There are documents for other mines in the area that bear the names of women as the ancestors of other troncos. These data are relevant for the examination of the development of the troncos as a cognatic descent system in which membership is defined by optional affiliation with a group through either a paternal of maternal link.

Most traditions on the Güelmambí also show the founders as brothers or brothers and sisters. When they were not siblings by descent, it is explained that they became siblings by settling on the same land to found the mine. Tradition tells of another instance when one of the founding brothers moved away and gave his rights to another individual who had arrived in the area. This person then became an ancestor and today is considered as having been a brother of the other founders.

Every aspect of life among the miners is in reference to the mina which includes the physical territory, the work, one's family and one's identity. The term mina applies to the complex formed by the settlement, the family chagra or agricultural plot, the family mine or comedero, and the communal mine or compañía where all the members of each tronco may exercise their rights to participate in communal mining activities (Figure 12.2).

IV Affiliation to troncos: active and latent rights

The actual recruitment of members in the tronco depends on the patrilateral or matrilateral affiliation that the individual chooses. Although a person can possess rights in several troncos at one time, the exercise of the right is maintained by constant participation in the communal work at the compania mining site on the property of the tronco of affiliation, by the use of lands for his chagra and comedero, and by the preferential use of the tronco's territory in establishing his residence. Consequently a person can maintain active rights in only one tronco at a time.

The founders and the second generations of miners along the river banks

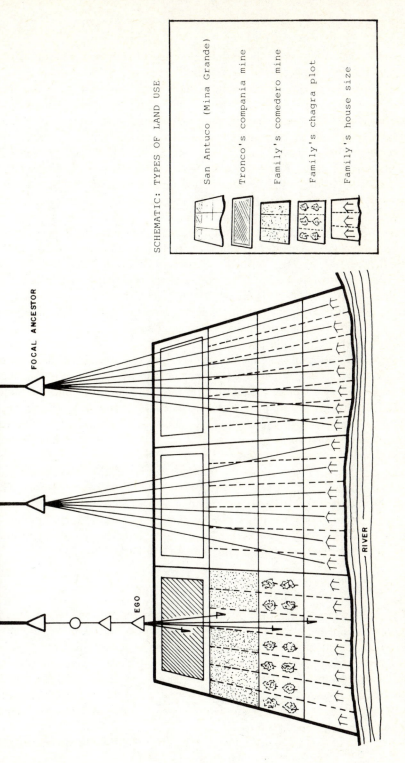

Fig. 12.1 Types of land use.

SCHEMATIC: TYPES OF LAND USE

San Antuco (Mina Grande)

Tronco's compania mine

Family's comedero mine

Family's chagra plot

Family's house size

FOCAL ANCESTOR

EGO

RIVER

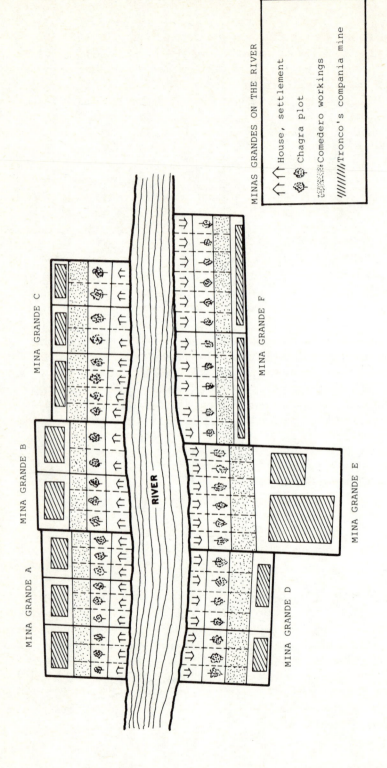

Fig. 12.2 Minas Grandes on the river.

maintained tronco affiliation and discrete, clearly defined parcels of land associated with each comedero, chagra and house site. This is not true today, although some people can still point out the old boundaries of the residential sites of each tronco. Some active members of San Antuco may and do, live in nearby villages. This occurs, for example, when a woman is married to, and residing with, a man of another mina grande along the river, and it has been agreed that the family unit would affiliate with the woman's tronco. Thus, the husband, wife and children go to work the chagra, the comedero and the compania sites using the wife's rights although their residence remains patrilocal.

As cognatic groups, the troncos can be conceived of as pragmatically restricted groups (Fox 1967: 156), since in practice the individual only exercises his active option in one mina at a time. However, this affiliation is not immutable, as seen by the maintenance of latent rights in other troncos, and their minas.

An individual can choose to affiliate with one tronco in preference to another through options provided by either a maternal or a paternal link. Each nuclear family unit also has the option of activating rights in any of the troncos to which either the man or the woman can trace ancestral lines. This optional affiliation creates a flexible system that can adjust to circumstances such as personal preference for residence, better mining opportunities, and cultivation on sites belonging to the woman's or man's tronco.

The ideal norm is that a man will have worked with his parents since adolescence on land received from them, and will continue to activate the rights received from his parental family unit. He will bring his wife to live in the territory and the tronco of his parents. In such case, the wife's rights in other troncos and often in other minas grandes remain latent in the family unit, to be exercised in case of need or to be invoked later by their grown-up children if they so desire. The preferred norm of patrilocality is changeable in cases of activation of the woman's rights by the family. When this occurs the rights of the man become the latent rights of the family unit but remain available to the husband, wife or children when circumstances require their activation. Maintenance of latent rights permits people to solve problems rapidly, as when house sites are flooded; flood victims can invoke their latent rights and acquire a new house site in a neighbouring village.

The miner conceptualises two categories of active rights within the family unit: (1) property rights to a house site, garden land (chagra), forest land and comedero mine as a member of a tronco; (2) work rights to participate in the communal mining work, the compania. The compania mine works on communal land that is never divided and that belongs to his tronco as a unit.

To exercise his first order rights the miner builds a house, generally overlooking the river, after having obtained approval of the group controlling the territory. He then prepares a chagra, by the slash and mulch system[4] where he plants mainly corn, beans, yuca, plantains and sugar cane. For his comedero he makes a small water reservoir for rain water, digs a sluice and washes sand each week, a process similar to his work at the compania level. The gold dust recovered may allow him to buy dry fish, salt, meat and staples to maintain the domestic group. The entire nuclear family cooperates in all these activities, thus exercising the rights associated with the tronco to which the unit is affiliated.

The father heads the domestic work group. He uses the iron bar to initiate the breaking up of the terrace. In domestic units of widows and children, a son is in charge of this work. If she does not have a son, the widow recruits a male worker amongst her male relatives. A rule in the mining world is 'there must be a man present'.

The second order of rights, those to work in the compania, are exercised by participating in the communal mine of the tronco, where its members work together three days a week. The whole family may go to the communal mine on these days, but it is also possible for only one person to exercise rights as a member of the larger unit. The compania names a captain to organise the workers and to assign the tasks along the trench or in the line that moves the large stones away from the cut. The captain keeps track of the days worked by each individual and at the end of the *picado*, or mining period, each worker receives a sum of money proportionate to the days he contributed. The captain and a tronco representative have the duty of exchanging the gold for money in Barbacoas. Every three months the profits are divided among the workers, who use their share primarily for the expenses of religious ceremonies, including clothing, drink and food.

In the beginning of the 1970s, the exercise of rights in one of the compania sites of San Antuco was practised largely by women, and male participation diminished noticeably. The same mine worked by seven men and nineteen women in 1969, had only three men and fourteen women in 1971. In recent years nearby oil fields and road construction projects have required unskilled labour, and many men have left the Güelmambí for brief periods to earn cash. While trying to get cash as workers in the larger society, they maintain their rights in their troncos and minas back in the forest and return to their families when lean times arise in the labour market. This is part of the adaptive function of the troncos, which permits the population to adjust to the boom and bust economic cycles of the Pacific Littoral.

In San Antuco, for instance, there was a period in the 1970s when the Leonco trunk had only three family units actively exercising their rights and consequently the compania works had to close for a long time. Notwithstanding, the members spoke of their compania mine as if the mining was being carried on regularly. By 1983, the communal mine had started operating again, with an average of seven men and ten women.

Finding gold is not easy for these miners. A group of twenty-five workers can wash the rocks and sands of a terrace for several months only to get one gram of gold dust. When this happens, the compania group agrees to dissolve for a while. It is at this moment that its members start activating their latent rights in other compania mines, but they must first obtain the consent of their own compania captain and the consensus of the members of his tronco, through its representative, who is the man or woman that keeps the document of land ownership. Another option is to relocate to another site within the communal land to try their luck. If yields are promising the group may then continue together.

The maintenance of latent rights is a constant feature in the daily inter-action of miners, expressed through mutual aid. A person can, and does, travel to nearby villages by canoe to help repair a house, clear land or mill sugar cane. Latent rights can also be maintained by helping in other troncos' compania mining operations.

The active and latent rights of each family unit in the troncos rooted in San Antuco are shown in the structural representation of the village (Fig. 12.3). This diagram could be extended to cover other minas grandes and their troncos where Otulios, Cristinos and Leoncos maintain latent rights and likewise with the latent rights that miners from other minas grandes hold in San Antuco. The result would be the same pattern of descent lines, spread over a larger area of the river. The system, linking people and rights across the physical boundaries of the minas grandes in a broad kinship network, leads the residents to correctly exclaim, 'on this river we are all relatives'.

Tables 12.1 and 12.2 are a summary of active and latent rights of each one of the family units within the troncos Otulio, Cristino and Leonco. Moreover, Table 12.1 shows eleven units actively affiliated with Otulio, ten with Cristino and three with Leonco. Also it shows that some of the twenty-four family units of San Antuco hold latent rights in other minas grandes B, C and D along the river Güelmambí.

Table 12.2 specifies the male or female links used by each one of the units to affiliate actively each one with the three troncos. Here it is shown that in the fourth generation six family units (1, 5, 7, 12, 18 and 21) activated rights received through the woman, and nine family units (3, 4, 8, 11, 13, 14, 17, 22 and 24) activated rights received through the man. In the third generation five family units (2, 10, 15, 20 and 23) activated rights received through women and four units (6, 16, 19 and 25) activated rights received through men. This means that of the twenty-four houses now occupied by members of the Otulio, Cristino and Leonco troncos, thirteen (or 54.1 per cent) of the families have affiliated by invoking rights received through men, and eleven of them (45.0 per cent) have affiliated by invoking rights received through women. As these data show, the descent system has a slight patrilineal bias, possibly induced by the ideal norm of patrilocal residence.

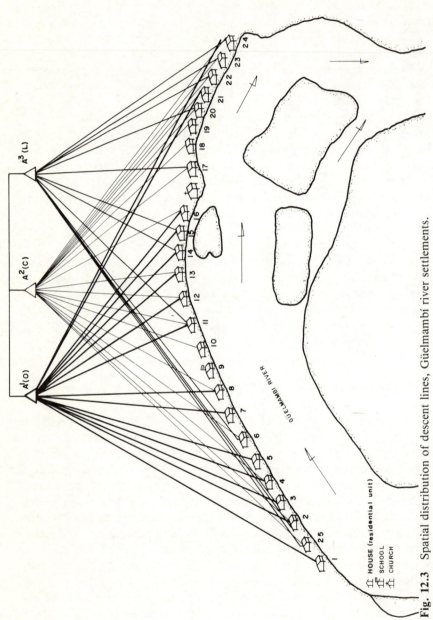

Fig. 12.3 Spatial distribution of descent lines, Güelmambi river settlements.

Table 12.1   Active and latent rights, San Antuco families

Family	Mine A₁(O) Active	Mine A₁(O) Latent	Mine A₂(C) Active	Mine A₂(C) Latent	Mine A₃(L) Active	Mine A₃(L) Latent	Mine B Latent	Mine C Latent	Mine D Latent
1	–								
2		–	–			–	–		
3		–			–				
4	–			–				–	
5	–								
6			–			–			
7	–								
8	–			–					
9									
10			–					–	–
11	–			–		–			
12	–			–					–
13	–			–					
14		–			–		–		
15	–				–			–	
16	–								
17			–			–			–
18			–				–	–	
19			–				–	–	
20			–			–	–	–	
21	–							–	–
22						–			
23			–						
24	–		–			–			
25			–			–	–		–

V Marriage

On the Güelmambí the basic domestic and economic unit is the nuclear family. A man ideally takes his wife to live in a house he builds when they begin their life together. If he as yet has no house, he may take her to his parents' house until he obtains the wood to 'raise' their own. Since both the man and woman have their own rights to a residence site, chagra and mining sites, a decision is made on the affiliation of the new family unit. The husband and wife may come from the same settlement, but each is likely to have an affiliation with a different tronco. A Cristino, for example, will marry a Leonco or an Otulio in the Mina Grande of San Antuco, or maybe someone from an entirely different tronco (here designated 'X') from another mina grande. These

Table 12.2   Ancestry, San Antuco troncos

	A_1(O)				A_2(C)				A_3(L)			
	G_1	G_2	G_3	G_4	G_1	G_2	G_3	G_4	G_1	G_2	G_3	G_4
1	F	F	M	D								
2					F	M	D					
3									F	M	F	S
4	F	M	F	S								
5	F	F	M	D								
6					F	F	S					
7	F	F	F	D								
8	F	F	F	S								
9												
10					F	F	D					
11	F	M	F	S								
12	F	M	F	D								
13	F	M	F	S								
14									F	M	M	S
15									F	M	D	
16	F	M	S									
17					F	F	F	S				
18					F	F	F	D				
19					F	F	S					
20					F	F	D					
21	F	M	F	D								
22					F	F	F	S				
23					F	F	D					
24	F	F	M	S								
25					F	M	S					

Key: G_1 Founding generation, G_2 Second generation, etc.
F = Father, M = Mother, S = Son, D = Daughter

cross-tronco marriages endow the new family unit with wider kinship links and, thus, its ability to call upon a larger set of people for reciprocal tasks.

A summary of this data is given in Fig. 12.4, showing that La Mina Grande, San Antuco, is not endogamous. It is also clear that troncos C, L and O maintain control of their territories. When a C, for instance, marries someone from another mina grande, tronco X, if this person joins tronco C he or she exercises the active rights of the C spouse. The person's own rights in X are left latent for the family unit CX. Conversely, when an X brings a spouse from C to his/her mine, C joins the group with his/her latent rights for the family unit XC.

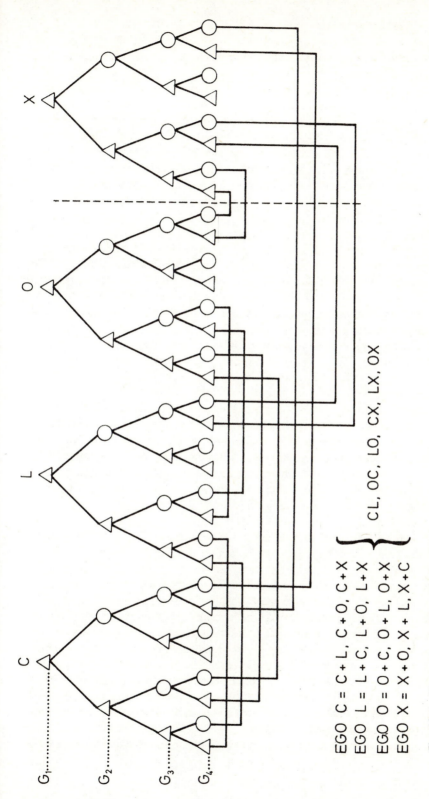

EGO C = C + L, C + O, C + X
EGO L = L + C, L + O, L + X
EGO O = O + C, O + L, O + X
EGO X = X + O, X + L, X + C

CL, OC, LO, CX, LX, OX

Fig. 12.4 Troncos: matrimonial linkages.

H

The descent groups of the Güelmambí are linked in a network over a large extension of territory.

The strong preference is for marriage between people from different troncos. Tronco exogamy increases (1) the breadth of latent rights in various troncos, and (2) the number of affinal relatives of Ego, to augment the potential groups of people who may be called to work. Marriage between children of siblings is prohibited as are premarital relations, which are transitory if they do occur. Kinship terminology for first cousins reflects this rule: Ego uses the term brother/sister with the children of his parents and also the children of his father's and mother's siblings. People on the Güelmambí are emphatic that 'marriage between cousins doesn't work'.

Marriages in San Antuco between people of the same tronco have not lasted long, as shown in Fig. 12.5, which also illustrates close adherence to the norm of tronco exogamy. Twenty of the twenty-four marriages follow this rule. On this point it is important to observe that only ten of these marriages are between members of troncos within La Mina Grande San Antuco. Each of the other ten marriages has one spouse (six women, four men) originating in a different mina grande along the river.

On the Güelmambí, premarital relations may or may not result in a permanent relationship. If a sexual coupling concludes with the birth of a child the man might only take responsibility for the costs of the birth, the midwife and clothing for the infant. A woman involved in such a brief relationship will remain in her natal family unit and may later marry someone else. These encounters may involve youths or a single woman and a married man. In the first instance the relationship may develop into a permanent union. Community recognition of these unions is not predicated upon a formal catholic religious ceremony. Although most couples will announce their intention of going through with the ceremony some day, this may not occur until several years later. Community recognition arises from the active exercise of rights by the man and the woman as a family unit in a mina grande and in a tronco.

When an unmarried mother marries a man other than her child's father, the child remains with her parents. He adopts the grandparents' surname and exercises the rights of a child in their tronco.

In cases of prolonged sexual unions between married men and unmarried women, the woman and children of such a union form a separate family unit. The man retains sexual access to the woman, as the father of her children, and is expected to provide gifts of clothing, small sums of money for the children, occasional food to the household, and to build a house in the territory of the woman's tronco. This house becomes the property of the woman. The woman and her children remain part of her natal household. They work in that group and participate in the circle of mutual aid corresponding to it. These women may live in the same settlement or in a different one than the man with whom they are involved.

Children of such unions are known as illegitimates (*bastardos*), a term

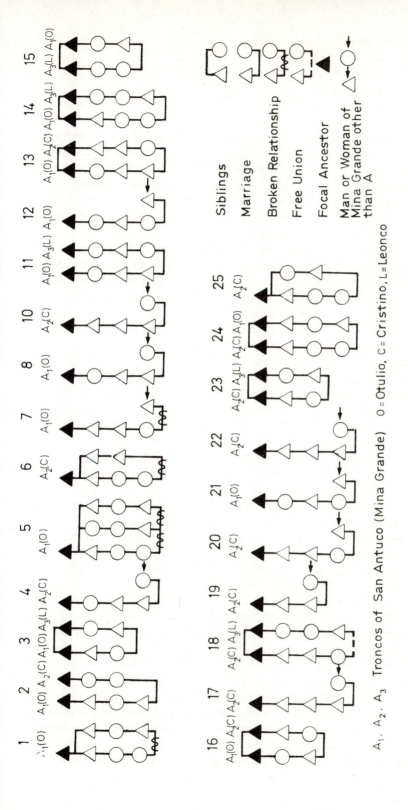

Fig. 12.5 Marriages of the twenty-four habitation units in San Antuco.

specifying the social illegality of their existence, and they identify themselves as 'the natural child of F' (name of father). In this way they maintain a latent paternal kinship link. During childhood and adolescence, exclusion from the descent group of the father is marked, but this may change significantly with increasing age. Today, when the emigration of young people to urban centres weakens the number of workers needed in the compania mining, the potential human energy offered by an illegitimate offspring is welcomed. His work enables him to claim rights in his father's tronco although his social status remains that of an illegitimate.

In San Antuco there was a unit formed by a semi-invalid old woman, her illegitimate daughter, and the children of this daughter, also illegitimate. Genealogical research showed that the old woman's mother had had children in extramarital couplings with an Otulio man. She had then left the region, taking her children. She thus forfeited her children's claim to affiliation with the tronco of father or mother. The daughter of this woman who had become a semi-invalid, later returned to claim rights as a natural child of an Otulio, but they were not granted to her. Her father's brother gave her a house site within his own settlement territory but she failed in her attempt to activate any tronco rights. Her own daughter, who accompanied her, thus had no rights to offer to anyone in marriage. Like her mother, she has had children from brief sexual unions. Through her illegitimate children with another Otulio man, a member of the same tronco in which her mother attempted to claim rights, it appears that her claim might eventually succeed. When they reach adolescence they can offer their physical labour for the communal mine and mutual aid obligations within their father's tronco. If they are accepted, they will be permitted to exercise their rights within the tronco.

Once an illegitimate child is allowed to activate descent rights by working he/she establishes an affiliation just as any other member of the tronco. One man, who had previously lived in another settlement affiliated to his mother's tronco, was able to exercise rights in tronco O as the illegitimate son of an Otulio. He then married a Cristino woman and came to live in San Antuco. A few years later the relations between this man and members of the tronco were still developing. He would show a ready availability to participate in the work of his mutual aid group, and was constantly being called on.

It can be concluded that illegitimates do not represent merely a deviance from marital and family norms that can be ignored; they constitute potential human resources that can be incorporated into the tronco. In San Antuco, the mina grande from which the two sample cases were taken, there is a distinct difference in the positions. In the first case of the semi-invalid woman, not only are they not accepted in the social system, but even their existence within the physical boundaries of the community is denied. Anyone asking where the settlement begins and ends will be shown a point that specifically excludes house no. 1 from the village limits. Correspondingly, none of the persons of this household have chagra land, nor can they work in any compania. All they can do is work sand in the river, whose waters are considered property of the national government. For their support the daughter obtains periodic work in other settlements and occasional help from her lover. In the second case of the man born of Otulio, the house is located at the other end of the settlements and is included *within* its recognised boundaries. The man in this case is the illegitimate son of an Otulio with a sister of the semi-invalid old woman heading household no. 1. The rule in this instance is modified by the fact that he lives with a Cristino woman. It also highlights the importance of active work in troncos and the need of human energy.

VI Transmission of property and work rights

To explain mining on the Güelmambí, the descent groups provide the miners with strategies of adaptation to a particular socio-economic, political and ecological environment. The potential resources and material products of their environment are secured and transmitted through tronco affiliation.

Possession of a comedero and chagra in a mina grande is subject to a process of 'registration', and active residence on the part of the family unit, similar to the original occupation of the mina grande by the ancestor/founders. To 'register' a plot means to obtain the approval of the tronco of affiliation to mark off within its territory a particular area for individual or family use. When the 'approval' is obtained, the interested parties clear the land, cut out the underbrush, and prepare the terrain for cultivating or mining. It can then be passed on as inheritance to the next generation. It may happen that the registered plot is used for only a few years. If another aspirant of the same tronco then wants it, he can obtain it on loan from the owner of the registration. There is, then, another process of consultation and approval from the tronco. When an owner abandons the settlements and/or land without notification that he will return, a new aspirant can register it for his own use and for transmission to his heirs. Growing emigration to urban centres has increased this secondary occupation through loan and re-registration.

Houses built by a man for his family belong to both spouses. If the man dies, his wife is the sole owner and vice versa. When both parents are dead the house and all its contents are divided equally among the children. These wooden houses are easily taken apart, so each child carries away a portion. It is interesting to locate the parts of houses that belonged to the parents or grandparents and to find that the inhabitants can show the pieces and identify from whom they were received. Don Inocencio, for example, one day showed his cousin the 'mother' beam of his house, inherited from his parents. His mother had, in turn, received it from the house of her parents.

House division is supervised by the eldest child, and his or her word is respected in this matter. If there are small children the eldest child also takes charge of them, receiving their portions of the house and accepting the responsibility of their care while they grow up. A child who wants to keep the entire house can have it appraised and divide its monetary value among the other children. The house can then be torn down and moved to this child's residence site, or it can be occupied where it stands. the house of an unmarried woman and her natural children is the sole property of the mother. When she dies it is divided among the children. Actions implicated in the division, appraisal, moving, or maintenance of a house in one place are matters for general discussion by everyone in the settlement and among the members of the involved troncos.

At age eighteen, children may receive land to work from their parents, and they are obligated, in return, to help the household with a portion of their harvest while they continue living in the parental house. This early inheritance of land, including trees and usable woods, induces young men and women to maintain their affiliation with their parents, tronco and to later bring their spouses there. It also permits the parents to keep their children nearby, helping in the chagra and comedero work. This redistribution of land may occur earlier if the parents are ill or are too old for active work. For example, in one of the units of San Antuco the father was old and becoming an invalid. Since most of his children were female, he gave them land to keep them and their husbands nearby when they married. In this way he managed to maintain a work force attached to the parental unit.

The lands given to children may not be used until they marry, but the new unit has to become affiliated to the parental tronco. Each tronco then makes room for new adults when they begin to exercise their rights independently of the parental domestic unit. Although each mina grande has limited land, availability of land has not been a problem, because the area is under-populated.

Both men and women work and their rights are transmitted to their children who begin exercising them with their parents from about the age of twelve. If, as an adult, a child affiliates with another tronco these rights become latent, as explained above. When members of a group leave for long period of time they can also loan out their lands. Those who do not return eventually lose their rights. 'Only the children who remain in the mina inherit', said a woman, who added 'The child that does not help, does not inherit'.

VII Discussion

This article has sought to describe relevant features of the tronco, a non-unilineal descent group, as an adaptive strategy for survival among blacks in a gold mining zone of the Pacific Littoral of Colombia.

Though the system has been traced back a century, evidence suggests that its roots reach even further, to well before abolition of slavery. While cognatic groups have been the subject of study and controversy since the 1950s, this case seems to be the first one to appear among blacks living in the rain forests of South America.

Some scholars maintain that cognatic systems result from the breakdown of patrilineal systems; others feel that they are the beginning of such systems. Some theorists consider them to be an independent type that, in some cases has resulted from a breakdown of a unilineal system (Fox 1967: 153). The Güelmambí evidence shows that a definite patrilateral bias exists, induced by the ideal norm of patrilocal residence. But only a study of the changes that have taken place within the socio-economic organisation of the earliest groups of freed miners, from before 1851 to the present, could help solve this question.

What this case provides is evidence of the tronco system as an adaptive strategy to cope with the use of land limited by the miners' documents, the squatter status established by the Colombian government, and also by the occurrence of such natural disasters as floods, depletion of gold and such external pressures as those posed by dredges of the multinational enclaves that rendered useless many kilometres of fertile soil along the rivers in the region. Tronco exogamy has provided the miners with a mechanism for spatial mobility, demographic adjustments and a distribution of labour force and resources within and outside the mina grande over considerable distances.

Adaptation for these people has to be understood as a process responding to fairly rapidly changing situations. It shows great efficiency at critical points, and, above all, the capability of drawing on an ample range of alternatives and resources from the physical and social environment. The role of ideology, especially that chartering kinship, marriage, and production groups, should be scrutinised in greater detail.

One notes that cognatic groups are often found among people that live in insular territories. The Güelmambí blacks, of course, are not island dwellers. Yet the pattern may still hold if we consider that these miners of continental rain forests suffer a type of isolation resultant from their historical background as forced migrants from Africa, turned slaves and then free, as well as the present social and racial discrimination they face. Furthermore, the respect given by the miners to the borders of their property surrounded by public lands may indicate their conceptualisation of their territory as an island.

One might also observe that an analysis of their creativity in transforming and reinterpreting facets of the Indian society, features of the Spanish colonial society, and their African traditions into a new way of living adds a deeper perspective to the understanding of South American blacks.

Notes

1 The basic field work for this article was carried out under the sponsorship of the Instituto Colombiano de Antropología. Further research was done under the sponsorship of CIID, IAF, FORD and FES. I would like to thank Dr Nancy Morey for her comments, as well for her help translating an early version of this article. Dr Jaime Arocha and Dr Ronald J. Duncan, having worked in the area, have contributed with discussions and suggestions. Dr Tom Greaves, who extended an invitation to join the 1982 symposium,

'Miners in the Americas', in Manchester, has made valuable recommendations for this chapter. I wish to thank them all for their greatly appreciated collaboration.

2 The Colombian government has legally transferred surface soil property to other interested parties and has extended sub-soil concessions to several industrial mining enterprises. In 1937, the South American Gold and Platinum Company began exploitation in the Telembi River. In 1963 this company merged with the International Mining Corporation. The multinational enclave obtained concession rights over 1,045 hectares and requested an additional 3,168 hectares. Moreover, it rented and acquired 2,332 hectares making a total of 6,545 hectares, or nearly 25,000 acres of river bottom land (Melo 1975: 68).

3 *Cf.* Firth (1936, 1957), Lambert (1966) and Hanson (1970).

4 Robert C. West (1957: 129) describes the system as evolved of Indian techniques: 'seeds are broadcast and rhizomes and cuttings are planted in an uncleared plot; then the bush is cut: decay of cut vegetable matter is rapid, forming a thick mulch through which the sprouts from the seeds and cuttings appear wtihin a week or ten days. Weeds are surprisingly few, and the crops grow rapidly, the decaying mulch affording sufficient fertilizer even on infertile hillside soils.'

References   * *See annotated bibliography for full details of all asterisked titles*

Buchler, Ira and Henry A. Selby
1968 *Kinship and social organization. An introduction to theory and method.* New York: The Macmillan Company.

Davenport, William
1959 'Non-unilinear descent and descent groups'. *American Anthropologist* 61: 557–72.

Firth, Raymond
1936 *We, the Tikopia; a Sociological Study of Kinship in Primitive Polynesia.* London: G. Allen.
1957 'A note on descent groups in Polynesia'. *Man* 57: 4–8.

Fox, Robin
1967 *Kinship and Marriage, an Anthropological Perspective.* Middlesex: Penguin Books.

Friedemann, Nina S. de
1966–9 'Contextos religiosos en un area de Barbacoas (Nariño, Colombia)'. *Revista Colombiana de Folklore* (Bogota) 4(18): 61–83.
1974 *Minería, Descendencia y Orfebrería, Litoral Pacífico (Colombia).* Bogota: Universidad Nacional de Colombia.
1980*

Friedemann, Nina S. de, and Jaime Arocha
1982 'Contribucion al etnodesarrollo de groupos negros en Colombia'. Manuscript.

Friedemann, Nina S. de, and Ronald J. Duncan
1974 *Güelmambí: A River of Gold.* Film, b/w, twenty-four minutes.

Goodenough, Ward H.
1961 'Review of G. P. Murdock (ed.) *Social Structure in Southeast Asia*'. *American Anthropologist* 63: 1341–7.
1970 *Description and Comparison in Cultural Anthropology.* Chicago: Alding Publishing.

Hanson, Allan F.
1970 *Rapan Lifeways, Society and History on a Polynesian Island.* Boston: Little, Brown and Company.

Keesing, Roger M.

1975 *Kin Groups and Social Structure.* New York: Holt, Rinehart and Winston.

Lambert, Bernd

1966 'Ambilineal descent groups in the Northern Gilbert Islands'. *American Anthropologist* 68(3): 641–64.

Melo, Hector

1975 *La Maniobra del Oro en Colombia.* Medellín: Editorial La Pulga Ltda.

Sahlins, Marshall.

1962 *Moala: Culture and Nature on a Fijian Island.* Ann Arbor: University of Michigan Press.

West, Robert C.

1952 *Colonial Placer Mining in Colombia.* Baton Rouge, Lousiana: Louisiana State University Press.

1957 *The Pacific Lowlands of Colombia.* Baton Rouge, Louisiana: Lousiana State University Press.

Whitten, Norman E.

1974 *Black Frontiersmen, a South American Case.* New York: John Wiley & Sons.

Whitten, Norman E. and Nina S. de Friedemann

1974 'La cultura negra del litoral Ecuatoriano y Colombiano: un modelo de adaptacion etnica'. *Revista Colombiana de Antropología* (Bogota) 16: 75–115.

CHAPTER 13

The workers of the modern mines in southern Peru: socio-economic change and trade union militancy in the rise of a labour elite[1]

DAVID G. BECKER

Mine labour movements in every country where miners are numerous have been the subject of sustained scrutiny, especially from theorists interested in class formation and class struggle. While in many instances the struggles of the miners have lent confirmation to these perspectives, the process of proletarianisation is apparently far from minutely unilinear, far from predictably straightforward.

In the following essay David Becker presents a critical, historical review of Peruvian labour in the mines, seen within the context of a national union movement that has now emerged as the strongest independent case in Latin America. He argues that Peruvian miners may be developing 'economistic' goals which will sustain effective political power and an advantaged economic share without, for the present, challenging the institutionalised capitalist structure.

David G. Becker (Department of Government, Dartmouth College, Hanover, NH 03755, USA) is currently researching a study of international capital, the state, and class formation in Latin America, as well as the theory of post-imperialist capitalist development. A related publication is: 'Development, democracy, and dependency in Latin America: a post-imperialist view'. *Third World Quarterly* 6(2): 411–31, 1984.
 — *The editors*

Those of us who study power and social control in developing societies have an abiding interest in the mining proletariat. Wherever there are mines in the capitalist world, the miner has made his entrance not only as one of the first proletarian elements but also as one of the most active politically. The argument for radicalism as the natural political condition of the mine worker is an old and familiar one. Taking account of factors tending toward a high degree of group cohesion, it posits that a socially cohesive work force will more readily make the transition from a class *an sich* to a self-conscious, politically aware class *für sich* (Marx 1963). Miners are, of course, just one occupational group and cannot be expected to make a proletarian revolution by themselves. They may, however, point the way for the larger working class that arises as capitalist industrialisation proceeds. Some maintain that in today's developing countries miners may further hasten the advent of socialism by transferring a revolutionary consciousness to other oppressed groups, notably peasants, whose socio-economic situations would not otherwise lead them to it on their own (Petras and Zeitlin 1967).

The notion that miners spearhead wider processes of proletarian class formation has much to recommend it. But the hope that miners' political action will lead promptly to socialism has not been realised. In Bolivia alone miners have stood in the forefront of a successful revolutionary movement; and the outcome there has hardly been of a kind to bring cheer to progressives. It may be that too much weight has been given to the rhetoric of union *pronunciamientos* and to occasional outbreaks of violence, not enough to the less spectacular daily practice of the ordinary worker.

Thus there has appeared a second school of developing-country class analysis, one which defines the industrial proletariat (including miners) as a 'labour aristocracy'. This pejorative term connotes far more than a class element which is privileged in comparison to the masses of peasants and urban poor. It implies that proletarian radicalism is wholly illusory, that its class practice is not a force for constructive change. Although this school's roots go back to nineteenth-century radical populism, its later reincarnation was due to the writings of Fanon (1963) and, with much more theoretical structure, Amin (1976, 1980). (Elements of it are also present in Lenin's (1969) attack on 'economistic' class practice.) Within the last couple of years it has been subjected to devastating criticism (e.g., Brewer 1980, Smith 1980) and has lost popularity. So we are left adrift in midstream in our effort to comprehend class formation and action in the mines of developing countries, a situation that plainly invites a new approach.

What is needed, I believe, is to bring to the empirical investigation of the actual class practice of proletarian groups the focus provided by the concept of ideological hegemony (Gramsci 1971: 57–8 and *passim*; see also Adamson 1980, Boggs 1976, and Jessop 1982: 148–9). This concept allows us to capture the interplay between a class's practice and its world view, or 'common sense'. Ideology, let us remember, is not 'false consciousness'. It consists, rather, in a moral explanation of political reality that works, in the sense that action based on it confirms the image it portrays of the nature of the social world (Mepham 1978).

The issues at stake here are more than academic ones. For, as Gramsci insisted, the task confronting the proletariat is dependent on the bourgeoisie's ability to incorporate into *its* world view both the autochthonous organisational initiatives and the most pressing material aspirations of the working class, and thereby to generalise that world view so that it animates the practice of all classes. If the bourgeoisie can accomplish this, the proletariat cannot seriously contemplate a breakthrough toward socialism until it has won a lengthy 'war of position' in the ideological realm – until, that is, it has entirely liberated itself from bourgeois 'common sense' in order to forge and propagate its own *weltanschauung* based wholly on its own interests and those of other subordinate classes which are its natural allies. Paradoxical though it may seem, 'economistic' class action can win battles in such a 'war of position' even as it produces immediate benefits for workers and their families – provided that in the long run it strengthens the class's political experience, and provided that the bourgeoisie is in fact potentially or actually hegemonic.[2]

Peru, a semi-industrialised country in which the mining industry looms large, presents conditions that are conducive to a study of mine workers' class practice (I now define it more precisely, as embracing class formation, ideology and ideologically motivated action) oriented around the issue of hegemony. As is true of all developing societies, miners originate in the peasantry and continue to exist in intimate association with it. There is also a numerous industrial proletariat (more than half of the economically active population is employed in the urban-industrial sector) with which miners interact. The modern mining industry and its working class developed in two discontinuous episodes – one beginning in the early twentieth century and the other in the 1950s – which permits us to examine class formation in two different epochs of world capitalism with many local variables held constant. Finally, Peru has been passing through a period of rapid socio-political change. A reformist military regime took power in 1968 and conducted an 'ambiguous revolution' whose measures included a far-reaching agrarian reform, alterations in the institutions of industrial property, restrictions on foreign investment together with expropriation of a number of foreign firms, and wholesale reconstruction of the state apparatus (Lowenthal 1975). It was succeeded in 1975 by a Thermidorian military regime (styling itself as the 'second phase' of the 'revolution') that arrested the reform process while still in a pre-socialist stage and, because faced with an economic crisis, took a harsher line *vis-à-vis* the labour movement. In 1980 it, in turn, was succeeded by a bourgeois democracy, one in which organised labour and the marxist left have played a prominent and legitimate role (Woy-Hazleton 1982). Mine labour's class practice has had a heavy impact on all of these political events.

The argument is structured as follows: First, the legal-institutional environment of the working class is specified, and historical processes of class formation are described. Mine workers' militancy is then evaluated on the basis of strike behaviour and rank-and-file expressions of politically relevant opinion. Next, the question is raised as to whether class action can be properly qualified as 'economistic' and, if so, why; it is answered by examining the evolution of wage rates and other material benefits, as well as the use that workers have made of the co-management mechanism that the military regime's enterprise reform put at their disposal. Last to be taken up is the political power of organised mine labour on the national front and in the local environment of the mining town or camp. In the concluding section I shall interpret the findings in terms of their meaning for hegemonic bourgeois domination and the probable near-term future of civilian governance in Peru and elsewhere.

I Labour law and working-class institutions

As we shall see, the working class has on occasion demonstrated real political power; but until the last few years it was excluded from formal legislative and bureaucratic-regulatory institutions and had to exercise its power extra-legally. Hence, it is futile to seek in the texts of labour laws and regulations an accurate indication of the balance of class forces (nor are the laws always enforced). We study them because they have shaped the institutional development of the proletariat and, thereby, major aspects of its class practice. Most of the laws currently on the books predate the 'revolution' and are characterised by efforts to retard the growth of centralised, multi-plant or multi-industry labour unions; and by surprising generosity, considering the era, in matters of wages, hours, benefits and bargaining rights with respect to individual employers. The apparent contradiction is readily explained. This legislation was the product of a dominant agro-export oligarchy with few direct interests in systems of industrial labour control. (In mining, oligarchic investment, though sizable, was limited to the domestic *mediana minería*, or medium-scale sub-sector, and was of a rentier character oriented toward long-term capital gains; the value of mining equities is determined over the long haul by reserves and world market conditions, not labour costs.) It was enacted, therefore, not for economic but for political reasons, to detour potential political challenges from below to the system of domination. If political militancy in working-class organisations could be headed off by combining repression with cooptative accommodation over immediate economic matters, so much the better – since the costs would be borne by foreign firms and local entrepreneurs who were not organic parts of the oligarchic elite.

In consequence, mine workers are entitled to the usual panoply of welfare state benefits (except unemployment insurance) and must be given free housing, medical care, education and other social services. The reforms of the military regime introduced just four innovations of note: physical standards for mining camp housing; national control of camp schools to prevent foreign companies from indoctrinating workers' children with 'alien' values; the Workers' Community system, containing elements of profit sharing and co-management; and *estabilidad laboral*, or guaranteed job tenure.

Formalised in the law is the usual Latin American distinction between *obreros* (ordinary workers) and *empleados*, who originally had to have some secondary eduation. In mining however, empleados are simply better-paid skilled workers, 'lead men' and foremen. Promotion to empleado status on the basis of on-the-job experience is routine.

Although founded on an outmoded rationale, this distinction affects workers' preceptions of their class situation and mobility prospects; it is a disarticulating factor within the proletariat and impedes the evolution of class consciousness. Those whom it coopts are presumably the most ambitious, the natural leadership of the class. The disarticulation has been deliberately nurtured: it is illegal for empleados, though they may unionise, to belong to the same local union as obreros. In recent years there has been some erosion of the economic differential between the two strata: the military regime not only united their once separate social security systems but also ordered that they no longer be segregated from each other in government- and employer-run medical facilities. These changes imply an advance in class cohesion.

Labour centrals cannot bargain and are limited by law to advising their member locals but, nonetheless, are vehicles for coordinated class action. Consequently, the labour movement is relatively free of centralised bureaucratic control. The other side of the coin is that the centrals' potential as coordinators is not fully realised, thanks to a profusion of competing organisations. There are 'federations' for trades and industries – the miners' federation is known as the FTMMP – and some locals belong to several at once. The federations are grouped into four nationwide 'confederations'. They are: the CGTP, since the early 1970s the largest and an affiliate of the orthodox (pro-Soviet) Communist Party (PCP); the CTP, a dependency of the moderate-reformist APRA party; the CNT, formed by the now-moribund Christian Democrats (much more radical than their European counterparts); and the CTRP, a creature of the military regime that escaped from state control (it is now politically independent but cooperates with the left). Since control from the top is almost nonexistent, affiliations by the base units are loose and can shift as momentary advantage dictates. The FTMMP began life as a CTP affiliate, switched to the CGTP around 1970, declared its independence in 1974–5 (a rump group split off and remained with the confederation) and is currently moving toward reaffiliation with the CGTP.

All forms of compulsory unionism are illegal; not even members can be compelled to pay dues. Save for a few federation and confederation function-aries supported from political party treasuries, union officialdom is unsalaried. Instead, officers remain after election in their regular jobs, from which they are granted fully-paid leaves (*licencia*) in order to attend to union business. (The law, strictly observed in this case, prohibits employers from discrimin-ating against union officials in matters of pay, seniority and promotion.) The amount of licencia above a statutory minimum is a matter for collective bargaining. Since 1970 the mine unions have forced employers to extend it to the point where released time approaches or equals 100 per cent; employers also now pay travel expenses if the official has to attend to union business off-site. In other words, a professional leadership cadre *subsidised by employers* has come into being.

The right to strike, too, has traditionally been subject to various restrictions – loosened somewhat by the military regime but since retightened by the Belaunde administration's latest labour legislation. These restrictions make it possible for the state to outlaw a strike on procedural grounds if it wants to, although the penalty – ineligibility for legally mandated back pay once the strike is over – is not much of a deterrent for well-paid miners. A more effective anti-strike weapon is the state's right to intervene in any labour dispute and impose a settlement at its discretion on any terms at all, without reference to the position of either party. This discretionary power of intervention has been exercised repeatedly in mine strikes, where the stakes for the national economy are great. This is still true today, despite the civilian government's preference for leaving industrial relations to the free play of the contending forces wherever possible.

Except when under extreme economic duress, recent governments have trodden lightly in dealing with mine strikes: in addition to possessing the economic wherewithal to withstand long periods without pay, mine workers have scarce skills and cannot easily be replaced if their *estabilidad laboral* is withdrawn. (For that reason, Belaunde's move to abolish tenure guarantees will have little impact in the mining sector.) The sole alternative to conciliation, therefore, is all-out repression. One task of this study is to understand why the military regime, which, presumably, had the means available to it, did not make much use of this alternative.

II Origins of the mining proletariat

Modern, industrial-type domestic mining companies have existed in Peru since the early years of this century, although they did not become a significant economic force until after the 1950s. The domestic mediana mineria accounts for 30 per cent of total mining employment, but only two locally owned firms have work forces as large as 1,000. With few exceptions, these firms were not unionised until the early 1970s, and their workers have yet to demonstrate the degree of militancy routinely observed in the bigger mines. In the domestic sub-sector it is not possible to utilise nationalism as an organising ideology, nor are profit margins so great as to hold out the hope that employers would accede to major improvements in wages and benefits rather than face production losses. Perhaps because it is geographically dispersed and its individual work units are small, the mediana mineria proletariat has not attracted researchers' attention to a degree commensurate with its importance.

The segment of the proletariat which has always been most visible politically, and which is the primary subject of this study, is employed in the *gran mineria* — the subsector of large-scale mining and refining whose enterprises were, before 1970, all foreign owned. But this subsector's proletariat is not homogeneous. Each of the two historical episodes of mine development referred to above involved the gran mineria in a different part of the country and gave rise to a different subgroup of proletarians.

The first episode began in 1901, with the establishment of the Cerro de Pasco Corporation in the central Andes (*sierra*) east of Lima. The second was launched in 1954, when the Southern Peru Copper Corporation (SPCC), a joint venture under the majority ownership and management of New York-based Asarco, Inc.,[3] started work in the southern sierra near the Chilean frontier. The operations of these two firms are not at all alike. Prior to the mid-1950s, Cerro's ores were extracted (some still are) from a chain of smaller underground mines exploited by traditional pick-and-drill methods; the company differed from the mediana mineria only in the number of mines it controlled and in the fact that it also operated smelters and refineries.[4] Even today the Centromin complex (Centromin is the para-statal enterprise which replaced Cerro after its expropriation on 1 January 1974) is relatively labour-intensive and employs around 16,500 workers — two-thirds as many as the whole mediana minería — of whom half or more are semi-skilled.

SPCC, in contrast, has always engaged in very large-scale open-pit mining of the most modern, capital-intensive sort. Its original Toquepala mine and associated smelter on the coast at Ilo produce well over twice Cerro's copper output with just 5,600 workers. Its second mine, Cuajone, which opened in 1977, is highly automated and produces even more than Toquepala with but 1,800 workers.

There are two other members of the gran mineria. MineroPeru is a para-statal enterprise founded *ex novo* in 1970 to develop and operate mines and refineries; it currently has in production zinc and copper refineries as well as a moderate sized but very modern open-pit copper mine, Cerro Verde, and its total employment is about 2,600. Marcona is a vast iron mine on the south-central coast that began producing in 1952 and was expropriated in 1975 to form HierroPeru; its total employment is about 3,500.

Class formation in the central sierra

The rise of a proletariat in the zone impacted by Cerro has received extensive treatment in the literature (DeWind 1977, Flores Galindo 1974, Kruijt and Vellinga 1979, Morello 1976). The work force has been drawn from the local peasantry, which initially resisted for historical reasons (memories of coerced labour in the mines of the colonial epoch) and religio-cultural ones. Cerro resorted to fraud in order to get labourers under contract and to brutal super-vision to keep them working — practices within the capitalist mores of the time and further justified in this case by the oligarchic elite's distaste for all things 'Indian'. Workers were also recruited among peasants driven from the land by smelter pollution, which, however, was brought under control in the late 1920s after Cerro had acquired most of the affected holdings. In this way the company became the area's preeminent *hacendado* and a formal ally of the landowning oligarchy; the effect on its reputation in the eyes of the peasantry and sympathetic intellectuals does not need to be spelled out. Physical conditions in the mining camps and the refinery town of La Oroya were abysmal, and basic services were nonexistent. Few workers stayed on the job for more than a few years. (Regarding Cerro's victims see Bonilla 1974, Flores Galindo 1974, and Kapsoli 1975.)

In the 1950s, with machinery becoming more complex and the average job tenure approaching ten years, Cerro finally instituted an up-to-date system of personnel administration and training and started to upgrade somewhat the residential infrastructure. But there was little fundamental change until the 1974 nationalisation, after which the state set out to improve management practices and to furnish new homes, schools and other facilities.

The outcome of these processes is a semiproletariat that retains important ties to a now-transformed smallholder agriculture in the zone. Its members have been motivated in their class action not only by the usual wage demands but also by intensely nationalistic opposition to their foreign oppressor; during the 'revolution' they insisted that the 'social debt' owed to them be redeemed by major improvement in their quality of life. They have resorted to extreme means – violence, land invasions, *marchas de sacrificio* down the Central Highway to the capital – in pursuit of their objectives; and in all such instances they have enjoyed the active support not only of wives and children but also of the peasantry and even the independent business people of the area. On the other hand, many workers still centre their aspirations on village life away from the mines and, using their earnings to purchase land or small businesses in their places of origin, elevate themselves into the village elites (DeWind 1977, Morello 1976). Laite (1981) finds that workers' value orientations often tend toward upward mobility and other aspects of a *petit-bourgeois* world view, and that marxist labour leaders have not been very successful in mobilising them on behalf of purely political objectives.

Details of class formation under the pressures of these struggles are discussed by Kruijt and Vellinga (1979) and by Sulmont (1974, 1975, 1977). Unionisation began in 1928–30 under communist aegis and then, after a period of severe repression, resumed in 1944–5 under APRA-CTP sponsorship. This was the first instance of fully-fledged industrial unionism in Peru; inasmuch as it ended by forcing the government and the country's largest employer, Cerro, to come to terms with the labour union principle, the miners of the central sierra can be said to have spearheaded class formation among the proletariat at large.

The CTP insisted that the various Cerro unions (there had to be one each for obreros and for empleados at each of the installations) moderate their activities in support of the APRA's national goals.[5] However, neither moderation nor discipline could be imposed on a group of workers still so close to their peasant origins and much given to spontaneous, protoproletarian forms of violent protest (Hobsbawm 1959). A more radical leadership group challenged *aprista* control during the 1960s, when general peasant unrest in the zone erupted into guerrilla warfare. The radicals were successful for a short time but lost rank-and-file support when their costly confrontational tactics failed to wring greater material benefits out of an intransigent government and employer. There was a resurgence of labour radicalism at the leadership level from 1969 to 1974 because, for reasons that we shall examine, a changed political environment made radicalism economically efficacious. Since the 1974 takeover of the Cerro installations by the state, which led to a complete indigenisation of supervisory personnel and better management communication with workers[6] in addition to improved infrastructure, radicalism — but not union solidarity on behalf of economic ends — has been on the wane.

Class formation in the south

By the 1950s, neither the prevailing ideology of international capital nor the political forces with which the Peruvian state had to contend would have permitted a repetition of the Cerro experience of fifty years before. Moreover, the new companies' interests were fundamentally divergent from Cerro's. Their capital-intensive installations needed well-disciplined workers, who had to be trained at considerable expense to operate costly and complex equipment. Employers could not let this training investment leak away in high turnover and knew that a few disaffected operatives could wreak enormous damage through inattention or sabotage. A stable, contented work force could not be dragooned into the mines, only attracted. The attractions were high wages and benefits by local standards; clean, well-appointed townsites offering a full range of social services; and, in matters of labour relations, a policy of benevolent paternalism.[7] There was also the fact that open-pit mining mostly involves less dangerous outdoor work and teaches generalised machinery skills whose value was widely appreciated.

Furthermore, these firms encountered a favourable surplus of supply over demand in the local labour market. By the middle of the 1950s a major out-migration from the south-eastern department of Puno was in progress. This department had been one of the country's most backward, its peasantry subjected to the harsh rule of the *latifundistas* and the rural power brokers, or *gamonales*, associated with them. There was a strong local tradition of peasant unrest and rebellion. Now the region's tenuous social equilibrium was being upset by population growth, by the spread of modern ideologies, and by the belated efforts of latifundistas to capitalise their holdings. Under these pressures waves of *puneños* flooded out of the department into cities and towns all over the southern half of the republic, swelling the *barriadas* (migrant settlements, since rechristened *pueblos jóvenes*, or 'new towns') on their outskirts. When they heard of the mine construction getting under way nearby, they flocked there in search of jobs. SPCC's and Marcona's problems were limited to deciding whom to hire and pacifying those who were rejected.

Despite their lack of formal education and industrial experience, the puneños turned out to be hard workers who learned quickly and adapted speedily to industrial discipline. Since for them the option of returning after a time to a life in the countryside was foreclosed, they were far more ready than their more northerly confrères to redefine their social situation in proletarian terms. Three indicators can be adduced: (1) The percentage of Toquepala and Ilo workers who claim twenty years' tenure, meaning that they were there when the facilities opened, is large; and it is becoming common for sons to fill positions left vacant by their fathers' retirement. (2) SPCC workers evince a strong preoccupation with job categorisation, promotion and status – far more, assuredly, than would have been detected at Cerro twenty years after its foundation. (3) Whereas Cerro workers did not begin to include camp infrastructure improvement in their *pliegos de reclamos* (lists of bargaining demands) until the 1970s, by which time the proletariat was seventy years old, at SPCC the proletariat was less than ten years old when its unions began to press for the same thing.

But the Puno tradition of resistance to authority, rather than dying out, nourished itself on opposition to SPCC's benevolent paternalism even as it transformed itself into new modalities more appropriate to an industrial setting. A chief source of resentment was the company's attempt rigidly to control many non-work-related aspects of life in the townsites, including physical movement in and out (partly to exclude 'agitators', partly to prevent the 'unsophisticated' work force from being cheated by unscrupulous local tradespeople and then blaming the company for their fate). A second was SPCC's highly intrusive acculturation programme, which, though its intent may have been laudable, was regarded – properly – as an invasion of privacy. (For example, the Townsites Division would conduct surprise inspections of workers' apartments and issue demerits for such health and safety violations as unauthorised construction, overcrowding, improper care and use of sanitary facilities, raising food animals on the premises, etc.)[8]

Resentment against overweening corporate authority soon spread to the issue of work rules. Workers have never disputed the company's right to discipline them for infractions such as unexcused absence or tardiness (SPCC does not use time-clock controls); but they have increasingly demanded a voice in shift assignments, task organisation, job definition, promotion and appointment of supervisors. As is true of workers in developed countries, it makes no difference if the company's unilateral decisions in these matters have the effect of lightening the work load or making it more pleasant.

However, strikes at SPCC have almost invariably been orderly. There are few pickets or demonstrations; workers simply go home or go off on vacation. Correspondingly, the compact network of residential organisations and dependents' support groups that Zapata (1980) finds to be typical of mining camp life – it is at Cerro – is absent here. In fact, SPCC social workers report that Toquepala and Ilo residential life is atomistic and that the company's efforts to form apartment block improvement associations, sports leagues, women's clubs etc. have met with scant success.

There is yet one more side to the story: SPCC workers are much more adept than Cerro's at using the sit-down, the slow-down, and the tactic of paralysing the installations by calling out only a small but crucial work group while the rest of the workers continue collecting their pay. This skill in the use of advanced struggle techniques attests that class formation is correspondingly advanced.

SPCC also differs from Cerro in the process of unionisation that it has undergone. Outside organisers sent by political parties were never a factor.

Instead, the newly hired workers displayed from the start a genius for planned, coordinated action that has indigenous roots and that has been much observed in the organisation of the *pueblos jovenes* (Collier 1976, Dietz 1969). Unions were initially formed to protest the company's refusal to hire more people during the construction of Toquepala. The unions organised their first strike when the police, acting without having consulted company officials, arrested their leaders. SPCC did not wait upon the Ministry of Labour but itself recognised the new unions and proceeded to bargain amicably with them.

During their first decade the SPCC locals accepted CTP advisement (the CGTP, although founded in 1928, had gone out of existence with the repression of the 1930s and was not reconstituted until 1968); nevertheless, they jealously guarded their autonomy. Strikes were occasional; relations with the employer were mutually satisfactory on the whole. Bargaining resulted in a tripling of the basic obrero wage and a gamut of unprecedented benefits.[9]

Radical ideological ferment reached SPCC *after* the 1968 'revolution' had unleashed it elsewhere in Peruvian society – indeed, almost nine years later than Cerro. The rank and file remained nonviolent, but union leadership was gradually assumed by a new generation of younger, better educated, more politically committed officers. They discovered how, lacking the more obvious grievances at Cerro, to exploit the issues of paternalism and authority to build a following. They discovered, too, that well-off workers with savings in the bank and free housing would tolerate lengthy strikes if they ended in economic payoffs. And they soon intuited that, since SPCC had come to account for the lion's share of Peruvian export revenues and was earning very high profits, and since the military regime's legitimacy in power depended on its plan of rapid industrialisation financed by mineral export revenues – 'bonanza development' (Becker 1982) – the regime stood ready to end such strikes by intervening in them and imposing settlements on terms close to union positions.

The SPCC locals furnished much of the impetus behind the formation of the FTMMP and, since then, a good part of the federation's officialdom. They have also supplied the federation with the majority of its funding, without which it could not have sought to become *the* nationwide miners' organisation. FTMMP organising efforts met with success in the mediana minería (but not at Cerro, where the thirty-eight local unions have their own federation and claim independence on the basis of their long and unique history of struggle); the majority of pre-existing locals affiliated with it, and it founded new ones in a number of previously non-union mines. Were SPCC workers a 'labour aristocracy', they would not have devoted their organisational resources to the unionisation of a separate subsector of the industry, whose fortunes do not affect them in the least.

The newest facilities of the gran minería – MineroPeru's, also SPCC's Cuajone mine – have recruited their workers from the *pueblos jovenes* of the nearby cities and towns (and even from Lima, when specialised skills were required). Most of these workers had resided for some time in urban environments and already possessed some industrial skills. Most belonged to labour unions before coming to the mines. Hence, they promptly organised the new facilities without meaningful opposition from the employers, who seemingly regarded this as the natural course of events. MineroPeru's unions have affiliated with the FTMMP; Cuajone's have not, due to rivalries with Toquepala. Labour relations at Cuajone and the MineroPeru installations are thought by both sides to be very good, in spite of the fact that there have been strikes at each of them.

Possibly the most noticeable difference between post-1970 gran mineria development and that of earlier periods is the virtual disappearance of the camp or townsite as a key factor in class formation. MineroPeru's copper and zinc refineries and its Cerro Verde mine are located near population centres (Ilo, Lima and Arequipa, respectively), obviating the need for any sort of special residential construction; workers receive a housing allowance in cash but otherwise are just part of the generalised urban proletariat. Cuajone has its own townsite, but SPCC learned from its previous mistakes in designing its infrastructure: paternalistic controls do not exist, local tradespeople are encouraged to set up shop in stalls supplied by the company, traditional popular entertainments have been vigorously promoted, and many services have been contracted out to national providers so as to minimise the 'company town' aspect of the place. Through these means and by virtue of its proximity to the city of Moquegua, Cuajone has taken on the air of a 'normal' suburban-industrial town with a multidimensional economic and social life.

III The militancy of the Peruvian mine worker

The foregoing suggests that mine workers' class practice contains significant elements of 'economism'. Since 1968, miners have been the targets of a steady barrage of marxist-leninist interpretations of their class situation. Moreover, it has been delivered to them by a home-grown union leadership that is admired and respected and that has not hesitated to risk its mandate at regular intervals in open, highly democratic local assemblies and elections. If it should turn out that, radical leadership and indoctrination notwithstanding, these miners hold fast to the principal elements of a bourgeois world view and ideology, we will have obtained powerful confirmation of the strength of the latter and of the potential for bourgeois hegemony in Peru.

Political opinion survey data

Unfortunately, no one has yet conducted a comprehensive survey of miners' political attitudes and beliefs, one that would sample all subsectors and geographical regions with the identical instrument. All that can be done for now is to draw upon two partial surveys – one performed in 1974 on a sample of medium mines designed to reflect accurately the whole subsector, the other taken at Cerro Verde in 1977.[10] The mediana mineria sample can serve very roughly as a surrogate for Cerro; although the miner respondents have not passed through the same formative experiences, they come from the same peasantry, relate to mining camp and home village life in a similar way, share in the same oral tradition (which has kept alive and propagated the memory of the Cerro workers' oppression), and have been part of the broad current of radicalism − most of it specifically marxist-leninist − that has washed over and through the peasantry during the last fifteen to twenty years. In like manner, the respondents in the Cerro Verde survey (the entire work force) can stand in for the workers of SPCC if one interprets the results tentatively and with caution.

Medium mine workers were requested to select from a list the group or institution most responsible for recent increases in the cost of living. Both obreros and empleados ranked the president and council of ministers first, followed in descending order of culpability by 'property owners and the wealthy' and 'middlemen'. Obrero responses differed from those of empleados in just one respect: a notably larger proportion of the former chose to rest the blame on the political authorities. The data therefore imply that all workers, but especially the *least* privileged, perceive their enemy as the political 'superstructure' and not the structure of class domination. In view of the radical nationalism espoused by all mine unions at this time, it is noteworthy that insignificant percentages of obreros and of empleados chose to blame 'US imperialism'.

Empleados alone were given an opportunity to describe their own class position and the basis on which they determined it. Just under half used salary as the criterion, while 28 per cent used education. More than half − 57 per cent − placed themselves in the 'middle class'; a surprising 12 per cent defined themselves as 'upper middle class'; whereas 22 per cent responded with 'lower middle class'. Empleados also placed more emphasis on noneconomic, status aspects of their jobs. Asked to state what single improvement they would most like their employer to make, one-third of them requested 'better workplace organisation and employee training', and a similar fraction expressed a preference for more employer-paid educational programmes. Salary was not mentioned in this context at all! In contrast, almost one-half of the obreros opted for better residential conditions, and about one-fourth asked for wage increases.

In the Cerro Verde survey, respondents were queried as to whether they claimed some understanding of the differences between 'capitalism', 'socialism' and 'communism'; those who did − let us call them the 'knowledgeables' − were then asked to indicate if they had a preference and to state it if they wished. Two-thirds of the empleados and 31 per cent of the obreros answered the first question in the affirmative. Among the 'knowledgeables', many more empleados than obreros − 76 versus 57 per cent − admitted to a preference. Of all those with preferences, 34 per cent of the empleados and 8 per cent of the obreros chose 'capitalism'; 34 and 69 per cent respectively, 'socialism';

zero and 19 per cent respectively, 'communism'; and the balance declined to state. Relatively few obreros, then, claim even superficial knowledge of the principal ideologies in contention; for that reason their preference for alternatives to capitalism more likely represents a diffuse alienation from the status quo than it does a firm ideological commitment to something different. Empleados, being on the average better educated, are more knowledgeable and more willing to make a choice. They are, however, less alienated and less attracted to noncapitalist alternatives – a reflection, one would assume, of experiences that ratify the existence of mobility chances and thus validate as 'common sense' the ideology of equal opportunity and individual achievement.

Impressionistic evidence on the political orientation of SPCC workers will have to serve in lieu of direct survey data, which does not currently exist and whose acquisition the company is unlikely to permit. It can be of some value if it confirms the Cerro Verde findings – and it does. My method was a simple one: I merely asked the many foremen and line supervisors with whom I conversed to characterise the political attitudes of the workers under them as best they could.[11] Not one believed that more than a tiny minority of their operatives, whom they usually knew rather well on a personal level, was committed to radical political viewpoints: most characterised the typical worker as apolitical, a probable left voter but not an enthusiastic one. Interestingly, a group of marxist-leninist intellectuals serving in SPCC's medical corps had come to the same conclusion.

Militancy as assessed from strike data

Table 13.1 summarises strike activity from 1967 to 1979. We see that there is no secular trend of increase or decrease in the ratio of mine strikes to the national total: except for a surge in 1970–3 and another in 1978 (to be discussed subsequently), the ratio has been constant at 6–7 per cent. Miners' contribution to the national total has been disproportionately large, inasmuch as only 2 per cent of the economically active population works in the mines and only 1 per cent of all local unions is based there. The table shows that this disproportion has become less in recent years: but all indications are that the change has to do with an increase in the frequency of strikes in the rest of the industrial sector. The time sequence suggests that the success of the mine strikes of the early 1970s may have emboldened the rest of the industrial proletariat to take similar action (at least it made clear that in the new political environment strikes would not be met with severe repression) – another example of a 'spearhead effect' at work. This should be regarded as an unproven hypothesis, however, until we learn more about the relationship.

Three measures of strike size and intensity are presented in Table 13.2. Prior to the latter half of the 1970s, mine strikes usually involved more workers than did strikes in other industries. Mine work forces were, after all, larger on the average than those in any other industry. What has altered the picture is the rise of strong federations outside of mining that have not hesitated to organise

Table 13.1   Strikes in Peru, 1967–79

	Number of strikes			Number of strikers			Man–hours lost to strikes		
	A	B	A/(A+B)	C	D	C/(C+D)	E	F	E/(E+F)
Year	Mine	Other	Per cent	Mine	Other	Per cent	Mine	Other	Per cent
1967	32	382	7.7	17,818	124,464	12.5	5,269,664	3,103,108	62.9
1968	21	343	5.8	9,426	98,383	8.7	2,825,376	552,425	83.6
1969	24	348	6.5	17,803	73,728	19.5	1,900,748	1,988,552	48.9
1970	66	279	19.1	56,205	54,785	50.6	4,325,853	1,456,003	74.8
1971	76	301	20.2	58,454	102,961	36.2	6,270,632	4,611,320	57.6
1972	33	376	8.1	16,657	113,986	12.7	958,008	5,373,008	15.1
1973	80	708	10.2	59,471	356,780	14.3	3,831,888	11,856,800	24.4
1974	38	532	6.7	27,433	335,304	7.6	1,878,148	11,534,892	14.0
1975	57	722	7.3	50,387	566,733	8.2	2,652,609	17,616,799	13.1
1976	29	411	6.6	31,505	226,596	12.2	572,228	6,249,996	8.4
1977	n/a	234[a]	–	n/a	406,461[a]	–	n/a	6,543,352[a]	–
1978	53	311	14.5	48,596	1,349,791[b]	3.5	4,680,388	31,464,348[b]	12.9
1979	40	537	6.9	25,342	678,141	3.6	1,187,288	9,364,064	11.3

Source: International Labour Organisation 1980; Ministerio de Trabajo 1973; unpublished data supplied by the Sociedad Nacional de Minería y Petróleo.

Notes: a. Data for all strikes in the country, mine plus 'other'.
   b. Figures are inflated due to two-day general strike.

Table 13.2   Indicators of strike intensity, 1967–79

	No. of strikers per strike		Man-hours lost per strike[a]		Man-hours lost per striker	
Year	Mine	Other	Mine	Other	Mine	Other
1967	556.8	325.8	164.7	8.1	296	25
1968	448.9	286.8	134.5	1.6	300	6
1969	741.8	211.9	79.2	5.7	107	27
1970	851.6	196.4	65.5	5.2	77	27
1971	769.1	342.1	82.5	15.3	107	45
1972	504.8	303.2	29.0	14.3	58	47
1973	743.4	503.9	47.9	16.7	64	33
1974	721.9	630.3	49.4	21.7	68	34
1975	884.0	784.9	46.5	24.4	53	31
1976	1086.4	551.3	19.7	15.2	18	27
1977	n/a	1737.0[b]	n/a	28.0[b]	n/a	16[b]
1978	916.9	4340.2	88.3	101.2	96	23
1979	633.6	1262.8	29.7	17.4	47	14

Source: Computed by the author from data in Table 13.1.
Notes: a. In thousands.
   b. Data are for all strikes, mine plus 'other'.

industry-wide work stoppages; prominent among them are the very militant 'middle-class' federations of teachers and bank employees. Contrariwise, the FTMMP's effort to become a true national miners' union has no reflection in these data.

Mine strikes have always entailed greater than average losses of labour time. This used to be due to differences in the size of work forces but now must be ascribed to the fact that mine strikes tend to last longer than others. The tabulated data for time lost per striker confirm this. Note that these data confirm Zapata's (1980) observation that militancy has been declining of late; but they show that the trend has been in evidence for some time and is not confined to the mining sector. In other words, observed increases in the aggregate number of strikes and number of workers affected per strike are due to development: a larger industrial proletariat and better industry-wide organisation will raise aggregate strike totals even as the average length and bitterness of individual strikes decline.

Table 13.3, which gives a partial breakdown of two strike-intensity indicators by enterprise, permits us to correlate militancy with the worker groups defined earlier. The trend toward decreasing militancy (it appears most vividly in the right half of the table) is here seen to be especially noticeable in the Cerro group. Indeed, Cerro/Centromin workers have become less militant on the average than those of the mediana mineria, who account for the bulk of the 'rest of sector' category. Since the trend first became visible in 1973, by which time the impending expropriation of the firm was public knowledge, it would appear that the transfer to national control has resulted in reduced

Table 13.3   Strike intensity indicators by enterprise, 1969−79

	Man-hours lost per strike[a]				Man-hours lost per striker			
Year	Cerro/ Centromin	Southern	Marcona/ HierroP.	Rest of sector	Cerro/ Centromin	Southern	Marcona/ HierroP.	Rest of sector
1969	169.8	315.3	0.0	25.3	95	146	0	93
1970	92.9	58.2	76.1	34.9	81	66	81	70
1971	82.1	68.8	215.5	73.1	99	69	107	162
1972	4.1	48.2	0.0	16.5	17	86	0	31
1973	131.0	72.9	110.8	35.0	16	91	81	75
1974	282.5	186.2	0.0	30.5	24	216	0	79
1975	59.8	105.9	0.0	25.8	46	75	0	56
1976	147.2	26.7	0.0	11.3	16	22	0	17
1977		92.6				54		
1978		101.6		85.2[b]		117		92[b]
1979		50.6		21.7[b]		96		32[b]

Source: Same as Table 13.1 plus SPCC consolidated declarations for 1977−9, filed with the Ministerio de Energía y Minas.
Notes: a. In thousands.
b. Data for all mining enterprises except SPCC.

strike intensity. Sceptics might argue that the state, as employer, will have repressed the labour movement more strenuously than before, or at least will have reduced the payoff to militancy by more firmly resisting wage demands. Wage data, to be examined shortly, refute the second argument. The first contains a grain of truth, but post-1976 repression, while extant, was never extreme and was applied across the mining sector. More important, I feel, has been the quick action by para-statal management to clear the 'social debt' and to improve its channels of communication.

SPCC workers are more militant; their actions are cyclical, peaking at four- to five-year intervals with little apparent long-term change. A comparison with the near-passivity at Marcona/HierroPeru, where workers' backgrounds and conditions are similar, suggests that the employer's profit level and nationality are key explanatory factors.

I conclude that labour militancy in the mines of Peru no longer has much to do with the national strategies of political parties or other external elites, with populism (i.e., peasant reaction to the disruptive advance of capitalist relations of production), or with the grosser forms of oppression. It has come to represent the kind of class action that is to be expected from a more mature proletariat.

IV The issue of 'economism'

'Economism' refers to the object of proletarian class action and need not imply a tendency of specific groups or strata to pursue their immediate economic interests to the exclusion of those of other subordinate groups. That is, 'economistic' class action in no way confirms the presence of a 'labour aristocracy' and can be undertaken − as it frequently is − by a broad alliance of subordinate classes. 'Economism' is a necessary but not a sufficient condition for bourgeois hegemony. If it typifies proletarian class practice, the bourgeoisie may be able to attain hegemony if it can manage to: (1) meet a satisfying portion of the 'economistic' demands without sacrifice of its own overriding interest in maintenance of its social control; and (2) also without such sacrifice, accommodate within its world view the existence and activities of labour unions, labour parties, and other institutions constructed by the working class and deemed essential for the latter's self-protection. The actions of corporate mining enterprise in Peru since 1968 seem to meet the second requirement. What of the first?

Advances in wages and benefits

Table 13.4 portrays the recent evolution of real wages[12] in the mining industry. (Also shown, for purposes of comparison, is the standard Peruvian index of industrial wages.) Most miners experienced a steady increase in real income until the late 1970s. Wages thereafter came under pressure as inflation accelerated. Nevertheless, miners came closer to holding their own than did

Table 13.4   Index of real wages in Peruvian mining and other industry, 1967–79[a] (100 = average wage for all nonferrous metal-mining in 1967)

	Index				
Year	All mining	Southern	Cerro/ Centromin	Medium mining	Other industry[b]
1967	100	219	118	67	
1968	100	217	117	67	76
1969	105	216	128	72	83
1970	122	264	150	87	82
1971	133	280	154	100	82
1972	148	335	164	114	96
1973	157	402	187	109	108
1974	153	412	190	102	106
1975	164	353	189	123	102
1976	126	424	255	104	92
1977	126	335	239	109	81
1978		221	201	107	70
1979		207			61

Source: Compiled by the author from data contained in International Labour Organisation 1980; annual reports of the mining companies; consolidated declarations of the mining companies, filed with the Ministerio de Energía y Minas; and private documents supplied by the Sociedad Nacional de Minería y Petróleo.
Notes: a. Money wages (note 12) deflated by the consumer price index for metropolitan Lima.
   b. Average non-agricultural wage for metropolitan Lima.

other industrial workers: the average mining-sector remuneration, 32 per cent higher than the industrial in 1968, was 62 per cent higher in 1971 and remained 54 per cent higher in 1977. Since nationalisation Cerro/Centromin miners have finally attained one of their cherished goals – wage parity with the more capital-intensive SPCC. Even miners in the mediana mineria have done better on the whole than have industrial workers.

Note that the reversals recently suffered by the SPCC labour force are not as severe as they seem from these data. SPCC employees receive more high-quality free services than does any other work group, and their cost (or, from a standpoint more sympathetic to workers' interests, the excess cost over and above the average cost of services in the gran mineria) is in truth a form of remuneration. As costs of services have escalated as fast as or faster than the general rate of inflation, figuring them in would result in a lesser inflation-induced erosion of worker purchasing power.

Worker purchasing power has also benefited from direct and indirect company contributions to local unions, absorbing costs that would otherwise have to come out of pay packets. Such contributions, mandated by collective bargaining agreements, are becoming widespread, although SPCC has been, as usual, the pacesetter. SPCC union officers receive a total of 2,140 man-days of unrestricted annual licencia, up from 340 man-days with restrictions as of 1970. Federation and confederation officers belonging to the empleado locals are on permanent leave with full pay. All officers receive a company-paid health and accident insurance policy. The company has donated construction materials, labour and $326,000 in cash to the union building fund; it also supports union-run libraries to the tune of $19,500 per year. And the company has had to underwrite a union donation of $142,000 to Tacna and San Augustín (Arequipa) universities; this donation was made, interestingly, in order to disprove charges of 'labour aristocracy' which had inhibited the kind of student–labour alliance found everywhere else in Peru.

These wage data are consistent with aggregate income distribution data showing the principal beneficiaries of the military regime's redistributive policies to have been the four middle deciles of the national income distribution (FitzGerald 1979). While the 'revolution' did little in economic terms for the poorest, it accomplished some reduction of upper-class privilege and allowed for the advance of well-organised groups of proletarians. What is more, a breakdown of the wage data reveals that mining obreros have gained proportionally more than have empleados (as also seems to be the case for industry at large), thus narrowing the wage gap between the two strata. Therefore, wage trends have proceeded in a manner that enhances intraproletarian cohesion, even though they have not led to erasing the differential between industrial proletarians and all other subordinate classes.

The proletariat and control of enterprise

A truly class-conscious proletariat would take action, were the opportunity presented, to erode as much as possible the authority of capital within economic enterprise and to seize it for itself. Such an opportunity appeared with the 1971 institution of the Workers' Community system. The Workers' Community was a property reform that required employers to grant workers immediate minority co-management rights and to distribute annually a share of profits, paid out both in cash and in capital stock. The latter, held by all workers in common, constituted an ownership interest that would increase year by year; as it did, workers' representation on the firm's board of directors would increase in proportion until it reached 50 per cent, or full co-management. In mining, profit distributions were partially apportioned out over the entire sector, so as to correct for the wide difference between SPCC's profit per employee and that of other firms and to promote worker solidarity throughout the industry. (For additional detail see Stephens 1980.)

Although manufacturing firms frequently resorted to quasi-legal and illegal devices in order to get around these requirements, such was not the case in mining. In fact, some mining companies have provided courses in accounting and business management for their worker-directors. But the mine unions regarded the system, which workers had never sought for themselves, as a cooptative device intended to promote labour – management harmony at their expense – an accurate reading of the regime's intent (see e.g., Alberti *et al.* 1977). They therefore exhorted their members to boycott Workers' Community assemblies and elections; the boycott was widely observed for two years. Then, in 1974, workers started to receive very large distributions from the high industry profits of the previous year (world metals prices reached an all-time peak in 1973). Interest in Community affairs at once increased. The unions, wisely, decided not to buck the trend. Instead, they began to play an active behind-the-scenes role in the operation of the Community. Workers' loyalty to organised labour did not diminish at all; the Community simply became a parallel avenue for advancing the same class concerns.

These concerns had nothing to do with management *per se*, i.e., with the power of decision over the internal structures of the corporation and over the uses to which capital and labour are to be put. None of the many corporate executives whom I interviewed in the course of this research complained that the real authority of capital had been undercut. Rather, their main complaint was that worker-directors complicated and slowed the transaction of directorial business by submitting repeated requests for detailed financial information – information that was destined for the unions, in order to help them bolster their pliegos de reclamos. A second complaint was that worker-directors used board meetings as forums for discussing individual grievances – input that management is glad to have but would prefer to acquire at lower administrative levels. There is no doubt that the utilisation of board representation primarily as a grievance mechanism responds to rank-and-file preferences expressed in democratic assemblies of the Community.

With the loss of the government's reformist vocation has gone, *pari passu*, a progressive watering-down of the system's communitarian and co-management features. Capital stock is now distributed as individual property in the form of *acciones laborales* ('labour shares'), a sort of preference stock that carries no voting rights; naturally, most workers sell it (it can be traded on the Lima stock exchange) as soon as it is received). Workers' institutional presence in the management process has been limited to minority board representation; or, if the company prefers, it may be implemented instead through labour – management *comités de gestión* with less power than the board of directors. *Workers have not protested these changes in any way.* They seem quite content with cash profit sharing and with the possession of a forum guaranteeing that their personal grievances will reach the ears of upper management. It must be assumed either that their perception of the character of entrepreneurial authority remains dim, or that they perceive it but choose not to challenge it. This does not suggest a revolutionary proletariat.

V Organised mine labour and political power

In its reformist phase the military regime was motivated by a perceived need to foster more rapid industrialisation; by nationalism; by a desire to capture for itself the popular support that had in the past gone to the APRA;[13] and by a sincere concern for social justice and for the plight of the poor (North and Korovkin 1981).

Included in the 'bonanza development' strategy of industrialisation was the notion that workers' purchasing power should be improved in order to broaden the domestic market for manufactured goods. Nationalism counselled that more of the earnings of the large foreign-owned mining firms should remain in Peru. Since both objectives could be met by increasing miners' wages, the ministry of labour was instructed to intervene in and settle labour disputes on terms favourable to the miners. But favouritism was to be shown only to locals unaffiliated with the CTP, thus encouraging them to desert the APRA's labour central while putting them in debt to the government. The generals' moral concerns, together with their awareness that poverty and deprivation were the chief threats to public order, worked to the further advantage of the Cerro miners in particular (Cerro, due to its past behaviour and friendly relations with the oligarchy, was additionally deemed an epitom- isation of the old order's excessive liberalism toward foreign investment).

Other Latin American authoritarian regimes – the *loci classici* being Perón's Argentina, Vargas's Brazil, and Calles's Mexico – have used their power to reward labour for the purpose of gaining control over it through the medium of *oficialista* (state-sponsored) unionism. The oligarchic preference for *laissez-faire*, however, had prevented the establishment of such a tradition in Peru; and by 1968–9 the autochthonous basis of labour organisation was already too strong to be overcome by cooptation or by repression within the limits of the available political resources. So the regime adopted a strategy of favouring the CGTP and its affiliates; labour's national power, it reasoned, could be kept in check by competition between centrals, exploiting the com- munists' antagonism toward APRA. The strategy was put into effect in dealings with mine labour; as we have seen, the mine unions did transfer their allegiance from the CTP to the CGTP, helping the latter to attain first place in total membership, number of affiliated locals, and financial resources. In return the CGTP and its parent, the PCP, lent the regime their 'critical support'.[14]

But the strategy had the unintended effect of exacerbating rather than mitigating mine union militancy. As the regime could not intervene on the miners' behalf unless there were a strike in progress, a premium was placed on fomenting disputes; as usual, the grease went, *ceteris paribus*, to the squeakiest wheels. And with the breakdown of centralised control over the labour movement came a competition among leaders for rank-and-file backing, a competition in which calls for moderation were a sure road to defeat.

Militancy was further spurred by ideological divisions within the ruling military. Their existence was public knowledge and stimulated societal groups across the spectrum to mobilise politically and to apply pressure in the hope that the regime's ideological balance might be tipped in the desired direction (North and Korovkin 1981). Hence, the mine unions could not rest secure in the cooperative attitude of the labour ministry but had to couple their material demands to public ideological appeals. They also had to counter a bourgeois ideological offensive that had become vehement by 1972–3.

Though powerful economically and, at the leadership level, committed ideologically, the mine unions were at a disadvantage in the rough-and-tumble of the national political arena – due in part to their lack of experience, in part to their geographical isolation from the capital. If they wished an effective national voice, they would have to link themselves to a political party. (Parties were not abolished by the military; despite having been stripped of their formal representational function, they remained as spokesmen for ideological positions and for key economic interests.) The problem was the persistent tendency of party cadres to subsume union goals to wider electoral strategies: the reason that the FTMMP broke with the CGTP was resentment over the confederation's call for wage moderation in order to demonstrate support for the 'revolution'. Wanted was a party organisation without a wide base of its own (thus susceptible to union domination) but which had an ideological affinity with mine labour.

The marxist intelligentsia's numerous sectarian vanguard groups were the obvious targets of this search. Most had split off from the PCP or the APRA some years earlier and had fissioned again and again over minor issues of doctrine and/or personality. They nevertheless shared a consensus on a 'maoist' approach to socialism modified by the thought of the prominent Peruvian marxist, José Carlos Mariátegui (d. 1930). In essence, this approach foresaw a socialist revolution headed by a comprehensive coalition of proletarians, peasants, the urban poor and radicalised middle-class elements; it posited, however, that the objective conditions for revolution did not yet exist and would have to be created via appropriate class action. The revolutionary classes were to press forward simultaneously on a number of fronts: organisation, economic struggle, mass protest, ideological propagation and electoral activity should the opportunity arise. Once FTMMP had made known its availability as a partner, it became possible for the 'maoists' to set aside some of their doctrinal quibbles in order to forge a practical political unity. In the end the FTMMP joined with no less than thirteen such groups to constitute the Popular Democratic Union (UDP; Woy-Hazleton and Gorman 1982). The UDP was the first true Peruvian labour party – if by that term we understand a party founded and largely controlled by the labour movement with the intent of carrying on its struggle in the political plane.[15]

Mine labour and direct political action

Beginning in 1975, Peru was gripped by an economic crisis occasioned by the collapse of international metals prices and the profligate borrowing habits of the military regime – a crisis that has not yet abated. It brought about the ousting of the reformist generals and the installation of the 'second phase' of military rule. It also brought galloping inflation and unemployment; while the miners did not suffer from the second, they saw the first eroding the gains that they had won since 1970 and redoubled their strike activity in order not to lose ground. The new rulers were not wholly unsympathetic but felt obliged to call for restraint. Initially they appealed to the miners' national loyalties; when that appeal failed they were prepared, unlike their predecessors, to resort to coercion. In April 1976 a state of emergency and a general prohibition of strikes were imposed on the mining sector. In July and August they were made universal.

These events had the result of 'nationalising' mine labour's economic struggle and turning it into a defensive one. No longer could miners' incomes advance by wringing higher wages out of individual employers; instead, the target had become the state and its economic austerity polices. Inasmuch as those policies contemplated higher prices for essential consumer goods, other subordinate classes besides the wage-earning proletariat had an incentive to join in closing ranks against the common adversary. The UDP, pursuing its broad front strategy, argued so convincingly for united direct action that the more reticent working-class parties, such as the PCP, had to go along; the argument was clinched by the promise of full cooperation from the mine workers, which also served (especially in the central sierra) to assure the adherence of the peasantry. Thus the successful general strikes of July 1977 and May 1978 were mounted. General strikes have continued sporadically into the current era of civilian governance.

Direct action by way of general strikes has not been Sorelian; i.e., it has had the 'economistic' aim of modifying economic policy and does not contemplate the overthrow of the government. Its aim has not been achieved, since austerity programmes have been imposed at IMF insistence under circumstances which left Peru with no alternative. However direct action has stiffened the government's spine in its negotiations with the IMF and has probably helped the country by making clear to the international agency that there are limits beyond which it is wiser not to press. Moreover it has induced official evasion of IMF strictures after agreements were signed (see *Quarterly Economic Report for Peru-Bolivia*, various issues for 1978–83). Most of all, it has contributed mightily to the political unity and effectiveness of the popular sectors *in toto*. For

[t]he organization of the unions combines with widespread popular involvement as the causes of strikes reflect issues of deep concern to the urban and rural poor who are not organized in unions. ... [S]uch a widespread popular coalition is difficult to coerce or to co-opt precisely because it is so widespread and draws its strength from a variety of sources. (Angell and Thorp 1980)

Paradoxically, popular mobilisation for direct action may, in the end, promote bourgeois hegemony by ruling off the agenda a quasi-authoritarian political option *à la* Mexico. As direct action in Peru has economic rather than revolutionary objectives, its threat potential is limited; the bourgeoisie may find it advantageous to provide institutional channels for peaceable popular involvement in politics, as opposed to outright dictatorship (Therborn 1979). In fact, it was *both* the rising tide of popular direct action *and* pressure from the bourgeoisie that convinced the military regime in late 1977 to relinquish power and that kept the process of transition on track from 1978 to 1980.

Mine labour and electoral politics

Peru's transition to civilian governance extended from the 1978 election of a Constituent Assembly charged with revising the constitution to the 1980 elections for national and, later in the year, municipal offices. The UDP participated actively in all three campaigns.

The 1978 election was mainly a preliminary test of strength for an assemblage of political parties some of whose members were new and others of whom had not contested an election for twelve years; there would be little incentive for the left to deal with its ever present factionalism until this trial had been completed. The UDP entered the lists as an independent entity and garnered 4.6 per cent of the vote, which translated into four of the 100 seats in the assembly; it was outpolled by other leftist coalitions that had the advantage of well-known, charismatic personalities such as Hugo Blanco (a radical peasant leader of the 1960s) at the head of their tickets. A degree of left unity was achieved during the assembly's deliberations, enough to write a bill of socio-economic rights into the new constitution and to extent the franchise to illiterates. It was not enough, however, to produce a single slate for 1980. Eschewing both a moderate–left coalition led by the PCP and a far-left grouping of trotskyist parties, the UDP ran its own candidates for president and congress.

Whereas the combined left had won over a third of the seats in the Constituent Assembly in spite of its internal divisions, its vote total fell to barely 20 per cent in 1980; the UDP's share was only 2.3 per cent of the presidential vote, 3.5 per cent of the nationwide senatorial vote, two senate seats, and three seats in the Chamber of Deputies (Woy-Hazleton 1982). Two lessons were learned: that if the mine unions wished to play a national political role, they would have to do so as part of a wider political grouping at the sacrifice of some of their prized independence; and that the left's electoral fortunes would henceforth depend on its ability to put forward a coherent programme of viable, concrete policy alternatives.

These lessons were applied in the November 1980 municipal contest: separating itself from the trotskyists and moving closer to the moderates whom it had previously spurned, the UDP joined the United Left (IU) electoral front. IU candidates did well in Ilo, Moquegua and other southern towns where the

FTMMP is present; they also demonstrated strength in the larger cities, capturing the mayoralty of Arequipa and placing second to the government party in Lima.

In sum, the mine unions have begun to engage in the kind of labour party politics that is found in the bourgeois democracies of western Europe. If IU evolves from a coalition into a single party, the miners will have to share it with other elements of organised labour. But the labour movement as a whole – in which the place of the miners will be central so long as the industry retains its prominence in the national economy – will control the party rather than being manipulated, as in the past, by middle-class cadres with their own purposes in mind. On the other hand, the evolving party will have to face a problem that plagues its European counterparts: workers' intense loyalty to their unions does not translate automatically into support for Labour at the polls. Cerro miners have been willing to cast their votes for conservative parties led by charismatic figures who project a concern for the welfare of the less privileged – in 1980, President Belaunde and his AP party. Even in areas of FTMMP strength, such as Tacna department, parties other than IU have triumphed on the basis of one or another form of charismatic appeal.

Mine labour and community power

It can be argued that a proletariat desiring to develop a capability for political leadership and hegemony, looking toward the eventual supersession of bourgeois domination, can best begin by involving itself in issues of community power. By assuming primary responsibility for the social and political organisation of the mining camp or townsite, workers free themselves from utter dependence on the employer; gain a greater sense of personal effectiveness; acquire habits of participation in a wider circle of affairs; fortify class cohesion and solidarity; and, in the process of creating socio-political facts, breed their own 'common sense' independent of bourgeois conceptual categories and ideologies. Such an effort does not encounter the same intense opposition as does a frontal challenge to the bourgeoisie at the national level and, thus, is more apt to yield successes given the current balance of class forces. Indeed, it would dovetail nicely with the decline of employer paternalism and the mining companies' desire to be rid of the trouble and expense of organising and running residential settlements. The mine unions, with their full treasuries and loyal base of rank-and-file backing, could take the lead in a community power movement.

Unfortunately, self-help activities of this sort are conspicuous by their absence. At Toquepala and Ilo, ordinary workers and union officials feel that townsite affairs are entirely SPCC's responsibility; many of those with whom I spoke complained of the lack of this or that facility but were astounded at the suggestion that they might take a direct hand in providing it. This finding is in keeping with SPCC's own observation, noted above, about the atomisation that characterises the social life of all of its townsites except Cuajone.

Studies of the Cerro work force suggest that its highly organised social life revolves around matters of economic concern and does not extend to camp organisation proper (Morello 1976).

It was mentioned previously that pueblo joven dwellers, drawn from the same peasant origins as miners, have displayed a genius for organising their community affairs. Interestingly, the people are usually employed in *petit-bourgeois* pursuits and frequently orient themselves around a dream of individual mobility (Palmer 1982). It may be that, at Peru's current level of development, community power and workplace economic power are alternate, competing outlets for proletarian class practice, not mutually reinforcing ones.

VI Conclusions

Miner labour's political gains since 1968 have been real. They were and are being won through the proletariat's own efforts. These efforts have often been dressed up in radical ideological rhetoric. They are, however, profoundly 'economistic' and do not represent a bid to displace the bourgeoisie or overthrow the capitalist system. Not even nationalism serves any more — since the expropriation of Cerro and Marcona — as a wellspring of class action in the mines; there have been few calls for the expropriation of SPCC, which, besides being utterly unfeasible, could well kill the goose of high profits per employee that has laid the golden egg of the highest wages paid anywhere in the country.

Yet, 'economism' *can* be useful for enlarging proletarian class power short of socialist revolution. Had the military regime perceived mine labour as politically ambitious in national terms and, thus, an obstacle to its plans for controlled 'revolution', it would not have responded to mine strikes in ways that served, even if unwittingly, labour's ends. Had the bourgeoisie felt more at risk from a revolutionary proletariat than from the inept economic management of the military regime, it would not have joined with the workers in demanding a return to civilian governance nor accommodated, as it has thus far, Latin America's strongest independent union movement and most active electoral left. If the outcome is bourgeois democracy rather than socialism, that is nonetheless real progress for a country that has never known such an open political order. The issue of socialism versus capitalism will remain alive for as long as the latter exists; there is no reason why the proletariat cannot consolidate its political gains and settle in for a hegemonic struggle — a 'war of position' — over the long haul.

For the immediate future, then, the mining proletariat can look forward to a hegemonic integration into a stable capitalist order. Hegemony is possible if the bourgeoisie, responding to the challenge of the labour movement, comes to rely on high productivity rather than depressed wages (relative rather than absolute surplus-value) in its strategy of accumulation, and admits unions to a role in the mediation of interclass conflict. The bourgeoisie in Peru, both native and foreign, appears to be doing so.

This study has been undertaken because the probability of a hegemonic outcome cannot be assessed by attending to bourgeois class structure alone. For a hegemonic 'pact' requires

a working class both strong and capable of stimulating change through wage demands and other types of pressure. ... In (hegemonic) systems there is a congruence between the economic and political power of the workers, in the sense that the latter's political strength grows out of and is coexistent with economic development. This is not so in ... countries where political mobilization is largely the work of revolutionary parties and other factors only indirectly related to the economy. (Graziano 1980)

It is the case, though, in Peru – and, possibly, in other countries whose industrialisation takes the form of 'bonanza development' and where proletarian class practice reflects that of the mine worker.

Notes

1 This article is based in part on field research conducted in Peru during 1977–8 and 1981. Financial assistance was provided by the US Department of Health, Education and Welfare, Office of Education in the form of a Fulbright-Hays Dissertation Research Abroad fellowship; by the UCLA-Meiji University Pan-Pacific Studies Program, administered by the Council on International and Comparative Studies at the University of California, Los Angeles, and by the School of International Relations, University of Southern California. Special thanks are also owed to the following: the Peruvian Fulbright Commission and its director, Marcia Koth de Paredes; Southern Peru Copper Corporation and its senior vice-president, Arthur G. Beers; Empresa Minera del Perú and its president, Luis Briceño Arata; Centromin-Perú and its president, Guillermo Flórez Pinedo; Compañía de Minas Buenaventura and its chief executive officer, Alberto Benavides de la Quintana; the Sociedad Nacional de Minería y Petróleo and the chief of its Department of Economic Studies and Statistics, Róger Arévalo Ramírez; and David Ballón Vera, president of Sudamericana de Metales SA and past president of the Banco Minero del Perú.

2 It is fashionable to assert that bourgeoisies in less-developed countries are structurally incapable of attaining hegemony. I dispute that contention and present a counterargument elsewhere (Becker 1983).

3 The other owners are the Marmon Group of Chicago, 20.8 per cent (this share belonged to Cerro until Marmon purchased the company in 1976); Phelps-Dodge, 16.2 per cent; and Newmont Mining, 10.7 per cent. Billiton NV, a Royal Dutch/Shell subsidiary, is not an integral partner in SPCC but holds an 11.5 per cent joint venture in its Cuajone mine. The shareholdings indicated here differ, but only slightly, from those at the moment of SPCC's foundation.

4 There are several of these, since the Cerro complex produces copper, lead, zinc, silver and a variety of minor nonferrous industrial metals.

5 The APRA has always been a predominantly middle-class party. However, from its origins in 1924 until the late 1940s it sought power by revolution and was periodically suppressed. Beginning in 1956 it changed tactics, seeking to develop a more moderate image so as to be better able to compete openly in electoral politics. After 1960 the party's principal concern became competition from Fernando Belaunde's Accion Popular (AP) movement, which also flew a banner of moderate reformism; in the 1963–8 congress the APRA allied itself with the oligarchic opposition to Belaunde so that the latter's reform programme would not succeed.

6 Centromin has set up its own television channel, whose signal is received throughout the enterprise's area of operations. In addition to re-broadcasting popular programmes from Lima it offers news about Centromin – its activities, finances, market conditions,

etc. – announces social and cultural events in the camps, and so on. Whenever labour disputes are brewing, it presents management's point of view. Guillermo Flórez Pinedo, Centromin's president (interviewed on 23 August 1981), believes that the latter presentations have been very effective in weakening the militancy of the work force, and especially that of miners' families.

7 The Guggenheim family, founders and long-time owners of Asarco, were among the early bourgeois critics of the brutalisation of labour. They practiced benevolent paternalism in their US and Mexican mines from their beginnings at the turn of the century. To my own knowledge no Asarco mine in Latin America has ever become, as did Cerro, a centre of labour violence (see also Marcosson 1949).

8 Interview with Peter Graves, Toquepala-Ilo townsite manager for SPCC, 22 June and 1 July 1978. Surprise inspections have since ceased. It may also be noted here that one of the early acts of the military regime was to disarm company police in the mining camps and to order free access to them.

9 Including: annual bonuses amounting to twenty days' basic wages; an allowance of 500 *soles* per month for those on the waiting list for family housing; workmen's compensation insurance, funded jointly by the company and the unions; company loans for purchases by workers of private houses; reimbursement of expenses for correspondence courses; college scholarships for children of workers.

10 These surveys were made available to me on condition that sources remain anonymous. However, I was shown in each case a full report and much of the raw data including the actual completed questionnaires, enabling me to vouch for both methodology and results.

11 I claim no great accuracy for this procedure: it was simply the best that could be done in the time available. It will be objected that workers will only tell their supervisors, who are, after all, representatives of their employer, what they think the latter wants or is willing to hear. However, most SPCC line supervisors were ordinary workers – and devoted union men – in US mines before signing on for stints in Peru and are characterised by top management as being rather too much pro-union in their personal attitudes and too sympathetic toward the views of the men under them. Union officials and radical intellectuals who volunteered their opinions during my visits to the SPCC work sites all agreed that workers would have no fear of their political views affecting their promotion possibilities and would speak freely about political matters to supervisors with whom they were friendly.

12 Defined to include: basic wage or salary; overtime pay; shift differentials; obreros' holiday pay and Sunday pay (*dominical*); vacation pay; and deferred payments – employer contributions to social security funds as well as amounts set aside for future payment of time-of-service indemnities.

13 The military had an intensely anti-aprista tradition that went back to 1931, when an APRA uprising in the city of Trujillo led to the murder of many of the officers of the local garrison. In 1948 and 1962 the military intervened to prevent the APRA from coming to power. It has been suggested that this was also a motive behind the *coup d'état* of 1968, as it was widely believed at the time that the eternal aprista leader, Víctor Raúl Haya de la Torre, would be the winner of the presidential election scheduled for 1969.

14 The PCP was following the Soviet strategy for Latin America, which is that communist parties should avoid revolutionary 'adventurism', strive for legality, and support bourgeois nationalist political forces.

15 Significantly, no labour leader arisen from the rank and file has ever become prominent in the directorate of the PCP or, until very recently, the APRA. But Víctor Cuadros Paredes, an SPCC obrero who became a local union official and then the first president of the FTMMP, is now a member of the UDP's delegation in the national congress. The new kind of union party nexus allows for upward mobility by workers in a way that the old arrangement never did.

References   * *See annotated bibliography for full details of all asterisked titles*

Adamson, Walter L.
1980 *Hegemony and Revolution: Antonio Gramsci's Political and Cultural Theory.*
Berkeley and Los Angeles: University of California Press.

Alberti, Giorgio, Jorge Santistevan, and Luis Pásara
1977 *Estado y Clase: la Comunidad Industrial en el Peru.* Lima: Instituto de Estudios
Peruanos.

Amin, Samir
1976 *Unequal Development: An Essay on the Social Formations of Peripheral Capitalism.* New York: Monthly Review Press.
1980 'The class structure of the contemporary imperialist system'. *Monthly Review*
31 (January): 9–26.

Angell, Alan, and Rosemary Thorp
1980 'Inflation, stabilization and attempted redemocratization in Peru, 1975–1979'.
World Development 8: 865–86.

Becker, David G.
1982 '"Bonanza development" and the "new bourgeoisie": Peru under military rule'.
Comparative Political Studies 15: 243–88.
1983*

Boggs, Carl
1976 *Gramsci's Marxism.* London: Pluto Press.

Bonilla, Heraclio
1974*

Brewer, Anthony
1980 *Marxist Theories of Imperialism: A Critical Survey.* London: Routledge & Kegan
Paul.

Collier, David
1976 *Squatters and Oligarchs: Authoritarian Rule and Policy Change in Peru.*
Baltimore: Johns Hopkins University Press.

DeWind, Adrian W., Jr.
1977*

Dietz, Henry A.
1969 'Urban squatter settlements in Peru: a case history and analysis'. *Journal of
Inter-American Studies and World Affairs* 11: 353–70.

Fanon, Frantz
1963 *The Wretched of the Earth.* New York: Grove Press.

FitzGerald, E.V.K.
1979 *The Political Economy of Peru 1956–78: Economic Development and the
Restructuring of Capital.* Cambridge: Cambridge University Press.

Flores Galindo, Alberto
1974*

Gorman, Stephen M., ed.
1982 *Post-Revolutionary Peru: The Politics of Transformation.* Boulder, Colorado:
Westview Press.

Gramsci, Antonio
1971 *Selections from the Prison Notebooks.* Quintin Hoare and Geoffrey Nowell
Smith, transl. and ed. New York: International Publishers.

Graziano, Luigi
1980 'The historic compromise and consociational democracy: toward a "new democracy"?' *International Political Science Review* 1: 345–68.

Hobsbawm, Eric J.
1959 *Primitive Rebels: Studies in Archaic Forms of Social Movement in the Nineteenth and Twentieth Centuries*. Manchester: Manchester University Press; New York: W. W. Norton & Co.

International Labour Organisation
1980 *Yearbook of Labour Statistics 1980*. Geneva.

Jessop, Bob
1982 *The Capitalist State*. New York: New York University Press.

Kapsoli E., Wilfredo
1975 *Los Movimientos Campesinos en Cerro de Pasco, 1800–1963*. Huancayo: Instituto de Estudios Andinos.

Kruijt, Dirk, and Menno Vellinga
1979 *Labor Relations and Multinational Corporations: The Cerro de Pasco Corporation in Peru (1920–1974)*. Assen, The Netherlands: Van Gorcum & Co.

Laite, Julian
1981*

Lenin, V.I.
1969 *What is to Be Done? Burning Questions of Our Movement*. New York: International Publishers.

Lowenthal, Abraham F., ed.
1975 *The Peruvian Experiment: Continuity and Change under Military Rule*. Princeton: Princeton University Press.

Marcosson, Isaac F.
1949 *Metal Magic*. New York: Farrar, Straus & Co.

Marx, Karl
1963 *The Poverty of Philosophy*. New York: International Publishers.

Mepham, John
1978 'The theory of ideology in Capital'. In *Marxist Philosophy*, 3: *Epistemology, Science, Ideology*. J. Mepham and D.-H. Ruben, ed. London: Harvester Press. Pp. 141–69.

Ministerio de Trabajo (Peru)
1973 *Las Huelgas en el Peru, 1957–1972*. Lima.

Morello, Gino
1976 'Los mineros de la Cerro de Pasco, 1940–1970'. Bachelor's thesis, Pontificia Universidad Catolica del Peru, Lima.

North, Liisa, and Tanya Korovkin
1981 *The Peruvian Revolution and the Officers in Power 1967–1976*. Montreal: Centre for Developing Areas Studies, McGill University.

Palmer, David Scott
1982 'The post-revolutionary political economy of Peru'. In *Post-Revolutionary Peru: The Politics of Transformation*. Stephen M. Gorman, ed. Boulder, Colorado: Westview Press. Pp. 217–36.

Petras, James, and Maurice Zeitlin
1967 'Miners and agrarian radicalism'. *American Sociological Review* 32: 578–86.

Smith, Sheila
1980 'The ideas of Samir Amin: theory or tautology?' *Journal of Development Studies* 17 (October): 5–21.

Stephens, Evelyne Huber
1980 *The Politics of Workers' Participation: The Peruvian Approach in Comparative Perspective*. New York: Academic Press.

Sulmont, Denis
1974 *El Desarrollo de la Clase Obrera en el Peru*, Lima: Publicaciones CISEPA, Pontificia Universidad Católica del Peru.
1975 *Sindicalismo y Política en el Peru*. Lima: Taller de Estudios Urbano-Industriales, Pontificia Universidad Católica del Peru. Mimeographed.
1977 *Historia del Movimiento Obrero Peruano (1890–1977)*. Lima: Tarea.

Therborn, Göran
1979 'The travail of Latin American democracy'. *New Left Review* 113–14: 71–109.

Woy-Hazleton, Sandra L.
1982 'The return of partisan politics in Peru'. In *Post-Revolutionary Peru: The Politics of Transformation*. Stephen M. Gorman, ed. Boulder, Colorado: Westview Press. Pp. 33–72.

Woy-Hazleton, Sandra L., and Stephen M. Gorman
1982 'The Peruvian left since 1977: ideology, programs and behavior'. Paper presented at the annual meeting of the American Political Science Association, Denver, September 2–5.

Zapata, Francisco S.
1980*

CHAPTER 14

Nationalisation, copper miners and the military government in Chile

FRANCISCO ZAPATA

Zapata has written an analysis of the dynamics of the miners' union at the Chuquicamata open pit copper mine during the 1970s and 1980s. His chapter reveals an extremely sophisticated working class, that while using a variety of ideological styles, is intent upon protecting its economic position. Even under the military dictatorship after 1973, the union has been able to remain strong due, primarily, to the strategic role of Chuquicamata in earning foreign exchange. To the extent that leaders were coopted or repressed, the ranks of the miners continued to supply fresh ideas and new people to continue their labour action, from negotiation to striking. Importantly, Zapata finds labour continuity from the years of Allende to those of Pinochet.

Francisco Zapata is with the Centro de Estudios Sociológicos, El Colegio de Mexico, Camino al Ajusco No. 20, Apdo. Postal 20–671, Mexico, DF, Mexico. A related recent publication is: *Enclave y Polos de Desarrollo en México*, El Colegio de México, 1984. Currently he is working on an analysis of temporal theories of strikes in Argentina, Chile, Mexico, Peru and Venezuela in relation to their economic and socio-political context.   *— The editors*

I Introduction

Labour action and political behaviour of Chilean copper miners is, in many respects, an open question. In spite of having data concerning electoral results both in union elections and in congressional and presidential elections, their results do not show consistency in terms of the political attitudes of the copper miners. Sudden changes in the direction of their support, revealing more of a clientelistic than a class approach to politics, indicate that miners are structurally conditioned by the characteristics of their spatial localisation and of their work within a company town (Zapata 1979). This is in contrast to the view that traditional ideological factors determine their political attitudes. Additionally, the strategic character of copper exports for the Chilean economy contributes to the bargaining power of miners in the industrial relations system of the country. Copper miners represent only 3.5 per cent of the economically active population, while their contribution to the national economy is much larger. Indeed, in the period 1971–77, mineral exports represented, on average, 70.5 per cent of total exports.

Mostly located in the north of the country, desert and sparsely populated, the copper industry possesses a highly combative work force if one considers striking activity or capacity to confront the political system to obtain demands. Interpretations of labour action in the copper mines fluctuates between those that argue in favour of characterising the miners as extremely trade unionistic in their overall strategy, and those who argue in favour of a more ideologically motivated work force, highly functional for the political strategy of the left (Barrera 1978; Barria 1974; Hunceus 1974). The history of labour action in the mines since the beginning of the century and especially since the legitim-isation of the trade union organisation in the late 1920s supports the character-isation of miners as combative in the pursuit of their demands, but this history is not conclusive in terms of the consistency of their political commitments.

One can say that at least up to the nationalisation of the great copper mines in 1971, copper miners, especially at Chuquicamata, supported the socialist and communist parties as they were represented by union leaders in their work-places. Between 1971 and 1973, after nationalisation, a majority supported a splinter socialist party in Chuquicamata (Union Socialista Popular – USOPO) and christian democrats in the rest of the mines. During this latter period both the socialist and communist parties had serious difficulties obtaining political support from the copper miners. The long strike of April–June 1973 proved to be one of the principal factors in the defeat of Allende. In addition, nationalisation implied that labour relations were to be admin-istered by the new authorities in the mines, named by the Chilean state. Unlike in the past it was not possible to blame the foreign companies for the labour problems that arose in the mines. And in the first two years after national-isation, new ways of exercising authority, high levels of politicisation of management decisions in labour relations and, in general terms, the develop-ment of a new power structure in the copper mining sector (including new relations between the government and the confederation of copper workers – Confederación de Trabajadores del Cobre, CTC) changed traditional schemes to which both miners and technical personnel were accustomed.

These elements contributed to an increase in social tensions in the mines, especially in the second semester of 1971 when confrontations arose between the 'supervisores' and the government. Later on, all through 1972 and 1973, up to the *coup d'état* of September, tensions erupted frequently between union leaders, workers and the new executives as a consequence, as we shall see, not only of strictly defined labour issues but especially as a result of different political commitments. When the military took over the government of the country in 1973, many labour leaders were fired or executed; collective con-tracts were declared void or not negotiable, wildcat strikes and legal strikes were prohibited; union meetings and union elections were suspended. Copper miners entered a new phase in their relationship with the management of the companies and with the government of the country. The situation in the mines was strictly controlled by the military authorities, reflecting the importance of copper mining to the overall economy. Nevertheless, at the end of 1977 and all through 1978, especially at Chuquicamata, Chile's largest mine, both the rank and file and the union leaders began to act against the military's strategy of suppressing workers' demands. These manifestations of discontent coincided with disagreements inside the government concerning policy toward labour which eventually were at the origin of the so-called Labour Plan implemented at the beginning of 1979. This process shows that miners had some capacity to influence the political decision making in the capital and obtain some of the demands they were asking for.

This chapter builds from this general image of what happened in Chilean copper in the last few years and details some considerations on the relationship between mining and politics in that country. The specific consideration of trade union action at the Chuquicamata mine will be utilised to discuss the more general problem of the consequences of repressive and authoritarian government on the development of labour action. On the basis of this analysis, some light will be thrown on the way Chuquicamata's miners have been able to give a remarkable continuity to their style of pressing demands in spite of the political changes that have occurred in the country.

II Chilean copper mining, Chuquicamata and miners: 1970–9

The Chuquicamata copper mine is located in the north of Chile, two hundred kilometres east of the port of Antofagasta, and about twenty kilometres north of the small town of Calama, the only urban centre relatively close to the company town that surrounds the mine (Manning 1975). The deposit is 2,895 metres high in the arid and mountainous northern Atacama desert. When American capitalists began to search for large scale deposits all over the world at the beginning of the twentieth century, taking off from favourable legislation approved in the United States and from easy access to mining property in Latin America (Culver 1980), the possibility of building a large-scale operation in Chuquicamata became real. At the beginning of 1911, Guggenheim, who also had large investments in Chilean nitrate, just a few miles from Chuquicamata, founded the Chile Exploration Company (CHILEX), incorporated in New York City, to develop the deposit. Systematic exploration began in 1912 under Guggenheim's direction, and plans were made for large-scale exploitation of the mine. Copper production started in 1915 with an initial output of 4,300 metric tons. Guggenheim managed the mine for eight years and in 1923 sold the concession and industrial installations to the Anaconda Mining Corporation. The Chile Exploration Company became a subsidiary of Anaconda until it was nationalised by a constitutional reform implemented on 11 July 1971 (*Mining Engineering* 1952, 1969; Sawyer 1960; Suseelman 1978).

By 1941 Chuquicamata was producing two hundred thousand metric tons of refined copper a year and employment reached 7,406 workers, including white- and blue-collar employees. During the period 1923–41 unionism appeared in Chuqui. Indeed, as a consequence of the passing of the social laws in 1924 and of the labour code in 1930–1, unions were legally permitted and white- and blue-collar workers became organised, white-collar employees in a 'sindicato profesional' and blue-collar in a 'sindicato industrial'.

Working conditions in the mine were not very favourable; Gutierrez and Figueroa demonstrate (1920) accidents were common. The labour force in the mine came from the south of Chile and had rural origins. Some of the man-power had been engaged in small mining operations in the so-called 'Norte Chico' (small north) around the towns of Andacollo, Copiapo and Vallenar where gold mining was predominant. Labour turnover was frequent and the policies of the company assured that 'agitators', especially from the nitrate mines, could not enter the mines. Nevertheless, at the end of the thirties, politics entered Chuquicamata.

As a consequence of the refusal of Chilex to allow Pedro Aguirre Cerda to carry his Popular Front presidential campaign inside the company town, the government of President Aguirre Cerda tried to limit the extra-territorial rights that the company defended within the Chuqui compound. As the series of governments presided over by politicians belonging to the Radical Party succeeded each other (1938–52), a process of gradual movement to the political right was taking place within the country; a process not unrelated to inter-national events and the beginning of the Cold War. In 1948, the Law for the Defense of Democracy was passed and led to political persecution particularly in those productive centres where communist labour leaders represented workers. In this way, both in the coal mines of the southern part of the country (Lota-Coronel) and in the copper and nitrate mines of the north, dismissals and disappearances began to occur. This legalised persecution did not end until the Law for the Defense of Democracy was repealed in 1958.

During this period, employment in the copper mines decreased as a result of the political dismissals and political persecution. Employment at Chuquica-mata went down from 8,174 workers in 1949 to 5,945 workers in 1953, a 27.3 per cent decrease. The employment level of 1949 was not reached again until 1970 as a result of various expansion projects: the construction of a sulphur smelter (1948–52), a concentration plant, a refinery and maintenance shops (1964–8). Notwithstanding the decrease in personnel, production levels continued to rise.

Chilex sold all of its production to the United States where it was refined and manufactured. It put pressure on the United States government to establish agreements with Chile on the price level of the copper exports (Moran 1974). As a result of these agreements the price of copper did not reflect international demand. In 1951, the so-called Agreement of Washington (Convenio de Washington) fixed a price of 27.5 cents a pound for Chilean copper production and supposedly extracted a commitment from Anaconda to expand production at Chuquicamata. The experience with this agreement was extremely negative for Chile and was not continued in 1952. Chile lost great amounts of revenue as a consequence of these agreements which reflected the interests of the companies more than those of the country. An indirect indication of this is demonstrated by the fact that Anaconda's investment in Chile, equal to 19 per cent of its global investment commitment, produced more than 52 per cent of the total profits of the company in the period 1935–68 (calculations of the nationalisation decree, 1971).

In 1955 new regulations were established for the relations between the companies and the Chilean state including a new special labour code for copper workers (Estatuto de Trabajadores del Cobre) and the creation of a state bureau for copper affairs.[1] With the so-called 'chilenisation' agreements of 1965 and especially in the El Teniente mine, property of Kennecott Copper Corporation, new arrangements in production matters were reached but, in general terms, all the policies that were implemented in the pre-nationalisation period did not succeed in giving Chile all the benefits it expected from the copper companies (Moran 1974; Sigmund 1981). This explains in part the unanimity that was achieved in 1971 when a joint session of the Chilean Congress approved the nationalisation process.

In many ways, the nationalisation process reflected a long history of frustration on the part of the Chilean government concerning the copper mines. Both Anaconda and Kennecott used political pressures to obtain higher prices, fix labour problems and influence elections, without giving back what many conservative governments expected in exchange, that is, increases in production and revenue for the country (Moran 1974). Conservative governments assumed that production increases would increase the percentage of Chilean participation in the overall value of that production: they forgot that the corporations were more interested in stable levels of profits at given levels of production than in increasing levels of production and unstable levels of profits. In addition, as the case of the price agreement of 1965 illustrates very well (Berteau 1982), instead of assuring economic aid from the United States government in exchange for price restrictions, Chile was made the fool of the game by losing on both the price level and on the economic aid level.

Nationalisation became then the only solution that could establish new bases for the development of the industry and assure that copper would be the locus of the Chilean development process. The experience of the period 1970–9, to which we will refer now, shows that, in overall terms, the decision to nationalise the copper industry was correct as production increased significantly (by more than one third), revenues doubled, and productivity increased dramatically, although this may be as much a result of political repression by the military than as a reflection of technological innovation and managerial capacity. In any case, the copper industry has been conceived by all governments as central to the development of the country, and especially by the military government which follows an economic policy of export generated internal growth. Copper mining remains strategic in terms of the hard currency obtained through exports and profitable in terms of the high levels of productivity that existed in the industry.

As the data in Table 14.1 shows, two periods can be distinguished in the recent evolution of copper production in Chile. On the one hand, between 1970 and 1973 total Chilean production increased very slowly while Chuquicamata decreased its contribution to total production. Total production reached an average of 716,000 metric tons between 1971 and 1973 while Chuqui's production averaged 249,900 metric tons during the same period, twenty thousand tons less than it had produced, on average, in the period 1965−70.

As can be seen in Table 14.2, production per employed person also decreased in the period 1971−3 continuing a tendency that had started in 1968 when productivity levels began to deteriorate in the 'Gran Minería del Cobre'. The overall correlation between production and employment for the period

Table 14.1   Basic data on the Chilean copper mining industry 1970−9

| | 1 | 2 | 3 | 4 | 5 | 6 | 7 |
| | Production (thousands) | | Production | Employ- | Earnings ($) | Dependency | Pro- |
Year	M.T.	Price	value	ment	(thousands)	index	ductivity
1970	692	1,415	979,2	20.766	312.639	77.1	22.81
1971	708	1,081	765,3	21.500	−	79.0	21.86
1972	717	1,070	767,2	23.932	313.304	87.0	20.31
1973	735	1,780	1.308,3	24.000	−	69.5	19.54
1974	902	2,058	1.856,3	27.863	196.895	59.0	23.18
1975	828	1,236	1.023,4	25.941	160.727	64.0	21.59
1976	1.005	1,410	1.417,0	28.526	222.746	58.0	27.41
1977	1.056	1,310	1.383,4	29.238	164.805	53.2	29.02
1978	1.036	1,362	1.411,0	29.504	234.082	50.8	28.36
1979	1.036	1,985	2.106,0	29.062	233.429		29.55

Sources: 1−3: Bureau of Metal Statistics, *World Metal Statistics*, New York, 1979, cited by Comercio Exterior (Mexico), June and August 1981, p.715 and 939 respectively.
4−5: Guillermo Campero 1980.
6: Fondo Monetario Internacional, *Estadísticas financieras*, Washington, Several numbers, 1970−9.
7: Corporación del Cobre (CODELCO), 'Inversiones proyectadas, quinquenio 1981− 1985', Santiago, 1980.

Definitions: Production: copper production in metric tons of copper content in the mines of the Gran Minería del Cobre (GMC) on a yearly basis.
Price: annual average for refined or electrolytical copper on a cash basis in American dollars for metric tons at the London metal exchange.
Production value: equals production by price.
Employment: supervisors, white-collar and blue-collar employees at the mines of the GMC for each calendar year.
Earnings: total median incomes on an annual basis in pesos of December of each year.
Dependency index: value of mining exports divided by value of total exports for each calendar year.
Productivity: average annual production by employed worker.

Table 14.2 Chuquicamata, 1940–76: production, employment, investment book value, profits, rate of return and taxes paid by Chile Exploration Company

	1	2	3	4	5	6	7	8
Year	Production (,000 M.T.)	Employment	Production per employed person	Investment (thousands)	Book value	Profits (millions)	Rate of return	Taxes (millions)
1940	151.0	7.406	20.4	–	–	–	–	–
1941	216.8	7.793	27.8	–	–	–	–	–
1942	225.7	8.855	25.5	–	–	–	–	–
1943	238.0	9.589	24.8	–	–	–	–	–
1944	241.2	9.670	24.9	–	–	–	–	–
1945	237.6	–	–	886	44.2	13.2	30.0	11.1
1946	210.4	8.716	24.1	907	44.2	16.3	38.3	12.2
1947	221.0	9.650	22.9	1166	43.8	22.4	51.1	23.5
1948	208.0	9.542	21.8	5588	50.6	24.6	46.6	24.3
1949	175.1	8.174	21.4	20641	57.2	15.4	26.8	15.3
1950	156.3	5.828	26.8	22446	63.6	17.0	26.8	22.0
1951	163.5	5.984	27.3	37716	73.6	20.6	28.0	23.9
1952	159.2	6.353	25.0	28521	82.7	14.7	17.8	16.3
1953	156.8	5.945	26.4	17584	90.0	4.7	5.2	5.4
1954	186.2	6.033	30.9	–	86.6	13.9	16.0	22.1
1955	209.3	6.496	32.2	1218	100.0	31.7	31.8	62.9
1956	241.3	6.731	35.8	11933	139.0	49.3	35.4	67.8
1957	239.0	6.783	35.2	14078	137.0	18.9	13.8	31.3
1958	212.8	7.005	30.4	10275	138.4	11.9	8.6	20.3
1959	278.0	6.816	40.8	14087	162.8	34.9	21.4	20.9
1960	231.1	6.990	33.0	10080	162.4	21.3	13.1	34.4
1961	249.7	7.013	35.6	5571	200.0	25.5	12.8	33.7
1962	275.8	7.137	38.6	8461	206.1	31.2	15.2	52.4
1963	274.8	7.206	38.1	9807	213.0	29.3	13.7	25.8
1964	288.1	7.252	39.7	7207	213.3	34.3	16.1	54.5
1965	252.7	7.288	3.7	9000	200.6	32.0	16.0	69.8
1966	303.5	8.090	27.5	22593	222.0	61.6	27.8	96.6
1967	276.9	8.353	33.1	43611	273.0	72.8	26.7	102.0
1968	278.9	8.373	33.3	62324	288.4	73.7	25.6	98.5
1969	283.4	8.711	32.5	–	279.4	61.6	22.2	24.9
1970	263.0	9.357	28.1	–	–	–	–	–
1971	250.2	10.013	25.0	–	–	–	–	–
1972	234.3	10.839	21.6	–	–	–	–	–
1973	265.3	11.428	23.2	–	–	–	–	–
1974	356.8	10.617	33.6	–	–	–	–	–
1975	304.6	10.355	29.4	–	–	–	–	–
1976	445.5	10.232	43.5	–	–	–	–	–

Sources: World Bank 1978 for columns 1, 2 and 3; 3–8: 1974.

1940–76 in Chuquicamata is very low (r = .5326; $r2$ = .2837) and confirms the impression that non-technical aspects such as political and union affairs continuously affect the level of total employment in the mine. In the last period, the correlation seems to be higher and employment levels keep pace with production, contrary to what had been the case in the period before 1960.

But, in general terms, one can conclude from this series that the two variables are not related very closely.

Indeed, if one examines data in Table 14.3 on production and employment by presidential periods, that is to say, on average terms, it is quite clear that given levels of production can be achieved with very different numbers of personnel. If we compare, for example, average production during Ibañez's presidency (1953–8) and González Videla's presidency (1947–52), we find a higher tonnage was reached with fewer personnel in the latter period. The same could be said in the recent period where high tonnages have been achieved with the same personnel that worked in the mine five or six years ago.

Table 14.3 Chuquicamata: production, employment, productivity, investment, book value, profits, rate of return, taxes (averages for political periods, 1940–76)

Political period	1 Production (thousands of metric tons)	2 Employment	3 P/E	4 Investment	5 Book value	6 Profits	7 Rate of return	8 Taxes
				[in millions of dollars]				
1940–6	217.2	8.671	24.6	–	61.9	–	–	–
1947–52	180.5	7.588	24.2	19.3	115.2	19.1	33.2	21.2
1953–8	207.6	6.498	31.8	9.1	192.9	21.7	18.5	34.9
1958–64	266.3	7.069	37.6	9.2	210.6	29.4	15.4	48.8
1965–70	276.4	8.362	33.2	22.9	–	50.3	19.7	63.6
1971–3	249.9	10.760	23.3	–	–	–	–	–
1973–6	368.9	10.760	35.5	–	–	–	–	–

Sources: 1ā3: World Bank 1978; 4–8: Moran 1974.

Notes: political periods: 1940–6: various presidents after death of Pedro Aguirre Cerda, among which Alfredo Duhalde, Juan Antonio Rios; 1947–52: Gabriel González Videla; 1953–8: Carlos Ibañez del Campo; 1959–64; Jorge Alessandri Rodriguez; 1965–70: Eduardo Frei Montalva; 1971–3: Salvador Allende Gossens; 1973–?: Augusto Pinochet Ugarte.

Another interesting fact concerns the high levels of variability in the value of total production (see Table 14.1). This reflects the close relationship that exists between copper production and the international economy. Copper is a raw material closely related to the dynamics of economic sectors whose rhythm of production reflects the overall rate of growth of the economy. This creates acute changes in price levels which preclude the formulation of development plans based on the assumption of fixed levels of foreign revenue. These variations have seriously hampered Chile's possibilities of development. In the period 1971–3, while production increased very slowly and actually stagnated, prices were at the same time very low; this generated losses that reached almost 200 million dollars which severely affected the Chilean economy. This pattern is confirmed by the dependency index which shows that 1972 mineral exports represented 87 per cent of the total value of the country's exports.

This vulnerability to the variations in copper prices was compounded by the intensification of political confrontation in the mines, and especially in Chuquicamata where illegal strike activity increased between 1971 and 1972. In 1973, both in El Teniente and in Chuquicamata, production progressed at half speed for almost four months (April to July) due to the political confrontation between miners and the Popular Unity government. During the military government, price variations have continued to affect the Chilean economy and in spite of the relative tranquillity in terms of labour relations, the increase in production that has taken place from 1974 onwards has not been reflected in dramatic increases in foreign revenues. In spite of the increase in production by 29.3 per cent between 1974 and 1975, its value did not increase in the same proportion, essentially because the price of copper in the international market, after having increased, experienced decreases which coincided with the world recession that started around 1979. Therefore, revenues from copper exports increased as a result of increases in production levels and productivity and not as a consequence of price increase.

All this has affected Chile's capacity to import from its own revenues and given way to foreign indebtedness which has risen to dramatic levels in the period 1979−81. This evolution of prices, production and revenues indicates that in many ways the country is dependent upon the copper miners and their capacity to increase production and productivity and thereby assure that some continuity in revenues can be ascertained in spite of price variations. Indeed, productivity increases, equal to 27.5 per cent between 1974 and 1979 in a context where employment has increased only by 4.3 per cent, support an argument that stresses the importance of increased efficiency in the operation of the mines as a factor in increased production. At the same time, this has resulted in increased exploitation of the miners because earnings have not kept pace with their effort reflected in increased productivity. On the contrary, it seems that after an absolute deterioration of the standard of living between 1974 and 1975, it has increased very gradually from 1976 onwards.

In addition to the fact that levels of income and fringe benefits that miners had achieved before 1973 were questioned by the economic policy implemented by the military, many rights judged by miners as irreversible were not recognised by the government-appointed managers. Even if this defined a radically new situation in terms of the relations system of the mines, it did not mean that miners had lost their strategic location in terms of the bargaining power they could exert on the government. They could continue to obtain better conditions of work, higher salaries and a higher living standard than the rest of the Chilean working class. Political changes did not affect this position very much and one can see that even during a military government such as the one that is in power in Chile today, copper miners remain in a better position than the rest of Chilean workers. But in spite of the fact that the level of salaries at Chuquicamata has not decreased in the same proportion as those of the rest of Chilean workers in the period 1973−80 and that they have not had to accept a policy of massive lay-offs, miners at Chuqui had to absorb important changes. The protections of company town life, in terms of subventions for food consumption, transportation allowances and scholarships for children, have been drastically curtailed. Social mobility expectations which are important in the explanation of the miner's type of worker consciousness have suffered with the labour policies of the military government. Concrete possibilities for social promotion have disappeared and miners have had to accept that they are, indeed, 'workers'.

This is substantiated when one looks at the structure of the company town in terms of housing facilities. It is certain that social segregation according to types of housing and types of furnishings for houses has always been typical of Chuquicamata's physical lay-out. This lay-out is even reinforced by the mountainous topography in the sense that social hierarchy corresponds to the slope of the mountain. Supervisors live higher up. White-collar workers inhabit houses built on the only plateau of the town where the union meeting hall as well as commerce and schools are located. Blue-collar workers live down in the so-called 'sunken' camp (Campamento Hundido). This structure goes together with size of houses, number of bathrooms, types of furniture, free use of appliances and also with possibilities of repairs by the quaint 'welfare' department of the company where historian Ricardo Latcham, back in the twenties, succeeded in getting employment and which was very useful for his insights into the workings of Chuqui at that time (Latcham 1928).

It seems that in recent years, according to demands made by the unions, excess privileges obtained by the supervisors have accentuated social segregation in the town. The unions have denounced both the increase in the proportion of supervisors in the total employment of the mines[2] and the better quality of housing given to them, while at the same time deploring the deterioration of the level of housing provided for white- and blue-collar workers. Housing privileges go together with travel allowances, longer vacations, trips to Santiago and other benefits that are not available to non-supervisors. Unionised workers have become more and more alienated from the social structure of Chuquicamata. When this situation combines with the relative losses in terms of salaries and fringe benefits and with political persecution of their union leaders, one can explain the process of mobilisation that began to take place after 1976 and which climaxed in 1978 in the so-called 'lunchbox strike' to which we will refer later on.

Miners are also worried about the absence of collective bargaining which could help to correct abuses such as those we have mentioned. The automatic readjustment of salaries that the government uses to compensate workers for inflationary pressures is questioned by the unions who judge the percentages insufficient. Also, they mention that deductions for taxes, social security, housing and other items cut gross salaries by 40 per cent and indicate that salary and readjustments do not increase the level of net salaries actually received by miners. Nevertheless, these statements by the union leaders overestimate the degree of loss because they do not mention that miners receive, in addition to their salaries, all kinds of supplements such as overtime payments, production bonuses, year-end compensations and family allowances which are considered as part of the salary. Therefore, the real problem of the miners is not strictly related to the level of net income but to the fact that they are not able to increase this income through collective bargaining.

Union leaders argue that they deserve this bargaining possibility because of the productivity increases we have mentioned before. Leaders of the CTC argue that while the total number of miners has increased since 1975 from 25,941 to 29,062 workers, (+ 12.0 per cent) productivity in the same period has increased from 21.59 tons per employed worker to 29.55 tons per employed worker, an increase of 36.9 per cent.

Table 14.4   Employment and earnings in the mines of the Gran Minería del Cobre (GMC) in 1977

Mine	Occupational category							
	Supervisors		White-collar		Blue-collar		Total	
	N	$	N	$N	N$	N	Average	
El Teniente	803	28.193	4.571	13.496	6.856	8.824	12.230	16.537
Chuquicamata	975	27.420	6.042	10.857	3.288	7.915	10.305	15.397
El Salvador	462	21.056	2.250	9.181	2.922	7.113	5.634	12.450
Andina	199	26.992	1.034	14.030	914	10.536	2.147	17.186
Total employment	2.439		13.897		13.980		30.316	
Average salary		$25.915		$11.891		$8.597		$15.467

Sources: El Mercurio (Santiago), '¿Cuanto ganan los trabajadores de la Gran Minería del Cobre?'. 11 November 1977.

Also, they mention the fact that profits of the Copper Corporation (Corporación del Cobre-CODELCO) have gone from 159 million dollars in 1977 to 178 million dollars in 1978, an increase of 12 per cent. All this supports their pressure for obtaining the right to bargain. While this right was granted in the Labour Plan approved in 1979 and some kind of collective bargaining took place in 1980 and 1981, Chuquicata's miners were refused the right to strike in support of their demands, in contrast to El Teniente's miners which were not refused this right and struck in both years.[3] It seems that the historical evidence concerning cohesion, solidarity and combativeness, as well as the fact that Chuqui represents almost half of the total production of the 'Gran Minería del Cobre' (due to the high content of copper in the Chuqui ore), motivated the military government to discriminate between the different mines in labour relations. In any case, we conclude that Chuqui miners, while

losing some of the privileges that they had before 1973, have not lost everything. They have succeeded, as we will see, in maintaining some leverage with
the government.

III The labour relations system in Chuquicamata

*Rank and file, union leadership and labour action during the
'lunchbox strike' of July–September 1978*

Perhaps the best way of focusing upon the leverage exerted in recent years
by Chuqui miners is to examine one specific conflict which can throw light
on the more general characteristics of the 1973–81 evolution. This conflict,
the so-called 'strike of the lunchboxes' (presion de las viandas), took place
between July and September of 1978.

The conflict started after Pedro Tapia, union leader at the Tocopilla electricity generating plant of Compañía de Cobre Chuquicamata (COBRE-
CHUQUI) and leader of the local chapter of the CTC, had asked on various
occasions for specific responses on the part of the company concerning payment of benefits to the miners. Overtime payments, food during work-hours,
extraordinary payments for three year services (*trienios*) were not being paid
as in the past, and Tapia had petitioned the company on these points. These
petitions had been mentioned to the Minister of Labour on many occasions
in the period May–June 1978. No answer was forthcoming from the company, nor from the minister. This gave rise to the miners' refusal to eat food
during work, yet to remain in the dining rooms without eating, and subsequently to begin a work slowdown.

At the beginning of August a union meeting took place where several miners
addressed their fellow workers and demanded satisfaction. On the following
day the company dismissed all miners that had taken the floor. From this
moment on tension began to rise in Chuquicamata and negotiations were
demanded by the CTC in Santiago. Also, El Teniente miners offered their
solidarity. The condition laid out by the CTC to start negotiations concerned
the reinstatement of the dismissed workers. At the end of the month a trade-
off was accepted by the leadership of the CTC whereby the dismissed miners
would be reinstated in exchange for the end of the lunchbox strike.

This trade-off was refused by a massive union meeting that also repudiated
the presence of the president of the CTC, Bernardino Castillo, who had to
leave the union hall. The miners gave an ultimatum of one week to solve the
problems, which in turn provoked the establishment of a state of siege by the
military in the department of El Loa where the mine is located. The military
accused various persons of communist agitation, provocation in the form of
inciting miners to slow work, slowdown and distribution of illegal propaganda.
Those arrested were sent to a far-off town in the 'cordillera'.

Tension in the mine did not subside and miners maintained their two demands: reinstatement of the dismissed miners and satisfaction to their economic petitions. In Santiago, things began to move and the president of CODELCO, pressured both by the President of the Republic and the Minister of Labour, had to meet with the CTC leadership. Public statements by both sides made the conflict known to the population. Notwithstanding the contacts between CODELCO and the CTC, the military continued with arrests and dismissals, with a special focus on christian democratic politicians in Calama and Chuquicamata. Various ministers, including the Minister of Finance, disqualified the petitions of the miners by exaggerating and ridiculing the miners' demands. The ministers also argued there would be very high costs if petitions were granted. This process continued all through the month of September, and the pressure of the miners continued.

Suddenly, on 22 September in Chuquicamata, the company agreed to sign an agreement which satisfied most of the miners' demands including payment of utility bills, supplement for national holidays and 'trienios'. One week later, the president of CODELCO and the vice presidents in charge of finance, labour relations and operations resigned, giving proof of the success of the pressures built up by the miners. The 'lunchbox strike' had come to an end. All through it many aspects of labour relations under military government stand out. We will try to specify these.

The first element that appears from the development of the strike is the leadership crisis that seems to have affected Chuqui's miners. It is interesting to note that instead of weakening the position of the rank and file, the strike had an inverse effect. Indeed, spontaneous leadership began to appear in different departments of the company. Among them were some leaders of wildcat strikes from the pre-1973 period, but there were also many leaders who emerged for the first time. Repressive policies had the consequence of developing a parallel leadership that had high levels of legitimacy among rank and file workers. Relations with the national leadership of the CTC became strained as it was unable to obtain better salaries and benefits for the workers. As we mentioned before, the president of the CTC was shouted out of the union meeting hall in 1978 because he dared suggest that miners should abandon their pressures and agree to negotiate with the government. Miners at Chuquicamata became progressively suspicious of any negotiation that took place outside of the mine and demanded that all talks take place in Chuquicamata instead of in Santiago.

It is possible to say that, at the local level, miners chose a new leadership that was more responsive to their concerns and that gradually they began to replace the old, politicised leadership that had controlled the process of demand-making and demand-bargaining at Chuquicamata. This process was also a consequence of the ambiguity that some leaders had in relation to the military government; ambiguity which reflected the positions of the political parties to which they belonged. In particular, the christian democratic position which, at least until 1976, did not question the legitimacy of the military government, affected the positions taken by the only remaining union leader of the elections of 1973, a christian democrat. When this ambiguity was replaced by outright opposition on the part of the christian democrats, and when some degree of mobilisation began to take place in the mine in 1977–8, workers began seriously to question management procedures and ask for the reestablishment of collective bargaining and the reinstatement of union elections. In 1979, some of these demands were granted through the implementation of the Labour Plan (Haworth and Roddick 1981), but at the same time limitations were established such as the prohibition of strikes in the mine.

All these events point to a general consideration: from 1976 onward a new system of labour representation began to appear in the mine. From that date miners began to structure some legitimate communications channels deriving from their newly found power. Union leaders became subordinated to the pressures of the rank and file and could not manipulate nor instrumentalise that base as they saw fit. On the contrary, union leaders had to tell the military leadership that they could not act without previous consultation of the rank and file. In turn, the military had to recognise this situation and grant some demands so that the union leaders could obtain some degree of legitimisation from the rank and file, and thereby diffuse the danger of having an anonymous partner as a negotiating agent.

In this way, from the events of 1978 one can conclude that the military in charge of negotiations had to deal with delegates who were more spokesmen of the rank and file than its leaders. This situation can hypothetically explain the small impact that outright repression had in diminishing labour mobilisation in 1978. In spite of arrests, dismissals and pressures on the wives of the miners, the military could not reach agreements to end the 'lunchbox strike'. A more substantive deal had to be structured. Later on, in 1981, the same would be demonstrated in El Teniente, when the strike there could not be ended without some concessions made to the miners. A direct consequence of this process was that those workers that were in charge of the 1978 movement obtained enough support from the rank and file to be elected as leaders by the miners in the first union elections that took place in June 1980.[4] One can say then that relations between union leaders and rank and file experienced a serious change from what was typical in pre-1973 Chuquicamata. Leaders became more responsive to their constituency and military leaders had to see their repressive capacity within the context of this relationship.

The military, management and the miners

Another angle at which to approach the restructuring of labour relations in Chuquicamata is to focus our attention upon the relations between the military and the managers of the mine, both at the corporate level of Santiago and in the operations part at the mine itself. After nationalisation in 1971, supervisors became a difficult group for the new state owners to manage (Zapata 1979). Supervisors had been accustomed to receiving their salaries in dollars and had a very dependent relationship with the American owners of the mines. At the same time, their relations with the national government were strained by their foreign perspective. In many ways they were Chilean only in name. Their attitudes, way of life and cultural frames of reference were shaped by their socialisation within the American companies. Their attitudes toward the miners had always taken a paternalistic bias, which in turn created resentment on the part of workers.

When nationalisation took place in 1971, this paternalistic situation did not change very much. Many supervisors left Chuquicamata and were replaced by a mixture of politically appointed managers and technicians that accepted to work for the nationalised companies. In many ways, the company town structure made its mark upon this new group and miners referred to them as 'the gringos del Mapocho'[5] to characterise the fact that they behaved just as paternalistically as the real 'gringos' without the benefits these brought along with them. Tensions rose in Chuquicamata and entered a crisis at the beginning of 1972 when the more technocratic management that had been collaborating with the administration resigned and a purely political administration took over (Gall, 1972). From this moment on, labour relations became seriously tainted by political considerations. As unions were not controlled by the Unidad Popular but by a small splinter socialist party, the USOPO which had its central support in Chuqui as revealed by national election results,[6] the management resorted to confrontation politics. This was especially true when the USOPO-controlled union leaders engaged in straightforward manipulation of the miners to provoke wildcat strikes that were really political moves. In the period 1971–2 this condition was behind most of the strike activity in the mine; strikes which led to serious production losses. But, interestingly enough, miners became tired of this eternal confrontation and instead of renewing USOPO's mandate in the union elections of February 1973, they gave their support to the UP candidates and to the christian democratic candidates, thus changing the political panorama of the mine for the rest of the UP government.

Management had to deal very closely with all these matters and often had to contradict national policies concerning the working classes as it had to take into account the fact that miners at Chuqui were a part of the opposition and not, as many supposed, the 'vanguard' of the revolution. The importance of political considerations in the management of Chuquicamata in the period 1971–3 revealed that strictly technical issues had to take a second place; this was a discovery of something that American management had always known and something that the military would find out when they took power after September 1973. Indeed, miners at Chuqui were extremely clear-headed about their bargaining power. Despite the fact that one could associate this power with manipulations or clientelistic relations under control of political parties such as USOPO or the traditional communist and socialist left, in fact miners revealed themselves to have a very high level of worker consciousness, capable of confronting adversaries as different as American managers, leftist leaders or military officers.

When the 'lunchbox strike' took place in mid-1978, the military government experienced this phenomenon very directly and the military-appointed management succumbed to the pressures of the Chuquicamata miners. All through this conflict, management played a secondary role. The miners succeeded in displacing the locus of the negotiations from the capital to Chuquicamata and thereby provoked many shifts in the negotiating position of the government. In the same way the UP managers consulted with political parties, military men in charge of CODELCO had to define a position in terms of the different pressure groups which make up the military coalition. And miners kept up the pressure all along, finally succeeding in obtaining most of the demands and in particular provoking the resignation of the vice presidents in charge of labour relations, finance and operations of CODELCO. This reflected that, at the last instance, the president of the Republic had to intervene in the negotiation.

The 'lunchbox strike' had catastrophic consequences for the military group in charge of CODELCO and Chuquicamata.[7] It showed very clearly that miners had a bargaining power that continued to exist and that transcended the type of political structure that existed in the country. In other words, one could say that there is a system of labour relations where the miners play a role within a structure that is more or less independent of the type of political leadership the country has. It is possible to suggest that the company town hypothesis[8] is useful to explain this situation because it shows that this type of social context generates specific behaviour on the part of workers, managers and political leaders, independently of the concrete differences which these last two can have in different historical periods.

In this respect, the three crucial elements contributing to the explanation of the intensity that social conflict has in Chuquicamata are: (a) the strategic character of the mining sector for the national economy, (b) the geographic isolation of the miners and (c) the instrumental character of the action of the miners. These three elements, more or less independent of the political context of the country at any one time, are useful to explain why the type of labour action does not vary in relation to the political changes that take place. On the contrary, it seems that even in a repressive situation such as the one that affected other sectors of the working class, in the case of the miners of Chuqui they did not have the results expected by the military. This supports the hypothesis that the interaction of the three elements mentioned is the foundation of a 'structure' of social relations that give strength to labour action.

This action can take various forms. Sometimes it can follow legal procedures or take the form of wildcat strikes; in other situations, as is the case during the present military government, it takes the form of conflict that stops short of outright strikes, and instead takes the form of passive protest, work to rule or work slowdowns. Any one of these forms affects production and the general working climate. It also modifies the relations between the rank and file and the union leadership because the latter is incapable of channelling demands in the 'usual' way and has to take a much more militant and enterprising role that the military has to recognise as legitimate if it wants to solve the problems that gave rise to the protest. All this generates high levels of cohesion among the rank and file and provokes the appearance of a new type of leader who has little in common with the one that prevailed in previous epochs. Not necessarily related to political parties, generally young and relatively unknown to management, these new leaders have a high level of legitimacy among their fellow workers and can lead conflicts such as the ones we have mentioned here, essentially characterised by direct demands. Politicisation takes a new dimension, more related to the rank and file than to political organisations.[9]

Union meetings take a new character as they become the only forum where miners can express their grievances and show their disagreements with the policies being implemented and with the way union leaders are negotiating their demands. Traditional ways of administering these demands are questioned in the union meetings, as some union leaders have found out to their surprise and dismay. This contributes to the development of new forms of labour action that counteract the repressive atmosphere that has been created by the military. Miners have succeeded in maintaining their identity and in facing acute repression to pursue and obtain some degree of autonomy in their bargaining position *vis-à-vis* the government. Resistance to repression as well as to the erosion of salary levels and their standard of living through simple demands such as reestablishment of collective bargaining and union elections have characterised the pressures exerted by miners on the company and on the government. They could not be accused of making political demands as they had been ardent supporters of the arrival of the military; so the military had to deal with the demands on their own merits.

The military had to accept and tolerate this resistance by accepting the existence of some form of labour relations that could assure the necessary peace in this strategic sector of the economy. As we have seen, this took quite a long time to make itself clear in the minds of those officers in charge of the sector. Only after the 'lunchbox strike' was it clearly perceived that it was indispensable to institutionalise some type of labour relations system both in the mine and in the country. That strike contributed significantly to the establishment of the Labour Plan implemented in 1979. In Chuquicamata itself, relations between the military government and the miners develop within a structural context where confrontation cannot be exacerbated at the risk of disturbing the rhythm of development of the country as a whole.

We can see some form of armed peace where both actors maintain their positions and use the specific conjuncture to advance or retreat. They thus establish a system of reciprocal expectations which is useful to solve problems. When those expectations are not fulfilled, straight conflict arises. There is a continuity in the social behaviour of miners. Both in the initial period of nationalisation, when the Popular Unity government had to deal with a very recalcitrant working class and in the recent period when a repressive and bloody military government has had to do the same, the miners have succeeded in generating enough cohesion and solidarity to nullify initiatives to domesticate them. At the same time miners have maintained an autonomy that does not imply political or ideological radicalism; they remain firmly attached to economistic and instrumental tactics in the implementation of a very trade unionistic strategy.

Notes

1 The 'Gran Minería del Cobre' is composed of five mines: Chuquicamata, El Teniente, El Salvador, Andiana and Exotica. The GMC is owned and managed by the Corporacion del Cobre (CODELCO), a state owned company with fixed assets amounting to 2.5 billion dollars in 1979. Capital and reserves equalled 1.9 billion dollars and net profits in 1979 amounted to 467 million dollars, out of a total sales figure of 2.0 billion dollars. CODELCO is the first enterprise in Chile with 22.6 per cent return on sales and 25 per cent return on capital and reserves. Production levels exceed one million metric tons of refined copper for the last five years. Employment fluctuates around 30,000 workers and productivity has increased to a figure of twenty-nine tons of refined copper per employed person (see, *Business Latin America*, 1 October 1980).

2 The increase in the number of 'supervisores' equalled almost one hundred per cent. There were 550 supervisores in 1973 and 975 in 1977. This increase weighed heavily in the total of Chuquicamata's employment in the period 1973–6, which almost did not increase in the blue- and white-collar worker categories.

3 For reasons that can only be assumed, such as Chuquicamata's worker combativeness and strike propensity, the Labour Plan prohibits strikes from taking place at the mine. Miners can obtain satisfaction from collective bargaining and if they do not agree with the company's offers they can ask for a government referee, which usually amounts to the same thing.

4 In the union elections of June 1980, 4,288 miners voted from a total of 5,611 that could vote. They elected seven new labour leaders of the 'sindicato profesional' (white-collar). None of the incumbents were reelected. The new leaders are: Santos Veas Zamorano (before 1973, a centrist socialist); Mario Gallardo Marin; Alfonso Gallardo Ugalde; Juan Ugarte Gonzalez; Ramiro Vargas Ortiz; Mario Meyer (before 1973), a leader of the local Partido Nacional); Roberto Guerra.

5 'Gringos del Mapocho': the new technical and political management of the mine. *Gringo* is a derogatory term for Americans all over Latin America. *Mapocho* is a river that runs through Santiago, capital of Chile. Thus the expression, created by Chuqui miners in 1971 is revealing of their cynical attitude towards the new management.

6 The Unión Socialista Popular (USOPO) received 29,123 votes in the 1971 municipal elections. From this national total, 7,899 votes were obtained in Antofagasta; from this provincial total, 4,828 votes were cast in Chuquicamata. We can conclude that most of USOPO's support came from the miners; this can easily be explained by the fact that USOPO's leadership, especially senator Ramon Silva Ulloa, had been labour leaders back in the fifties in the white-collar worker's union. In the period 1971–2, USOPO led a series of wildcat strikes which resulted in serious production losses; most of the strikes were politically motivated and centred on assuring the survival of the party in the national political arena. This calculation was wrong inasmuch as USOPO lost most of its labour leaders in the 1973 union elections and also its representatives in the March 1973 congressional elections. The miners changed allegiances to the christian democrats and to the Unidad Popular, both in the union elections and in the congressional elections of February and March 1973.

7 At the end of September 1978, Jorge Vial, Gabriel Vals and Danilo Rojic, vice presidents in charge of finance, labour relations and operations, resigned. Shortly, afterwards, the CODELCO president, General Orlando Urbina, also had to resign to be replaced by Colonel Gastoz Frez; these resignations were evidently related to the failure of these corporate officers to solve the 'lunchbox strike'.

8 The 'company town' hypothesis is related to the so-called Kerr–Siegel hypothesis concerning miners' high propensity to strike as a result of their geographic isolation. In the specific case of Chuquicamata, the isolated mass hypothesis seems to explain very well strikes as well as political activity: it represents an ideal test case for the hypothesis given the strategic character of the mining sector for the Chilean economy, the geographic isolation of the mine and the highly instrumental character of the actions of the miners. For a very adverse position, see Edwards, 1977.

9 During the 1983 protests against the Chilean military regime, copper miners have played an important role and new leaders such as Rodolfo Seguel have acquired a central position in the organisation and implementation of the so-called 'national days of protest which took place on 14 June, 12 July, 11 August, 8 September, 27 October and 18 November of this year. This situation only confirms one general hypothesis concerning the new CTC leadership.

References   * *See annotated bibliography for full details of all asterisked titles*

Barrera, Manuel
1978*

Barria, Jorge
1974 'Organización y políticas laborales en la Gran Minería del Cobre', in R. French-Davis and E. Tironi (eds.), *El Cobre en el Desarrollo Nacional*, Santiago, Ediciones Nueva Universidad.

Berteau, David
1982*

Campero, Guillermo
1980 'Transnacionalización de la economía y de la sociedad: su impacto en la estructura y la estrategia del movimiento sindical Chileno despues de 1973', Research report, Instituto de Estudios Transnacionales (ILET).

Corporacion del Cobre (CODELCO)
1979 *Cuarta Memoria Anual*, Santiago.
1980 'Inversiones Proyectadas, quinquenio 1981–4', Santiago.
1981 *CODELCO: a Profile*, Santiago de Chile.

Culver, W. W.
1980 'Illusions of capitalism and progress in nineteenth century South America: politics and the decline of the copper industry in Chile (1883–1888)', paper presented to the meeting of the Latin American Studies Association, Bloomington, Indiana.

Edwards, Paul K.
1977 'A critique of the Kerr–Siegel hypothesis and the isolated mass: a study of the falsification of sociological knowledge', *The Sociological Review*, 25(3).

El Mercurio
1974–82 Edición Internacional.

Gall, Norman
1972*

Gomez, Ramiro
1981 'Cobre', *Comercio Exterior* (Mexico).

Gutiérrez, Eulogio and Figueroa Marcial
1920 *Chuquicamata, sus Grandezas y sus Dolores*, Santiago, s/e.

Haworth, Nigel, and Roddick, Jackie
1981 'Labour and monetarism in Chile, 1975–1980', *Bulletin of Latin American Research*, 1(1).

Huneeus, Pablo
1974 'Estructura y dinámica social en los trabajadores del cobre', Instituto de Sociología, Universidad Católica de Chile, documento de trabajo no.25.

International Labour Office
1979 *Yearbook of Labour Statistics*, Geneva.

Latcham, Ricardo
1926 *Chuquicamata, Estado Yankee. (Vision de la Montaña Roja)*, Santiago, Nascimento.

Manning, Alice Elizabeth
1975 'Calama: patterns of interaction in a Chilean city'. PhD dissertation, Columbia University.

Mining Engineering
1952 '40 years old: Chuquicamata looks to the future', December.
1969 'Chuqui', November.

Moran, Theodore
1974*

Sawyer, Thorp
1960 'A portrait of Chuqui as a young mine', *Mining Engineering*, December.

Sigmund, Paul
1981 *Multinationals in Latin America: The Politics of Nationalization*, Madison, University of Wisconsin Press.

Suseelman, Robert
1978 'Chile's Chuquicamata: looking to stay number one in copper output', *Engineering and Mining Journal*, August.

World Bank
1978 *Chile: An Economy in Transition*, Washington.

Zapata, Francisco S.
1975*

PART THREE

Miners and mining in the Americas: an annotated bibliography of sources, 1970 – 84

SUE TURNER and TOM GREAVES

Sue Turner is a former editor of the *Anthropology of Work Review* and a founder of the Society for the Anthropology of Work. She has written on Mayan archaeology and is author of *Stone Artifacts of the Texas Indians* (with Thomas Hester) Austin: Texas Monthly Press, forthcoming. Tom Greaves is professor and dean of behavioural sciences at Trinity University and an Andeanist anthropologist.

I Why an annotated bibliography?

In this glistering new world of computers and on-line data banks, with your library trucking away the card catalogue and leaving keyboards and video terminals in its place, is there any need for a published bibliography? We think so, particularly for this book on this subject at this time.

On-line searches may eventually become the standard for defining our professional reading, but they are not so yet. What is generated from computer data banks is piecemeal, seriously incomplete and classified by highly variable criteria. These new tools are a useful supplement to standard techniques for bibliography building, but our profession is still some distance from being able to trust them. Perhaps with some comfort we can observe that the classic skills of the scholar are still required.

There is a second reason for an annotated bibliography of mining: it has to do with the early, developing phase of contemporary mining research. Beginning about 1970 the study of mining gained a sustained momentum within many of the behavioural sciences, belatedly joining our colleagues in history for whom mining topics have been a long-standing pursuit. This growing enterprise, nourished by the widening perception that miners could offer insight on many intriguing questions, has now led to an accumulation of important published work in a variety of disciplines – among them, anthropology, sociology, economics, political science, regional science and history. This bibliography puts these parallel literatures in one place; we are hopeful that the listings and annotations will introduce us to each other, reduce the partitions among disciplinary citation patterns and, perhaps, nourish convergence rather than divergence in the questions we ask and the answers we seek.

This bibliography begins in 1970 and continues through early 1984. We believe 1970 manages to catch most of the recent curve, although some important pre-1970 sources had to be wistfully put aside.

II Scope and limits

The central anchorage

The central anchorage of this collection of sources is *the social behaviour of the miner, past and present*. Our focus is on the people, not the machinery, not the geology, not the ore, not the market price. In selecting sources, and excluding others, we have asked ourselves, *'would this be helpful to someone researching the explanation to some facet of miner behaviour?'*

That judgement, in retrospect, came to be answered with liberality. We found ourselves including an occasional source on the physiology of sili-cosis, the flows of international capital or a new mining technology. When we came across an item which might serve as a beachhead for someone wanting to explore the impact of that factor on miners and their predicament, we put it in.

Disciplines represented

Responding to the ongoing miner research in various disciplines, we sought to look throughout the behavioural sciences and history. So, in addition to the standard fare from sociology, anthropology, history and political science, we reviewed economics, business administration, regional science, public health, and some of physiology, human engineering and psychology. Our sources in the latter disciplines are not extensive, but those included may assist the investigator as bridges into literatures beyond. We have also included a few journalistic items which we thought might be useful as public images of mining life.

Geographic limits

Without doubt the most outrageously arbitrary boundary to this bibliography is confining it to the western hemisphere. Quite obviously, the problems faced by Kentucky coal miners are more similar to those of the Yorkshire colliery-men than to men working drag-line machinery in the lignite beds of East Texas. Someone looking at the social pathologies of declining West Virginia mine communities should be seeking insights from Wales, not Wyoming. The western hemisphere criterion has another weakness: the miners themselves have crossed the oceans. Cornishmen, for instance shaped the nature of hard rock mining from Canada to Chile. Given these human and cultural links, to constrain our attention to the Americas is inept. And still a third reason: scholars investigating the miner in the Americas have colleagues specialising in European, African or Asian miners. Their research, not included here, is essential reading. There is nothing intellectually defensible about stopping at the beaches of the Americas.

We plead pragmatism. Just covering with reasonable liberality the literature of the Americas already abuses the patience and bank account of the publisher, not to mention our own available time and energy. Too, if our list encourages its users to work beyond their local region, their mineral, or their discipline, that will be progress enough.

Materials included

Given the present vigour of mining research we need to access materials in less established formats. Thus we have sought out doctoral dissertations and masters' theses. These frequently contain more first order data than do more mature sources, and introduce us to the many new and talented intellects now joining our enterprise of mining studies. By the same reasoning, we have included unpublished conference papers, limited circulation public documents, and private manuscripts. Lastly, we have included a smattering of novels of mining life; no one has yet demonstrated that the essence of the miner will not be put into print first by the novelist or poet. Perhaps it has already been done.

Materials not included

Aside from the geographic and time limits already noted, certain other exclusions were made. Some have to do with the resource being mined. For instance, we have not included petroleum extraction, although it is a form of mining sharing many features, good and bad, with other forms of mining. The extraction of sub-surface water and other liquids has similarly been left aside, as has off-shore mining. Although we didn't consciously eliminate sources on stone, gravel and sand quarrying, we find we have none, perhaps suggesting an unverbalised exclusion.

Remembering that our central anchorage is miners (rather than mining), we tended to abandon the trail when we got very far into literatures dealing with mine engineering, managerial models, geology, health and government policy reviews. All of these have their own large literatures.

A much more pragmatic limit was what we could put our hands on. We have long lists of citations which eluded our grasp, including many we know to be important. Indeed, many promising additional entries are found in the chapter bibliographies of this book. We decided early on, however, that our bibliography needed to be annotated, with the consequent, artifactual selection. For the user, however, the omissions will be transitory. We expect that anyone who begins with the sources on the present list will quickly be led to the sizeable pool of references and authors we have not been able to include.

280 Annotated bibliography

Some patterns observed

We could not have pursued this project without pausing to reflect on some patterns we observed. Two are perhaps worth a short comment here. First, comparing one mining region with another, the size of its scholarly literature departs noticeably from the relative importance of its mining. For instance, there are huge mine literatures for the United States and for the southern Andes (Peru, Bolivia and Chile). Other zones, Brazil and the Caribbean for example, have spawned quite meagre post-1970 literatures on their miners.

For the United States the sheer enormity of the scholarly community relative to other hemispheric nations probably means that there is more written about *everything* in the United States. Thus the US miner literature is large. The southern Andes may represent a synergistic process of scholars exciting each other and becoming intrigued with the dramatic political events that involve miners in those countries. A Kuhnian process is probably at work.

Second, Euro-American writers in many cases cite fewer materials with Latin American imprints than do their Latin American colleagues. We suggest that while this may stem partly from different problem agendas, it is mainly due to the economic and distributional constraints impeding the circulation of Latin American imprints beyond the borders of the country of origin. The present bibliography is, transparently, a victim of these same impediments. Surely this is doing major harm to our collective effort.

Apologies and homages

First, two apologia. If you are working in mining, you will immediately identify critical and favourite sources which do not appear here. Quite possibly these will include seminal self-authored pieces. Their omission may be due to the difficulties of finding copies to annotate in our city, or because we simply failed to take proper notice. For this, we apologise.

Too, your displeasure may be still further aroused when you read our annotation of a familiar source – including something from your own pen – and find that we have utterly missed the point. We shall indeed be grateful to be set straight on the matter and look forward to all those cards and letters.

Now, our thanks: librarians at Trinity University are surely numbered among the world's great human beings. We are the benefactors of their professional commitment and cheerful collegiality. Particular debts go to librarians Norma Carmack, Mary Clarkson, James O'Donnell and Maria McWilliams. We also owe thanks to the budget of Trinity University's dean of behavioural sciences which paid for a consequential amount of computer charges, postage, stationery and telephone bills, and provided untold amounts of word-processor time. To all go our sincere thanks.

1 **Abbe**, Donald Ray
  1982 'Austin and the Reese River mining district: Nevada's forgotten frontier'.
    PhD dissertation, Texas Tech University. Ann Arbor: University Microfilms.
Describes the establishment of super-charged silver boom towns in the 1860s, the
evolution into more organised, smaller-scale society over the next two decades, and
the withering of social commerce by the first world war, seen within the culture of
the America's western mining frontier.

2 **Ackerman**, John A.
  1979 The impact of the coal strike of 1977–1978. *Industrial and Labour Relations
    Review* 32(2): 175–88.
Reviews the events of the national coal strike of 1977–8, which resulted in the eventual
invocation of the Taft–Hartley Act, as a 'national emergency'. Concludes that no
national emergency developed, but that political factors made intervention imperative.

3 **Akin**, Gib
  1981 'Psycho-social aspects of productivity in underground coal mining'. Prepared
    for US Department of Energy through an agreement with National Aeronautics
    and Space Administration by the Jet Propulsion Laboratory, JPL Publi-
    cation 81–113. Pasadena, California: California Institute of Technology.
Status of the literature, concepts, experiments, new technology and clinical studies of
productivity result in a model with implications and recommendations for new
technology and management.

4 **Albarracín Millan**, Juan
  1972 *El Poder Minero en la Administración Liberal*. La Paz, Bolivia: Urquizo Ltda.
A highly detailed history of the development of Bolivian tin, tungsten and antimony
mining, principally covering the first two decades of the twentieth century. Focus is
on the interaction of foreign and domestic industrialists with government as national
mining policy is formed.

5 **Alcina Franch**, José
  1970 'La producción y el uso de metales en la América Precolombina'. In *La
    Minería Hispana e Iberoamericana*, Ponencias del 1er Coloquio Internacional
    Sobre Historia de la Minería, 1. A. Vinayo Gonzalez, ed. León, Spain: Cátedra
    de San Isidoro. Pp. 307–32.
Reviews and summarises geographic distribution of pre-Colombian mining activities
for the western hemisphere, with attention to gold, silver, copper, bronze, platinum,
lead and mercury. Includes very generalised distribution maps.

6 **Almaraz Paz**, Sergio
  1969 *El Poder y la Caída: el Estaño en la Historia de Bolivia* (2nd edition).
    Enciclopedia Boliviana. La Paz: Los Amigos del Libro.
One of the best histories of Bolivian mining, researched by a national of that country.
Excellent source of specific data on the major industrial, labour and governmental
figures whose decisions shaped Bolivia's mining history before, during and after the
1952 national revolution.

7 **Althouse**, Ronald
  1974 *Work, Safety, and Life Style Among Southern Appalachian Coal Miners, A
    Survey of the Men of Standard Mines*. Morgantown: West Virginia University.
    Office of Research and Development, Appalachian Center.
Monograph presenting a questionnaire study conducted in 1970 among 188 coal miners
and foremen in three West Virginia mines owned by Standard Mines (pseudonym).
Questionnaire covers broad sectors of miner attitudes, behaviour and personal history,
relating these to safety.

8 **Alurralde Anaya**, Antonio
    1973 *Cooperativas Mineras en Bolivia*. La Paz, Bolivia: Escuela Don Bosco.
Meant as a handbook to promote cooperativism in Bolivian mining and a description
of existing cooperatives. Reviews history of mining in Bolivia. Contains fairly extensive
section on the gold workings at Tipuani.

9 **Angle**, Barbara
    1979 *Rinker*. Washington, DC: Crossroads Press.
A short novel (74 pp.) on a day in the life of an American coal miner, Jake Rinker,
assigned to a supply job within the mine. The dialogue, which carries the book, is rich
and authentic, drawn from the author's personal experience as an underground miner.

10 Anonymous
    1981 'Importancia de la minería en la economía del país [Bolivia]'. La Paz:
         CEPROMIN.
Forty pages in length, paper presents and critiques the features of the Bolivian mining
economy, accompanied by numerous tables of production and financial figures.

11 **Antmann**, Frances Carol
    1983 'Sebastián Rodriguez's view from within: the work of an Andean photo-
         grapher in the mining town of Morococha, Peru, 1928–1968'. PhD disser-
         tation, New York University. Ann Arbor: University Microfilms.
Recounts life and vision of gifted photographer residing in a town owned by the Cerro
de Pasco Corporation. Author reassembled Rodriguez's scattered work and presented
it to the library of the Universidad Católica of Lima. Rodriguez's work shows alter-
nation between western and indigenous perceptions of mining life.

12 **Aquizap**, Roman B. and Ernest A. Vargas
    1970 'Technology, power and socialization in Appalachia'. *Social Casework* 51:
         131–9.
Written for change agents, article identifies three critical components of any social
system involved in the process of social change: the relationship among technology,
power and socialisation. The general setting in Appalachia; the technology is the coal
industry.

13 **Arble**, Meade
    1976 *The Long Tunnel: A Coal Miner's Journal*, New York: Atheneum.
Personal diary of a coal miner. Faced with no job, no money and no place to live,
he goes to northern Pennsylvania with his wife and children to find work in the coal
mines.

14 **Arce**, Roberto
    1978 'Small scale mining in Bolivia'. In *Small Scale Mining of the World* (con-
         ference proceedings, Jurica, Queretaro, Mexico, November). New York (?):
         United Nations Institute for Training and Research. Pp. 293–319.
Bolivian mining engineer describes small mining enterprises in Bolivia, arguing that
their potential contribution to national production is large, but that lack of capital,
technical assistance, national infrastructure and regressive tax laws cripple the potential.
Proposes various legislative and developmental changes.

15 Asamblea Permanente de los Derechos Humanos de Bolivia
    1978 *Estudio Sobre el Valor Adquisitivo del Salario de los Mineros*. La Paz:
         Asamblea Permanente de los Derechos Humanos de Bolivia.
Important source of statistics on the wage situation of Bolivian miners. Study is
based at Catavi-Siglo XX COMIBOL mine and compares data across the period
1971 to 1977. Data show that purchasing power of most miners, already among the
lowest in the world, declined across these years, while the cost of living for essentials

increased. Presents important information on how miners have responded to the absolute decline in purchasing power, and also contains helpful date on the incidence of silicosis.

16 1980 *La Heroica Resistencia de los Mineros de Bolivia*. Lima: Asamblea Permanente de los Derechos Humanos de Bolivia.
Documents and testimony covering the violent repression of organised miner resistance by the Bolivian military government between 17 July and 6 August 1980. Contains photograph series apparently taken just before massacre of a group of miner resisters.

17 **Aschmann**, Homer
1970 'The natural history of a mine'. *Economic Geography* 46: 172–89.
An essay in 'geographic economics'. The models presented are designed primarily to inform economic policy, specifically policies that would exploit local mineral resources to achieve economic development of underdeveloped countries or regions. Examples used are Chilean mines.

18 **Bailey**, Kenneth R.
1973 'A judicious mixture: negroes and immigrants in the West Virginia mines, 1880–1917'. *West Virginia History* 34(2): 141–61.
Examines polyglot population in Western Virginia that originated in countries of western and southern Europe, plus a sizeable negro minority. Comparative weakness of racial and ethnic prejudice is attributed to efforts to integrate working conditions in the mines and enlist all men in the UMW.

19 **Baird**, Wellesley A.
1982 *Guyana Gold, The Story of Wellesley A. Baird, Guyana's Greatest Miner*. Washington: Three Continents Press.
Autobiographical account of an independent life of gold mining in the upper Barama region, and barriers experienced for a black man seeking to enter the mainstream of the business capitalism. His story might be understood as a case study in personal aspirations.

20 **Baker**, Joe G. and Wayne L. Stevenson
1979 *Determinants of Coal Mine Labor Productivity Change*. US Department of Energy and US Department of Labor. Washington, DC: Government Printing Office.
Highly quantitative analysis of the decline of deep and surface mine coal production per unit of labour between 1971 and 1974. Conclusion drawn is that much of decline was due to increased production costs incurred from the Coal Mine Health and Safety Act of 1969, together with rising coal demand and coal prices, and coal strikes.

21 **Bakewell**, P.J.
1971 *Silver Mining and Society in Colonial Mexico, Zacatecas 1546–1700*. Cambridge: Cambridge University Press.
Much archival research is used to illuminate establishment, growth and eventual decline of the mining industry which underlay the political and social realities of Zacatecas. Good material on early mining technology and entrepreneurial styles. Helpful bibliographical essay.

22 **Baldwin**, William L.
1983 *The World Tin Market: Political Pricing and Economic Competition*. Durham, NC: Duke University Press.
Politics of tin through study of commodity pricing within international trade. Covers the history of tin industry with considerable emphasis on Bolivia and US, 1920 to present. Includes role of tin in economics of producing and consuming countries, supply and demand, and mining costs.

23 **Ballantyne**, Janet Campbell
    1975 'The political economy of Peruvian gran minería'. PhD dissertation, Latin American Studies Program. Cornell University. Ann Arbor: University Microfilms. Also, Ithaca, NY: Cornell University Latin American Studies Dissertation Series, 60 (1975).
Purpose is to analyse the political economy of the mining sector in Peru, 1950–73. Concern is Marcona, Cerro de Pasco and the Southern Peru Copper Corporation. Primarily oriented towards the non-ferrous mining sector.

24 **Ballesteros-Gaibrois**, Manuel
    1970 'Notas sobre el trabajo minero en los Andes, con especial referencia a Potosí (S. XVI y SS.)'. In *La Minería Hispana e Iberoamericana*, Ponencias del 1ᵉʳ Coloquio Internacional Sobre Historia de la Minería, 1. A. Vinayo Gonzalez, ed. León, Spain: Cátedra de San Isidoro. Pp. 529–57.
Review of the contents of several dozen particularly useful archive documents, dating from early 1500s to mid-1700s, from the collections of the Biblioteca Nacional in Madrid.

25 **Barendse**, Michael A.
    1981 *Social Expectations and Perception: the Case of the Slavic Anthracite Workers.* The Pennsylvania State University Studies, 47. University Park, Pennsylvania: Pennsylvania State University.
Study to demonstrate that culturally produced negative expectations concerning immigrants in American society generated perceptions that fulfilled those expectations. Hypothesis is applied to the case of slavic immigrants in the anthracite coal fields of north-eastern Pennsylvania, 1890–1902.

26 **Barnum**, Darold T.
    1969 'The negro in bituminous coal mining'. *Racial policies of American Industry* 14. Philadelphia: University of Pennsylvania, Wharton School of Finance and Commerce, Industrial Research Unit. (Distributed by University of Pennsylvania Press, Philadelphia.)
Part of series on the participation of American blacks in various industry sectors, funded by Ford Foundation. Concentrates in the coal producing sectors of southern Appalachia, showing blacks were employed extensively from 1890 to 1930 to divide workers, thwart unionisation and undercut coal prices from the northern regions. After 1930 participation of blacks declined, reflecting unionisation, prejudice among white miners and mechanisation eliminating the unskilled and semiskilled jobs held by blacks.

27 **Barrera**, Manuel
    1978 'El conflicto obrero en el enclave cuprifero Chileno'. *Revista Mexicana de Sociología* 40(2), 609–82.
Marxist analysis of labour conflict at Chile's three largest copper companies, 1956–66. Concludes that despite winning the nation's highest salaries, the continuing conflict demonstrated the enduring conflict of classes, exacerbated by enclave nature of mine communities.

28 **Barrios de Chungara**, Domitila (*see* Viezzer 1978)

29 **Barry**, Mary J.
    1973 *A History of Mining on the Kenai Peninsula.* Anchorage, Alaska: Alaska Northwest Publishing Co.
Good local history of mining on a prominent part of the Alaskan coastline adjacent to Cooke Inlet. Most of book focuses on 1880 to second world war, with coal, gold and oil providing the principal minerals.

30 **Basadre Ayulo**, Jorge
  1978 'El futuro de la pequeña minería en la ley Peruana'. In *Small Scale Mining of the World* (conference proceedings, Jurica, Queretaro, Mexico, November 1978). New York (?): United Nations Institute for Training and Research. Pp: 709–20.
Explains the national legislation of 1971 as it affects small mining enterprises. Offers brief analysis of likely effects on production levels.

31 **Becker**, David G.
  1983 *The New Bourgeoisie and the Limits of Dependency: Mining, Class and Power in 'Revolutionary' Peru*. Princeton, New Jersey: Princeton University Press.
Ambitious and important critique of structure of national elites in Peru. Argues that an in-country elite, principally based in the mining sector, maintains mutualist, if junior, relations with transnational capitalists, imposing its own will on Peru's social policy. An accommodation with Peru's proletariat has developed. Thus, usual dependency model fails. Implications also drawn for democratic stability, improving agrarian situation, and capitalist policy for Peru.

32 **Behrman**, Jack N.
  1982 'Taxation of extractive industries in Latin America and the impact of foreign investors'. In *Foreign Investment in the Petroleum and Mineral Industries*. Raymond F. Mikesell *et al.*, eds. Baltimore: Johns Hopkins Press. Pp. 56–80.
Good primer for those needing a general vocabulary and overview of in-country taxation of mining (and oil) in Latin America. Provides summary of each major mining nation's tax policies, description of taxation options for achieving various ends, and interplay between tax policies of host country and those of company's home country.

33 **Beith**, Harry H., Mark S. Sanders, and James P. Peay
  1981 'Using retroreflective material to enhance the conspicuity of coal miners'. *Human Factors* 24(6): 727–35.
Investigates various configurations of retroreflective material to enhance the visibility of mine workers. Results point to a cost-effective configuration, increasing detectability in the periphery approximately 100 per cent; could substantially improve worker safety.

34 **Bennett**, James Douglas
  1982 'Relationship between work place and worker characteristics and severity of injuries in U.S. underground bituminous coal mines, 1975–1981'. PhD dissertation, Pennsylvania State University. Ann Arbor: University Microfilms.
Correlates 83,297 injuries recorded in federal files, 1975–81, with characteristics of mine workers, job sites. Findings showed greatest incidence of severe injuries occur at coal face and intersections. Excavation method, age and length of experience had no relation to accident severity, but likelihood that an accident will be severe increases across the years studied.

35 **Benson**, H.W.
  1972 'Miners for democracy: a report from West Virginia'. *Dissent* 19(4): 632–8.
Story of the miners' revolt, major event in recent US labour history. For the first time since the 1930s, a movement of rank-and-file workers won effective public support in fight to reform union.

36 1973 'Labor leaders, intellectuals, and freedom in the unions'. *Dissent* 20(2): 206–10.
Discusses how the Miners for Democracy fight for UMW leadership was successful through collaboration of courageous local leaders and pro-labour professionals.

Access to courts and pressure on government agencies made effective action possible within the union.

37 **Bentivegna**, Joseph J.
  1974 'A study of vocational rehabilitation experiences of bituminous coal miners with pneumoconiosis'. PhD dissertation, University of Pittsburgh. Ann Arbor: University Microfilms.
Pilot study to document and form preliminary understanding of rehabilitation process for miners with pneumoconiosis. Twelve unemployed, disabled pneumoconiotics discuss rehabilitation experiences, as they perceived them, with the writer.

38 **Berney**, Barbara
  1978 'The rise and fall of the UMW fund'. *Southern Exposure* 6(2): 95–102.
Creation of the industry-financed, union-controlled health and retirement fund by J. L. Lewis in 1946 started a revolution in health care delivery in the coal fields. The fund ultimately failed because there was no effective movement demanding fundamental change in the national health care system.

39 **Bernstein**, Marvin D.
  1965 *The Mexican Mining Industry 1850–1950: A Study of the Interaction of Politics, Economics, and Technology*. Albany: State University of New York.
Examines four aspects of the Mexican mining industry: (1) economic and historical setting; (2) evolving technology and economic organisation of the production of minerals; (3) effects of the industry upon its economy; and (4) policies of the government to control the industry to its own ends.

40 **Berteau**, David
  1982 'The Harriman–Solomon mission and the 1966 Chilean Copper Agreement'. In *Economic Coercion and US Foreign Policy, Implications of Case Studies From the Johnson Administration*. Sidney Weintraub, ed. Westview Special Studies in International Relations. Boulder: Westview Press, Inc. (F. A. Praeger). Pp. 173–214.
Well researched and well written by a graduate student in Weintraub's advanced seminar at the LBJ School of Public Affairs (UT Austin). Shows in detail how US leverage was applied to Chile to secure and maintain price stability, favourable price and continued supply of copper from the Chilean Frei government, in order to enhance domestic political tranquility in the US and to augment supplies in conjunction with Vietnam war. Covers 1960 to 1965.

41 **Besserer**, Federico, Diaz, Jose and Raul Santana
  1980 'Formación y consolidación del sindicalismo minero en Cananea'. *Revista Mexicana de Sociología* 42(4): 1321–53.
Investigates roots of Mexican miners' struggle through historical examination of the workers of the Compañía Minera Cananea, SA, a subsidiary of Anaconda Copper located in Cananea, Sonora. Union local's association with national syndicate (SNTMMSRM), controlled by the PRI, has weakened its effectiveness for worker protection.

42 **Bethell**, Thomas N.
  1978 'The UMW: now more than ever'. *Washington Monthly* 10(1): 12–23.
A vigorous defence of the need for active unionism to generate more tolerance and sympathy from American general public. Written by former research director for the United Mine Workers.

43 **Bethell**, Tom and Bob Hall
  1976 'The Brookside strike'. *Southern Exposure* 4(1–2): 114–23.
Summary and historical recap of ongoing struggles in one of the larger coal mines in Harlan County, Kentucky. Interviews by Hall with local and national organisers and strategists involved in strike.

44 **Bilkey**, Warren J.
  1982 'Harmonizing economic and environmental needs via a consensus approach: the Wisconsin copper mining case'. *Review of Social Economy* 40(2): 133–46.
Analyses a consensus-building approach used in Wisconsin as means of dispute settlement through examining the public debate over a proposal to develop new copper-zinc deposits entailing sizeable sulphuric acid and cadmium wastes.

45 **Bilder**, Ernesto A.
  197? 'Países mineros en America Latina'. *Cuadernos de Marcha* (Montevideo) 3(15): 53–64.
Analysis primarily of Latin American mining economies in 1970s in order to project likely features of the 1980s in region. Uses dependency theory framework, coupled with sober and detailed analysis. Unusual in its balanced attention to Caribbean as well as mainland Latin America.

46 **Binder**, Federick Moore
  1974 *Coal Age Empire: Pennsylvania Coal and its Utilization to 1860*. Harrisburg: Pennsylvania Historical and Museum Commission.
Covers the period before civil war. Deals with utilisation of several kinds of coal found in Pennsylvania. Outlines development of fuel experimentation, introduction of mechanical improvements and subsequent markets for the wealth of the state.

47 **Bizberg**, Ilan and Leticia Barraza
  1980 'La acción obrera en Las Truchas'. *Revista Mexicana de Sociología* 42(4): 1405–41.
Interesting sociological analysis of internal dynamics of labour union which emerged for Mexico's state steel plant at Las Truchas/Lazaro Cardenas. Argues that changes in union's dominant social class and altered economic conditions at the plant underlay explosively radical posture of new union.

48 **Blackmore**, Harold
  1974 *British Nitrates and Chilean Politics, 1886–1896; Balmaceda and North*. London: University of London and Athlone Press.
General political history of the development of the nitrates industry in the Chilean north, focusing on the separate interests and strategies of Chilean president, Jose Manuel Balmaceda, and Cornish entrepreneur, John Thomas North.

49 **Blatz**, Perry K.
  1981 'The all-too-youthful proletarians'. *Pennsylvania Heritage* (winter): 13–17.
Article on young boys in the Pennsylvania anthracite coal industry who separated slate from coal in the mammoth industrial buildings called breakers, and comprised about one-sixth of work force in 1885. Based on oral history reviews of mine workers and WPA records.

50 **Bobo**, M., N.J. Bethea, M.M. Ayoub, and K. Intaranont
  1983 'Energy expenditure and aerobic fitness of male low seam coal miners'. *Human Factors* 25(1): 43–8.
Physiological responses of male low-seam miners measured both above and below ground to ascertain characteristics and task demands of mining in very restricted surroundings. Individual task analysis for oxygen uptake and kilocaloric expenditure are discussed.

51 **Bonilla**, Heraclio
  1974 *El Minero de los Andes, una Aproximación a su Estudio*. Coleccion Minima no.4. Lima: Instituto de Estudios Peruanos.
Small book of some controversy and substantial importance arguing that Andean miners, as demonstrated by those at the Cerro de Pasco mine, Morococha, are more

or less permanently in a state of 'transitional proletarianism' due to enduring ties with peasant backgrounds from which they stem. Contains useful quantitative data from Morococha, 1920–70.

52 **Bossio Rotondo**, Juan Carlos
   1976 'Cambios en la politica minero-metalurgica'. In *Cambios Estructurales en el Peru, 1968–1975*. Kerbusch, Ernst J., ed. Lima: Instituto Latinoamericano de Investigaciones Sociales and Fundacion Friedrich Ebert. Pp. 121–44.
Analytic critique of transformation of Peruvian mining sector under the revolutionary military government. While endorsing regime's goals, critique raises questions regarding the record of the regime in thoroughly cutting off capitalist structures of the past and adopting truly socialist, national, independent mining sector.

53 1979 'Internacionalización y regionalización del capital: el caso de la minería metálica Latinoamericana'. *Economía de America Latina* 2: 29–59.
Analyses economic and financial conditions under which integrated mineral and metal industries can be viable within Latin America as opposed to being partial dependencies on industrial centres in developed countries.

54 **Bowan**, Robert and Ananda Gunatilaka
   1977 *Copper: Its Geology and Economics*, New York: John Wiley and Sons, Inc., Halsted Press.
Authoritative source for technological and economic background surrounding copper production. Focuses on geology of copper deposits. Emphasises processes rather than descriptions. Attempts to combine data from many sources.

55 **Boyd**, Ernest Withers
   1982 'The miners' health program: a study of purposes in interorganizational planning'. PhD dissertation, University of North Carolina at Chapel Hill. Ann Arbor: University Microfilms.
Examines UMW's union-organised health care plan from founding in 1946 to decentralisation into single company plans in 1978. Boyd describes most of the first three decades of the plan as an effective, miner-oriented, personal health service, though constrained from industry-wide health problems by the union's alliances with industry. With the decentralisation of late '70s, plan loses its distinctive character and becomes a conventional third-party payment form of health insurance.

56 **Brading**, David A.
   1970 'Mexican silver-mining in the eighteenth century: The revival of Zacatecas'. *Hispanic American Historical Review* 50(4): 665–81.
How and why Mexico emerged as world's chief producer of silver by the eighteenth century. One mining camp was selected for examination, Zacatecas, 1767–1809.

57 1971 *Miners and Merchants in Bourbon Mexico, 1763–1810*. Cambridge: Cambridge University Press. Also, *Mineros y Comerciantes en el Mexico Borbónico, 1763–1810*. Mexico City: Fondo de Cultura Económica.
Trilogy of scholarly essays based on original sources: emergence of Bourbon-style colonial government, development of an active class of mine entrepreneurs and merchants, and detailed study of the major silver mining region, Guanajuato, especially Valenciana mine.

58 **Brading**, David A. and Harry E. Cross
   1972 'Colonial silver mining: Mexico and Peru'. *The Hispanic American Historical Review* 52(4): 545–79.
Careful comparison of development histories of Mexican and Peruvian silver industries during colonial epoch, applying such criteria as labour supply, mineral access, capital structure, crown policy, to account for national differences that depended critically

on a differing structure of production which subsequently influenced Peru's move away from the Toledoan system.

59 **Bravo**, Juan Alfonso
    1980 'United States' investments in Chile: 1904–1907'. MA thesis, American University.
Discussion of what Bravo considers a watershed period in the development of Chilean copper industry. Deals with willingness of Chilean landowning capitalists to allow American investors to monopolise completely and develop Chile's copper industry.

60 **Brennan**, Joseph
    1975 'Industry perspective on CWP and the black lung compensation program'. Paper presented at the Coal Workers' Pneumoconiosis Meetings (Legal and Social Aspects) held at the National Academy of Sciences, Washington, DC, 7 March.
Discussion of cost, retroactivity of benefits, eligibility, financing of programme, federal/state interface, occupational disease and role of the coal industry.

61 **Brestensky**, Dennis F.
    1982 'Pastimes and festive customs in early mine patches'. *Pennsylvania Heritage* (winter): 15–19.
Deals with Connellsville bituminous coal and coke region south-western Pennsylvania; describes weddings, holiday celebrations and baseball competitions characteristic of coal communities in early twentieth century.

62 **Brett**, Jeanne M. and Stephen B. Goldberg
    1979 'Wildcat strikes in bituminous coal mining'. *Industrial and Labor Relations Review* 32(4): 465–83.
Tests several hypotheses on causes of wildcat strikes in bituminous coal through examination of questionnaire and related data concerning 293 underground mines, and interview data gathered at two high-strike and two low-strike mines. Suggests grievance procedure to reduce problems.

63 **Bringhurst**, Newell G.
    1981 'The 'new' labor history and hard rock miners in Nevada and the West'. *Nevada History Quarterly* 24(2): 170–5.
Offers extended review of two recent books (Brown 1979 and Wyman 1979) dealing with Nevada and the West from 1860 to the early decades of the twentieth century. The critique centres on these books as illustrations of a newer historiography emphasising non-union and pre-union labour history.

64 **Brinley**, Jr., John Ervin
    1972 'The Western Federation of Miners'. PhD dissertation, The University of Utah. Ann Arbor: University Microfilms.
The Western Federation of Miners was founded as an association of local metal mining unions in Butte, Montana. Brinley disputes the stigmatisation of WFM as a radical and revolutionary union which has coloured many previous studies of WFM. Argues that WFM's image caused the American Federation of Labor to reject it, leading to the WFM's eventual demise.

65 **Brown**, Ronald C.
    1975 'Hard-rock miners of the Great Basin and Rocky Mountain West. PhD dissertation, University of Illinois at Urbana-Champaign. Ann Arbor: University Microfilms. (Subsequently published as Brown, 1979.)
66 1979 *Hard-Rock Miners, The Intermountain West, 1860–1920*. College Station and London: Texas A&M University Press.
Covers underground mineral mining in Wyoming, Colorado, New Mexico, Arizona,

Utah, Nevada. Historical treatment with attention to techniques, mines, towns, labour relations, accidents and health and sources of labour. Although provided with footnotes and sizeable bibliography, the treatment is generalised. Provides little attention to differences between different regions or mining on basis of differing minerals.

67 **Brundenius**, Claes
    1972 'The anatomy of imperialism: the case of the multinational mining corporations in Peru'. *Journal of Peace Research* (Sweden) 9(3): 189–207.
Investigates functioning and changing character of imperialism and how it affects people subjugated to it. Historical analysis of the methods of exploitation of modern multinational mining corporations. Intimates links between 'national' bourgeoisie in Peru and foreign capital.

68 **Brunstetter**, Maude Phillips
    1981 'The desperate enterprise: a case study of the democratization of the United Mine Workers in the 1970s'. PhD dissertation, Columbia University. Ann Arbor: University Microfilms.
Compares uncontrolled Miners for Democracy decentralisation movement in UMW during the 1970s, and eventual pressures for centralised union authority, using organisation theory of Robert Michels.

69 **Buechler**, Rose Marie
    1973 'Technical aid to Upper Peru: the Nordenflicht expedition'. *Journal of Latin American Studies* 5(1): 37–77.
Recounts in detail one of best recorded examples of early technical assistance, sent by the Spanish Crown to Potosi to improve silver mining efficiency, an effort, amid German ethnocentrism and trenchant local interests, which fell prey to many of same absurdities that plague such missions today.

70 **Buis**, Lois Slagle Cordell
    1974 'Kay Jay in its heyday'. *Appalachian Heritage* (summer): 61–9.
Author revisits the town in south-eastern Kentucky where she lived as a child. The town was a small mining camp, with Kentucky-Jellico Coal at its peak when she resided there. Now a 'ghost town', Buis recalls how it was in 1930s.

71 **Burke**, Gill
    1983 'The light infantry of capital: Cornish miners in the Latin Americas'. Paper presented at the 9th International Congress of Latin American Studies Association, 9 September – 1 October, Mexico City.
Discussion of the worldwide presence of Cornishmen in mining enterprises of nineteenth century, and analysis of the features and causes of the Cornish phenomenon. Uses Mexican Real del Monte/Pachuca mines as well documented case, and identifies the roving, unaccompanied Cornish miner as thematic of Cornish influence.

72 **Burke**, Melvin and James M. Mallow
    1972 'Del populismo nacional al corporativismo nacional (el caso de Bolivia, 1952–1970)'. *Aportes* 26 (October): 66–96.
Analyses process by which reformist-populist governing strategy came to be replaced by a national-corporativist strategy, beginning during first Paz Estensoro government, 1960, and intensifying during subsequent Barrientos period. Extensive longitudinal documentation provided within detailed summary of Bolivian political history.

73 **Bustamante**, Ciriaco Perez
    1970 'Las minas en los grandes geógrafos del período hispánico'. In *La Minería Hispana e Iberoamericana*, Ponencias del 1er Coloquio Internacional Sobre Historia de la Minería, 1. A. Vinayo Gonzalez, ed. León, Spain: Cátedra de San Isidoro. Pp. 295–306.

Reviews efforts to locate mineral deposits and mineral wealth in period of early conquest and exploration, as documented in chronicles and place names.

74 **Byrkit**, James W.
1982 *Forging the Copper Collar: Arizona's Labor-Management War, 1901–1921.* Tucson: University of Arizona Press.
Well-researched, detailed account of the Bisbee Deportation is used to probe Arizona's labour-management war, workers' unions, copper companies and politics. Investigates eastern power-sponsoring and manipulation of economic, political, social and cultural life of the American west.

75 **Cameron**, James M.
1974 *The Pictonian Colliers.* Halifax, NS (Canada): The Nova Scotia Museum.
Although the author asserts that 'geologists, engineers, sociologists, economists will not find this a volume to further their knowledge', his book contains much of value. Indeed, for descriptive history of coal mining, including technology, personalities, men, disasters and eventual withering of the Nova Scotia coal industry, this book will play a central part. Level of detail and depth of scholarship is high.

76 **Canelas Orellana**, Amado
1981 *Quiebra de la Minería Estatal Boliviana?* La Paz: Editorial Los Amigos del Libro.
A vigorous defence of the future of COMIBOL, the Bolivian state mining corporation, while offering biting criticism of its recent history and present condition, written by former COMIBOL director of public relations.

77 **Capriles Villazon**, Orlando
1977 *Historia de la Minería Boliviana.* La Paz, Bolivia: Biblioteca 'BAMIN'.
General history of mining from pre-colonial to post-revolution times, written by prominent Bolivian lawyer. Fairly comprehensive, though lacking in factual and statistical detail. Useful source for filling gaps, particularly in the historical record from the mid-nineteenth century to the early post revolutionary period (which begins in 1952). Account ends about 1967. No index.

78 **Cartajena Bakovic**, Manuel Francisco
1970 *La Pequeña Minería y las Cooperativas Mineras.* Santiago de Chile: Editorial Jurídica de Chile.
Although it is not clear that the author so intended, this is essentially a text on nature of cooperatives in small mining contexts, Chilean legislation which regulates these entities, and vigorous argument for extending the institution further.

79 **Cash**, Joseph H.
1973 *Working the Homestake.* Ames, Iowa: Iowa State University Press.
Examines lot of working man in one gold mine – the Homestake of Lead, South Dakota. The focus here is on a single mine, covering the period from 1877 to the 1940s. The salient feature of the Homestake was the extent of labour peace which the author ascribes to progressive company labour policies.

80 **Casteneda Delgado**, Paulino
1970 'El tema de las minas en la ética colonial Española'. In *La Minería Hispana e Iberoamericana*, Ponencias del 1er Coloquio Internacional Sobre Historia de la Minería, 1. A. Vinayo Gonzalez, ed. León, Spain: Cátedra de San Isidoro. Pp. 333–54.
Efforts of Spanish Crown to regulate use of New World Indians in mining enterprises, and to protect them from exploitation. Much necessarily applied to all use of Indian labour, but the specifics relating to mines can be found here. Author deals principally with sixteenth and early seventeenth centuries.

81   **del Castilla Crespo**, Alberto
    1970 'Las industrias extractivas en Bolivia'. *Estudios Andinos* 1(1): 93–113.
General description by Bolivian economist of his country's mining and oil sectors,
especially since 1952. Concludes production improvements since 1960 represent recovery
from previous period of economic depression.

82   **Caudill**, Harry M.
    1974 'The mountain, the miner and the Lord'. *Appalachian Heritage* (summer):
        29–39.
Personal story of black coal miner in Fleming, Kentucky.

83   **Cereceda**, Luis E. and Guillermo Wormald
    1975 *Genesis y Evolución de la Pequeña Minería del Cobre en Chile*. Documento
        de Trabajo no. 26. Santiago: Instituto de Sociología, Universidad Católica
        de Chile.
Those interested in small mining sector of Chile will find this useful. History of colonial
times to the present, giving particular attention to 1960–71 when government programme sought to encourage output, and 1971–3 when Allende government instituted
cooperativist programme.

84   **Chadwick**, Thomas Timothy
    1979 'Electoral patterns in coal mining areas: an examination of national elections
        in Great Britain and the United States, 1924–1970'. PhD dissertation,
        University of Virginia. Ann Arbor: University Microfilms.
Examines socio-economic environments and electoral patterns of coal mining areas
in two countries. Analyses impact of tradition, social class and union membership on
coal miners' support in parliamentary and presidential elections, 1924–1970.

85   **Cherder**, Michael
    1976 'Copper exploration restrained by resource nationalism and low metal prices'.
        Engineering and Mining Journal 177(8): 77–80.
Overview of economic and political forces likely to govern scarcity, investment and
production of world copper in 1980s, written from industry perspective.

86   **Christian**, R.T. *et al.*
    1979 'Coal workers' pneumoconiosis: *in vitro* study of the chemical composition
        and particle size as causes of the toxic effects of coal'. *Environmental
        Research* 20: 358–65.
Results demonstrate that difference between cellular responses of leachates of coal
particles from mine in Pennsylvania and one in Utah is not dependent on particle size,
but rather on chemicals leached from those particles. Pennsylvania coal more toxic
than Utah coal because respirable particles contain more leachable toxic chemicals.

87   **Church**, Ruth and Cathy Stanley
    1974 'Women miners – in the 40's; today'. *Mountain Life and Work* 50(11):
        14–15.
Personal interviews compare conditions of women coal miners in the '40s and today.
Two of the former, suffering with black lung, filed claims. Two of the latter say things
goind well.

88   **Clark**, Paul F.
    1981 *The Miners' Fight for Democracy: Arnold Miller and the Reform of the United
        Mine Workers*. Ithaca: New York State School of Industrial and Labour
        Relations.
Institutional history, organised thematically around efforts to reinstate democracy in the
UWM commencing with Miller, December 1972, and his call for district elections which
subsequently returned polarised International Executive Board which he had to control.

89 **Clark**, Paul F.
1983 'Democratic reform in the miners' union'. *Dissent* 30(2): 143−5.
Richard Trumka's election challenge, victory and opportunity to head UMW is legacy of the reform movement that took control of the UMW in 1972. Says legacy will ensure support, guiding the UMW through difficult times ahead.

90 **Cochran**, Alica Cowan
1980 *Miners, Merchants, and Missionaries: The Roles of Missionaries and Pioneer Churches in the Colorado Gold Rush and its Aftermath, 1858−1870.* ATLA Monograph Series, no. 15. Metuchen, New Jersey: The Scarecrow Press and The American Theological Library Association.
Thesis is that Colorado miners and those who followed them sought compulsively to implant the orderly structure of civilisation they brought will them, and within a decade had succeeded in dominating the frontier conditions of the earliest phase. Seen as an amendment to the Turner frontier thesis. Thorough bibliography.

91 **Colmenares**, German
1971−2 'Problemas de la estructura minera en la Nueva Granada, 1500−1700'. In *Anuario Colombiano de Historia Social y de la Cultura* (Bogota) 6−7: 5−55.
Extensive analysis of gold mining in first two centuries of colonial period, providing province-by-province accounts of production and labour, and discussing reasons for broad decline in production that characterised end of period. Role of Indians and black slaves receives detailed treatment.

92 Commission of National Resources
1976 *Mineral Resources and the Environment. Supplementary Report: Coal Workers' Pneumoconiosis−Medical Considerations, Some Social Implications.* A report prepared by the Committee on Mineral Resources and the Environment (COMRATE). Washington, DC: Commission on Natural Resources, National Research Council, Printing and Publishing Office, National Academy of Sciences.
Reviews and summarises current medical and scientific information on respiratory disease associated with coal mining. Clarifies ambiguities inherent in the Black Lung Benefits Programme, and focuses attention on whether this programme is appropriate model for other hazardous occupations.

93 *Congressional Digest*
1974 'Pro and con discussion: should the Federal Government assume a direct role in the regulation of surface mining of coal in the States?' *The Congressional Digest* 53(5): 131−60.
Documents relating to debate in US Congress on strip mining regulation. Notable is the statement of Helen Lewis relating to West Virginia strip mining experience, arguing for passage of regulatory authority (pp. 156, 158).

94 Consejo de Recursos Naturales no Renovables
1970 *Mineria Prehispánica en La Sierra de Queretaro.* Mexico City: Secretaria del Patrimonio Nacional.
Results of archaeological excavations of a prehispanic cinnabar mine appearing to have been worked from several centuries BC to about the tenth century AD.

95 **Corbin**, David Alan
1978 'Betrayal in the West Virginia coal fields: Eugene V. Debs and the Socialist Party of America, 1912−1914'. *Journal of American History* 44(4): 987−1009.

1976 Pelzer Award essay. Considers events surrounding Debs and the Socialist Party in terms of grass-root policies and actions. Addresses circumstances of compromises in the West Virginia coal fields, and their part in destroying the growing socialist movement, turning trade officials against the movement.

96 1981 *Life, Work, and Rebellion in the Coal Fields: The Southern West Virginia Miners 1880–1922*. Urbana: University of Illinois Press.
Well researched study of West Virginian labour movements. A highlight is the extensive treatment of the Logan/Mingo conflict (in which perhaps 15,000 to 20,000 miners confronted the United States army, and some 550 miners were indicted for treason). Author argues that class consciousness and class action characterised the later decades of the period, generated from harsh working and living conditions and exploitative practices of employers, abetted by collaborative local government. Bibliography and footnotes are extensive.

97 **Cortes Conde**, Roberto and Stanley Stein, eds.
1977 *Latin America: A Guide to Economic History, 1830–1930*. Berkeley: University of California Press.
General and national bibliographies, divided into ten major categories with appropriate subcategories, annotated by scholars from six nations of the Americas. Many mining citations, Library where entry was consulted appears at the end of most annotations.

98 Council of the Southern Mountains
1979 'Special issue: women coal miners'. *Mountain Life and Work* 55: 3–29.
200 participants attended First National Conference of Women Coal Miners. Focused on issues ranging from overcoming sex discrimination to training programmes for women miners and improving safety conditions in the mines.

99 **Couto**, Richard
1978 'Mud Creek, Kentucky: sick for clinics'. *Southern Exposure* 1 (2): 42–52.
Interview with an administrator and social worker at Mud Creek Health Project in Craynor, Kentucky. Focuses on dilemma they face if they are to keep their clinic open. The clinic faces financial ruin after providing for miners with no insurance.

100 **Crespo R.**, Alberto
1970 'El reclutamiento y los viajes en la 'mita' del Cerro de Potosí'. In *La Minería Hispana e Iberoamericana*, Ponencias del 1er Coloquio Internacional Sobre Historia de la Minería, 1. A. Vinayo Gonzalez, ed. León, Spain: Cátedra de San Isidoro. Pp. 467–82.
Brief study of the labour draft used at Potosí (and elsewhere) during colonial period. Statistics on sources and numbers of draftees from particular communities.

101 **Crooks**, W. H., *et al.*
1980 *Analysis of Work Areas and Tasks to Establish Illumination Needs in Underground Metal and Nonmetal Mines*. 2 vols. Final Report, Contract J0387230. Washington, DC: Department of the Interior, Bureau of Mines.
Technical study to determine necessary levels of illumination to reach job injury goals by type of activity in underground mining. On-site studies on ten mines ranging from 125 to 550 labour force, using various mine excavation techniques.

102 **Cross**, Harry Edward
1976 'The mining economy of Zacatecas, Mexico in the nineteenth century'. PhD dissertation, University of California at Berkeley. Ann Arbor: University Microfilms.
Helpful detail for examining validity of output statistics, labour supply, fodder prices and food sources, and mining-agrarian sector linkages.

103 **Curtis**, John Stephen
　　1982 'A study of the patterns of injuries in the mining industry following imple-
　　　　mentation of training requirements'. PhD dissertation, Catholic University
　　　　of America. Ann Arbor: University Microfilms.
Statistical study of mine accidents following implementation of federally mandated
training programme of 1977. Shows minor injuries reduced, but no effect on incidence
of serious ones.

104 **Daniel**, Pete
　　1975 'The Tennessee convict war'. *Tennessee Historical Quarterly* 34 (3): 273–92.
Considers events surrounding coal miners' insurrection of 1891 in Anderson County,
and the eventual overthrow of the convict lease system. Coal Creek rebellion, often
glorified in folklore, brought end to convict miners in Tennessee.

105 **David**, John Peter
　　1972 'Earnings, health, safety, and welfare of bituminous coal miners since the
　　　　encouragement of mechanization by the United Mine Workers of America'.
　　　　PhD dissertation, West Virginia University.
The effects of the United Mine Workers of America policy that encouraged mechanisation on union and on workers' earnings, health, safety and welfare in bituminous
coal mining industry, 1946–71.

106 **Dayton**, Stan, ed.
　　1975 'Brazil 75: for mining the best is yet to come'. *Engineering and Mining
　　　　Journal* 176: 89–176.
Issue devoted to developments in Brazilian mineral investment. Journalistic style, but
includes detailed minerals map of the country and account of then current investment
activities.

107 1977 'Mining in the Caribbean'. *Engineering and Mining Journal* 178: 49–198.
An extensive, industry-oriented survey of mining in the insular Caribbean and
surrounding Latin American mainland. Most constituent articles cover current major
workings, extent of known deposits, new initiatives and productivity figures. Separate
articles on Colombia, Venezuela, Guyana, Surinam, Trinidad/Tobago, Jamaica, Cuba,
Haiti, Dominican Republic, Puerto Rico, Guatemala, Honduras, Nicaragua, Panama,
Costa Rica and El Salvador. Fold-out geological map introduces the group of articles.

108 1979 'Chile: where major new copper output can materialize faster than anywhere
　　　　else'. *Engineering and Mining Journal* 180: 68–111.
General, well illustrated review of major copper workings in Chile, including
Chuquicamata, El Teniente, Salvador, Andina, Disputada, Pudahuel and Andacollo.
Attention to changing levels of profitability and future ore grade prospects.

109 **De Mesa**, José and Teresa Gisbert
　　1970 'Oruro, origen de una villa minera'. In *La Minería Hispana e Iberoamericana*,
　　　　Ponencias del 1er Coloquio Internacional Sobre Historia de la Minería,
　　　　1. A. Vinayo Gonzalez, ed. León, Spain: Cátedra de San Isidoro.
　　　　Pp. 559–90.
History of Oruro (Bolivia) during sixteenth, seventeenth centuries. Oruro, like Potosi,
survived exhaustion of its mining resources by becoming a regional administrative
capital and through provisioning mining centres at greater remove.

110 **del Castillo**, Guido
　　1978 'Development of small scale mining companies in Peru'. In *Small Scale
　　　　Mining of the World* Conference proceedings, Jurica, Queretaro, Mexico,
　　　　November 1978. New York (?): United Nations Institute for Training and
　　　　Research. Pp. 913–26.

Mainly consists of engineer's description of Carachaca, a small but relatively highly capitalised mine producing silver, lead, zinc near Puno. Author attributes the mine's viability to effects of financing legislation designed to promote enterprises of this sort.

111 **Delgado**, Guillermo
    1982 'A Mitmaq at Uncia; on cultural continuity in the Bolivian Andes'. Paper presented at the 44th International Congress of Americanists. 5–10 September, Manchester, UK.
A narrative description of agricultural and cooperative practices engaged in by the wife of a tin miner. Argues that the practices are derived from peasant origins.

112 **Delgado Gonzalez**, Trifonio
    1980 *Cien Años de Lucha Obrero en Bolivia*. Guillermo Delgado, ed. Oruro, Bolivia: Editorial Chaski.
A philosophic and personal account of major events in the Bolivian labour movement from the viewpoint of a union oraniser, veteran of the Chaco war, who worked in the Uncia tin mine as a *chapatero* (assistant to the mine foreman) until fired. Later, as a miner from the Huanuni mine, he rose to serve as worker representative during the Busch government.

113 **Denman**, William Neville
    1974 'The black lung movement: a study of contemporary agitation'. PhD dissertation, Ohio University. Ann Arbor: University Microfilms.
Examines and analyses the black lung movement to discover how it developed into a representative agitational campaign, the role communication played, and those elements in the movement that seem to account for its success.

114 **Densmore**, Raymond E.
    1977 *The Coal Miner of Appalachia*. Parsons, West Virginia: McClain Printing Company.
Personal narrative, sometimes rambling, of author's life as coal miner, eventually rising to more responsible positions and ending as federal mine inspector. Contains useful localisms and vernacular, and interesting accounts of the company store practices and other means by which management sought to increase income at expense of labour force.

115 **DeWind**, Jr, Adrian W.
    1974 'De campesinos a mineros: el origen de las huelgas en las minas Peruanas'. *Estudios Andinos* 4 (2): 1–31.
Traces roots of frequent strikes and work stoppages at Cerro de Pasco which characterised the 1969–71 period to historic changes in labour force, mechanisation of mining, rupture of links to peasant backgrounds, etc. Suggests nationalisation and leftist ideology neither instigated nor diminished proclivity to strike.

116 1975 'From peasants to miners – background to strikes in mines of Peru'. *Science and Society* 39(1): 44–72.
The transformation of mine labour force from part-time peasant miners into skilled full-time wage labourers. DeWind suggests present-day miners should be considered a proletariat in the classic marxist sense.

117 1977 'Peasants become miners: the evolution of industrial mining systems in Peru'. PhD dissertation, Columbia University. Ann Arbor: University Microfilms.
Examines mining operations of Cerro de Pasco Corporation with two questions: (1) has mining industry functioned as isolated enclave? and (2) if not an enclave, what relationship has existed between mines and rural regions in which they are located?

118 **Dilsaver**, Lary Michael
　　1982 'From boom to bust: post gold rush patterns of adjustment in a California
　　　　mining region' 2 vols. PhD dissertation, Louisiana State. Ann Arbor:
　　　　University Microfilms.
Analyses local socio-economic configuration of four-county California region in
transition from gold boom at mid-century to smaller, non-mining, economically
diversified, socially more homogeneous, politically stable society by 1920s.

119 **Dix**, Keith
　　1977 *Work Relations in the Coal Industry: The Hand-Loading Era, 1880–1930*.
　　　　West Virginia University Bulletin Series 78, no. 7–2. Morgantown: West
　　　　Virginia University, Institute for Labor Studies.
Labour process in bituminous underground coal mining in the United States from early
days of the industry to mid-1930s. Primary focus is on people and work relations at
work before mechanisation of coal loading.

120 **Dix**, Keith, Carol Fuller, Judy Linsky and Craig Robinson
　　1972 *Work Stoppages and the Grievance Procedure in the Appalachian Bitumi-
　　　　nous Coal Industry*. Morgantown: West Virginia University, Institute for
　　　　Labor Studies.
Attempts to ascertain the nature and extent of stoppages and why coal miners resort
to wildcat strikes.

121 **Doherty**, William T.
　　1971 *Minerals: Conservation in the United States, a Documentary History*. New
　　　　York: Chelsea House Publishers.
Documentary history, samples various sorts of federal and state conservation definitions. General and sectional introductions on conservation policy with respect to oil,
gas, coal, and gold and silver.

122 **Dore**, Elizabeth Wilkes
　　1981 'Accumulation and crisis in the Peruvian mining industry: 1900–1977'. PhD
　　　　dissertation, Columbia University. Ann Arbor: University Microfilms.
Dependency theory predictions fail to explain fluctuations and eventual economic crisis
of 1970s. Author's evidence suggests labour instability and variables related to cost
of labour account for much of variance. Beginning with 1950s labour becomes
increasing expensive relative to production, leading to industry crisis of the 1970s when
much capital transferred to more profitable sectors.

123 **Dougnac Rodriguez**, Antonio
　　1970 'Fuentes documentales Chilenas para el estudio de la historia de la minería
　　　　en el período Indiano'. In *La Minería Hispana e Iberoamericana*, Ponencias
　　　　del 1er Coloquio Internacional Sobre Historia de la Minería, 1. A. Vinayo
　　　　Gonzalez, ed. León, Spain: Cátedra de San Isidoro. Pp. 601–4.
Brief description of holdings of six archives in Chile dealing with the postconquest
period.

124 **Draper**, Theodore
　　1972 'Communists and miners'. *Dissent* 19(2): 371–92.
The story of the National Miners' Union, forerunner of Trade Union Unity League.
When formed, communists' hoped for a successful revolutionary trade union federation. Unexpected growth of UMW under John L. Lewis was death knell of NMU.

125 **Driever**, Steven L.
　　1977 'Mining and economic development: a case study of exploraciones y
　　　　Explotaciones Mineras de Izabal, SA of Guatemala'. PhD dissertation,
　　　　University of Georgia. Ann Arbor: University Microfilms.

Basic objective is to evaluate, through case study of Eximbal, the relationship between large-scale mining and economic development. The primary study area is El Estor, Guatemala.

126 1982 'Mining and the geography of cultural change in El Estor, Guatemala'. Paper presented at the 44th International Congress of Americanists, Manchester, UK, 9 September.

Study of negative and positive contributions to local economy of Eximbal, a large-scale nickel-mining operation. Evaluates Eximbal's alteration of local cultural environment.

127 **Dugi**, Don Thomas
    1981 'The political ideology of Kentucky coal producers'. PhD dissertation, Purdue University. Ann Arbor: University Microfilms.

Questionnaire study of political and economic attitudes of Kentucky coal operators. Finds operators generally subscribe to liberal capitalist philosophy of a fairly conservative type. Cross tabulations with demographic and political variance were generally nonsignificant.

128 **Dwiggins**, George Albert
    1981 'Variations in the properties of coal dusts, their significance to the health of occupationally exposed workers, and their relationships to parent material characteristics'. PhD dissertation, University of North Carolina at Chapel Hill. Ann Arbor: University Microfilms.

Although only peripherally related to mining *per se*, some readers may find this helpful in understanding complexities of dust composition in coal mine environments. Author finds that current regulations on coal dust and its respiration consequences are based on grossly simplistic assumptions which understate dangers of some dust environments.

129 **Eakin**, Marshall Craig
    1981 'Nova Lima: life, labor and technology in an Anglo-Brazilian mining community, 1882–1934'. PhD dissertation, University of California, Los Angeles. Ann Arbor: University Microfilms.

Twin roles of industrial technology and labour structure are examined for Brazil's largest gold mine, Morro Velho, from 1882 to 1934. The effort is to trace causality. Finds dialectic interaction between technology and human context to best account for the enduring development of this enormous mine centre.

130 1982 'The role of British capital in the development of Brazilian gold mining'. Paper presented at the 44th International Congress of Americanists, Manchester, UK, 9 September.

The role of the British in Brazilian gold mining boom viewed in larger context as a part of worldwide economic expansion of Great Britain in nineteenth century, especially as part of search for precious metals in Latin America. Detailed description of active British companies.

131 **Eberhart**, Thomas Milton
    1977 'The socio-political dimensions of Latin American peasant movements: a comparative analysis of the Bolivian syndicates following the 1952 revolution and the La Convencion Valley peasant movement of Peru'. PhD dissertation. The American University. Ann Arbor: University Microfilms.

Focus described in title.

132 **Edwards**, P.K.
    1977 'A critique of the Kerr–Siegel hypothesis of strikes and the isolated mass: a study of the falsification of sociological knowledge'. *The Sociological Review* 25(3): 551–74.

Methodological, logical and empirical critiques of weaknesses in Kerr–Siegel hypothesis are considered. Hypothesis is widely used in analysis of industrial conflict, but author suggests that there is no necessary relationship between correctness of a theory and status it attains.

133 **Eller**, Ronald D.
  1982 'Miners, millhands, and mountaineers: industrialization of the Appalachian South, 1880–1930'. Knoxville: University of Tennessee Press.
Regional history examining coming of bituminous coal mining along with capitalists and adventurers and their variety of 'New South' boosterism. Eller chronicles their coming and the results thereof, giving vivid portrait of the dark side of American economic progress.

134 **Enman**, John A.
  1974 'Coal company store prices questioned: a case study of the Union Supply Company, 1905–1906'. *Pennsylvania History* 41(1): 53–62.
Study to determine accuracy of prices charged by company stores. Uses actual accounts of a store in Buffington, Pennsylvania served by a supply company of H. C. Frick Coke Co. which indicate that coal company store prices were comparable to neighbouring stores in that area.

135 **Ervin**, Osbin L.
  1978 'Local fiscal effects of coal resource development: a framework for analysis and management'. *Policy Studies Journal* 7(1): 9–17.
Discusses research on potential community fiscal impacts of coal development in Illinois. Using applied policy analysis, a local 'fiscal system' is conceptualised, composed of economy; land-use service demand; and revenues. Preliminary research on impacts of a coal-gasification facility.

136 **Everling**, Arthur Clark
  1976 'Tactics over strategy in the United Mine Workers of America: internal politics and the question of the nationalization of the mines, 1908–1923'. PhD dissertation, Pennsylvania State University. Ann Arbor: University Microfilms.
Argues that internal political history of UMW between 1908 and 1923 can be explained in terms of competition between conservative and radical proposals for stabilising industry.

137 **Ewen**, Lynda Ann
  1979 *Which Side Are You On? The Brookside Strike in Harlan, Kentucky, 1973–1974*. Chicago: Vanguard Books.
Interesting, readable story of the Brookside strike. Highlights strikers and their supporters, and the Brookside miners who ultimately paid blood price for right to unionise.

138 **Ezquerra Abadia**, Ramón
  1970 'Problemas de la mita de Potosí en el siglo XVIII'. In *La Minería Hispana e Iberoamericana*, Ponencias del 1er Coloquio Internacional Sobre Historia de la Minería, 1. A. Vinayo Gonzalez, ed. León, Spain: Cátedra de San Isidoro. Pp.483–511.
Covers fairly evenly the labour policies of seventeenth and eighteenth centuries, with attention not only to Potosí but also to other areas of the Peruvian vice-royalty. Attention is well-balanced between analysis of legislation and the practical economics of labour supply.

139 **Farber**, Henry Stuart
  1977 'The United Mine Workers and the demand for coal: an econometric

analysis of union behavior'. PhD dissertation, Princeton University. Ann
Arbor: University Microfilms.
Farber develops a model of union behaviour applicable to United Mine Workers in
order to analyse response of union to shifts in demand for coal. Model is estimated
by Full Information Maximum Likelihood using time series data 1947–73.

140 **Fetherling**, Dale
1974 *Mother Jones, the Miner's Angel: A Portrait.* Carbondale: Southern
Illinois University Press.
Biography of vagabond agitator famous for her continuing fight against the oppression
of working people in the face of great power. Her extraordinary effect on workers
made her effective as an organiser in a strike.

141 **Filippelli**, Ronald L.
1976 'Diary of a strike: George Medrick and the coal strike of 1927 in western
Pennsylvania'. *Pennsylvania History* 43(3): 253–66.
Brief descriptive introduction precedes excerpts from diary of UMW labour organiser
during a major strike.

142 **Finley**, Joseph E.
1972 *The Corrupt Kingdom: The Rise and Fall of the United Mine Workers.* New
York: Simon and Schuster.
The Boyle–Yablonski campaign, the murder and its aftermath, and the rise of the
rebel movement inside the UMW. Examines how men behave inside their own organisations when powerful leaders deify themselves.

143 **Fisher**, John R.
1975 'Silver production in the vice royalty of Peru, 1776–1824'. *Hispanic
American Historical Review* 55 (1): 25–43.
Counters general assumption that loss of Upper Peru to Lima's vice-royalty in 1776
entailed drastic decline in silver production exported through Lima. Fisher's statistics
show a sustained rise. Remainder of article describes post-partition mining region by
region.

144 1977 *Silver Mines and Silver Miners in Colonial Peru, 1776–1824.* Monograph
Series no. 7. Liverpool, University of Liverpool Centre for Latin-American
Studies.
Changes in Peruvian silver industry spanning the years between the separation of Upper
Peru sector from Peru vice-royalty in 1776 to the end of civil war, 1824. Fisher evaluates
effects of administrative policies, political turmoil, technological changes and annual
silver production. Analysis of Nordenflicht mission is particularly worthwhile.

145 **Fitzpatrick**, John Steven
1974 'Underground mining: a case study of an occupational subculture of danger'.
PhD dissertation, Ohio State University. Ann Arbor: University Microfilms.
Constructs theoretical framework to explain how people adapt to stresses, losses and
injuries to which they are exposed in their work. A detailed case study of the social
setting of underground mining.

146 1982 'The social correlates of mine safety'. Paper presented at the 44th International Congress of Americanists, Manchester, UK, 5–10 September.
Paper describes how social interaction among miners leads to the formation of distinctive subculture which allows them to identify danger, manage its threat and work safely.

147 **Flores Galindo**, Albert
1974 *Los mineros de la Cerro de Pasco, 1900–1930; un Intento de Caracterización
Social.* Serie de Sociología. Lima: Pontificia Universidad Católica del Peru,
Departmento de Ciencias Sociales.

Important historical study of the evolution of a labour class among miners of the largest, highly capitalised mine operator in Peru. Establishes links between an evolving labour policy and capitalisation strategy and the degree to which labour defended its rights in an organised and often violent manner.

148 **Forman**, Pat
  1977 *Scandal at Gauley Bridge*. Health/PAC (Health Policy Advisory Center) Bulletin no. 79: 9–16.
Description of a hydroelectric tunnel project built in southern West Virginia in 1930–2 in which 500 died and another 1,500 were disabled by silicosis. Most of the victims were black, and the author's account shows project was conducted with reckless disregard for the known effects of silicosis in order to minimise construction costs. Ghastly.

149 **Fox**, Daniel M. and Judith F. Stone
  1980 'Black lung: miner's militancy and medical uncertainty, 1968–1972'. Bulletin of the History of Medicine 54 (1): 43–63.
Analytical essay concerning history of medicine, labour unions, people of the Southern Mountains and legislative politics in the United States.

150 **Fox**, David J.
  1970 *Tin and the Bolivian Economy*. London: Latin American Publication Fund.
As of 1970, some of attempts to help Bolivia during prior 10–15 years were bearing fruit. Tin still played principal role in the Bolivian scene, but did so with help of stronger and more active supporting cast drawn from a wider spectrum of national talents.

151 1971 'Tin mining in Bolivia'. *Mining Magazine* (London) 124 (1): 594–606.
Underscores Bolivia's reliance upon its minerals, and dependence upon metal prices, especially tin. Describes mining conditions and three distinct legal categories of industry, each with different set of rules – large, medium and small mines.

152 **Fox**, Jr., Harry Donald
  1975 'Thomas T. Haggerty and the formative years of the United Mine Workers of America'. PhD dissertation, West Virginia University.
Study of early union pioneer to illustrate what services may be rendered by men of ability and purpose in the more obscure positions. Details militant years of the UMW and the efforts of Haggerty to secure for the American miner some measure of respect.

153 **French-Davis**, Ricardo and Ernesto Tironi, eds.
  1974 *El Cobre en el Desarrollo Nacional*. Centro de Estudios de Planificación Nacional (CEPLAN). Santiago, Chile: Ediciones Nueva, Universidad Católica de Chile.
Series of coordinated papers reporting research on Chile's copper mining industries under nationalisation, after end of the Allende government. Studies are quite technical, done to provide an analytic and policy base for government decisions in managing copper mining sector of economy.

154 **Friedemann**, Nina S. de
  1980 'Negros: presencia indígena en un complejo orfebre Colombiano'. *Pluma* (Bogota) 4 (22): 50–7.
Describes gold working in pre-conquest Colombia and its persistence among black placer miners now indigenous to the Barbacoas (Pacific littoral) region. These artisans supply decorative and ritual needs both regionally and nationally within a complex university of technological arts, linguistic symbols and kinship.

155 **Galeano**, Eduardo
  1973 *Open Veins of Latin America: Five Centuries of the Pillage of a Continent*. Cedric Belfrage, transl. New York and London: Monthly Review Press. (Original: Las Venas Abiertas de America Latina, Mexico, 1971.)

Well-documented marxist-oriented history of Latin America. Focuses on the despoliation of the land, people and culture. Gold, silver, oil, iron, tin, copper and nitrates among 'open veins' whose drainage Galeano traces and analyses.

156 **Gall**, Norman
  1972 *Copper is the Wage of Chile*. Field Staff Reports, West Coast South America Series, South America, 19, 3. Hanover, New Hampshire: American Universities Field Staff.
On-site political analysis of conflicts and organisational problems which accompanied Allende nationalisation of Chuquicamata, El Teniente and other major copper mines in Chile. Includes short historical overview and useful statistical data in addition to interviews with on-site personnel.

157 1973 'Chile: the struggle in the copper mines'. *Dissent* 20 (1): 99–109.
Discusses the industrial accident that polarised the struggle for power between 'communist' management and USOPO union leaders at Chuquicamata following Allende's 'violent nationalisation' of mines in 1971.

158 1974 *Bolivia: The Price of Tin*. Part I, *Patino Mines and Enterprises*; Part II, *The Crisis of Nationalization*. South America: West Coast South America Series, 21, 1 and 2. Hanover, New Hampshire: American Universities Field Staff.
Both parts are essentially historical pieces centring on Catavi-Siglo XX-Llallagua tin mine complex built by Simon Patino. First part runs to 1952 revolution, the second until author's visit of 1973. Some demographic description stems from questionnaire administered to 183 families by author and helpers in 1973. Commentary is very gentle on those in both periods who could have improved lot of miners, but material helpful as rapid summary of history of major Bolivian tin complex. Contains helpful citations, photographs.

159 **Gamm**, Sara
  1979 'The election base of national union executive boards'. *Industrial and Labor Relations Review* 32 (3): 295–311.
Analyses the effect of the election base of executive boards on internal political life of national unions, especially capacity of such unions to marshall opposition to incumbent members of board.

160 **Garcia Morales**, Justo
  1970 'La bibliografía en España'. In *La Minería Hispana e Iberoamericana*, Ponencias del 1er Coloquio Internacional Sobre Historia de la Minería, 1. A. Vinayo Gonzalez, ed. León, Spain: Cátedra de San Isidoro. Pp. 671–7.
Brief description of major archival sources in Spain on mining, not only of New World, but also of Spain from fifteenth to twentieth centuries.

161 **Gartin**, Edwin V.
  1973 'The West Virginia mine war of 1912–1913: the progressive response'. *North Dakota Quarterly* 41 (4): 13–27.
Recounts efforts of two West Virginia governors to resolve violent dispute in Kanawha coal fields in which unionism, interregional competition and the railways were actors.

162 **Gaventa**, John
  1973 'In Appalachia: property is theft'. *Southern Exposure* 1 (2): 42–52.
Gaventa addresses corporate property holder in Appalachia as 'slave holder' possessing the power to deprive 'will and personality', a power of life and death. Combination of dependency and ideology makes support for energy industry an emotional issue with overtones of loyalty to home and country.

163 **Gavin**, James F. and Robert E. Kelley
   1978 'The psychological climate and reported well-being of underground miners: an exploratory study'. *Human Relations* 31 (7): 567–81.
Explores dimensions of underground miners' work environment perceptions, how these dimensions relate to measures of well-being, and to what extent such factors as age, seniority and work assignment account for perceptions of environment and well-being.

164 **Gavin**, James F. and Wendy L. Axelrod
   1977 'Managerial stress and strain in a mining organization'. *Journal of Vocational Behavior* 11 (1): 66–74.
Measures of job stress and strain obtained from ninety-five management-level employees in an underground mine. Thirteen potential moderators of stress-strain relations were assessed. Explanations and comparisons to other investigators' findings offered in conclusion.

165 **Geddes**, Charles F.
   1972 *Patino: The Tin King*. London: Robert Hale & Co.
Full-length study of South American industrialist Simon Iturri Patino. Acclaimed 'tin king' of the world, his interests, particularly in tin mining and smelting, stretched across the continents from the Americas to Europe and to the Far East.

166 **Gibson**, Anell M.
   1972 *Wilderness Bonanza: The Tri-State District of Missouri, Kansas, and Oklahoma*. Norman: University of Oklahoma Press.
History of Joplin and its economic region where great quantities of lead and zinc were exhausted through mining. Written on the centennial of Joplin.

167 **Gillespie**, Angus Kress
   1975 'The contributions of George Korson to American intellectual life'. PhD dissertation, University of Pennsylvania. Ann Arbor: University Microfilms.
Examines data available on the life of George Korson, newspaperman and folklorist who collected the stories and songs of coal miners. By selection and arrangement of material, Gillespie had tried to communicate his view of Korson's essential contribution to American intellectual life.

168 **Gillis**, Malcolm
   1982 'Allocative and x-efficiency in state-owned mining enterprises: comparisons between Bolivia and Indonesia, *Journal of Comparative Economics* 6 (1): 1–23.
Applying economics to the miner sectors, efficiency/inefficiency is measured against two indices. Countries are shown to be taking sizeable losses in the traditional measures of corporate efficiency in order to augment welfare costs. Probably a good source of investigating the contemporary economist's effort to quantify these national decisions.

169 **Gillis**, Malcolm, *et al.*
   1975 *Taxation and Mining: Nonfuel Minerals in Bolivia and Other Countries*. Cambridge, Massachusetts: Ballinger Publishing Company.
Seeks a clearer understanding of the complicated issues involved in the design and implementation of taxes on non-fuel minerals in less developed countries. Case study based upon Bolivian mining sector where minerals furnish a substantial share of export earnings.

170 **Goldman**, Marion Sherman
   1977 *Gold Diggers and Silver Mines: Prostitution and the Fabric of Social Life on the Comstock Lode*. Ann Arbor: The University of Michigan Press.
Incorporates sexual commerce within the social life of a mining town. Unsentimental perspective is maintained throughout – prostitution was a business in which women

worked for pay. Goldman argues that it was integral to understanding social attitudes toward all women in mining towns.

171 **Goldstein**, Joyce
 1978 'Cedar Grove, West Virginia: who will pay the bill?' *Southern Exposure* 6(2): 89–92.
Problems involved in establishing health clinic. The crisis of the miners' community clinic represents the crisis being experienced in health care throughout the country.

172 **Gomez**, Henry
 1971 'Venezuela's iron ore industry'. In *Foreign Investment in the Petroleum and Mineral Industries*. Raymond F. Mikesell, ed. Baltimore: Johns Hopkins Press. Pp. 312–44.
Reviews the historic directions of Venezuelan policy toward two giant US iron ore concessionaries in the period since 1951 when iron ore exports first became significant.

173 **Gomez-D'Angelo**, Walter
 1973 'Mining in the economic development of Bolivia, 1900–1970'. PhD disertation, Vanderbilt University. Ann Arbor: University Microfilms.
Strongly nationalist economic analysis of Bolivian economy, arguing that nationalisation of mines overcame parasitic pre-1952 mining sector. Nationalisation, though resulting in costly mining industry, funded agrarian sector, supply of needed imports, and contributed to a more just society.

174 1978 *La Minería en el Desarrollo Economico de Bolivia, 1900–1970*. La Paz: Editorial Los Amigos del Libro.
Spanish edition of Gomez-D'Angelo 1973.

175 **Goodman**, Paul S.
 1979 *Assessing Organizational Change: The Rushton Quality of Work Experiment*. New York: John Wiley and Sons, Inc.
The quality of work experiment at the Rushton Mining Company located near Osceola Mills, Pennsylvania. Historical account of the change and description of evaluation design and methodology is followed by analysis of results.

176 **Goodman**, Walter
 1972 'The sad legacy of John L. Lewis'. *Dissent*, 19(1): 91–8.
Article considers main concerns of the UMW hierarchy over prior decade and its abuse of power. Lewis' shadow still lies heavily on the UMW; challenge before the Miners for Democracy is to reawaken rank-and-file participation in governing of union.

177 **Goodsell**, Charles T.
 1974 *American Corporations and Peruvian Politics*. Cambridge: Harvard University Press.
Political implications for Peru of presence of sizeable American business investments. Addresses two questions as they relate to US investment in Peru: to what extent is national independence compromised by foreign capital? Are there lasting effects on the political system?

178 **Gordon**, Suzanne
 1973 *Black Mesa, The Angel of Death*. New York: John Day.
Careful and impassioned study of the impact of Black Mesa coal strip mining project and Four-Corners power plant on the Navajo and Hopi peoples who have been forced to host them. Excellent photographs by Alan Copeland.

179 **Goss**, Rosemary Carucci
 1982 'Housing for the Appalachian coal miner: conditions, satisfactions and aspirations'. PhD dissertation, Florida State University. Ann Arbor: University Microfilms.

Analyses housing survey among sample of mining families employed at mines in McDowell Co., West Virginia: though miners living outside county averaged somewhat better housing than those within, both groups aspired to better housing, defining it much as do Americans in general.

180 **Gower**, Abel
1979 'Morro Velho: a gold mine old but not over the hill'. *Optima* 28 (1): 22–35.
Although a journalistic description of history and current condition of Morro Velho mine (Minas Gerais), readers will find the piece well informed and a source of useful items on Brazilian mining in the interior.

181 **Graham**, Susan Brandt
1980 'Community, conformity, and career: patterns of social interaction in two mining towns'. *Urban Anthropology* 9 (1): 1–20.
Compares social networks of two Arizona mining communities, one controlled by mine officials, other by elected town council, to see how variation in community administration is reflected in lives of community residents.

182 **Granda**, German
1970 'Onomástica y procedencia Africana de esclavos negros en las minas del sur de la gobernación de Popayan (siglo XVIII)'. In *La Minería Hispana e Iberoamericana*, Ponencias del 1er Coloquio Internacional Sobre Historia de la Minería, 1. A. Vinayo Gonzalez, ed. León, Spain: Cátedra de San Isidoro. Pp. 605–38.
Contains introduction and commentary to three censuses conducted in 1716, 1717 and 1725 of slaves used in mines on south-west coast of Colombia. Data include ethnicity and origin, allowing interesting analysis of African origins for negroes. Complete transcriptions of documents are provided.

183 **Greaves**, Thomas C.
1972 'The Andean rural proletarians'. *Anthropological Quarterly*, 45 (2): 65–83.
Brings together the basic areal literature and proposes reinstatement of standard definition for rural proletarians. Discusses distribution, features and syndicates of Andean peasants and miners, suggesting that rural Andean world is rapidly becoming domain of the post-peasant.

184 **Greaves**, Thomas C. and Xavier Albo
1979 'An anatomy of dependency: a Bolivian tin miners' strike'. In *Political Participation in Latin America*, II *Politics and the Poor*. Mitchell A. Seligson and John A. Booth, eds. New York: Holmes and Meier. Pp. 169–82.
Recounts and analyses a miners' strike at a major COMIBOL mine, 'Bocasa', identifying factors weakening the miners' ability to win. Illustrates how transitional dependency is sustained at the local level.

185 **Green**, Archie
1972 *Only a Miner: Studies in Recorded Coal-Mining Songs*. Urbana: University of Illinois Press.
Comprehensive studies of songs which portray American coal-mining life and reveal miners' values. Ballad scholarship, labour history and popular culture studies used to explore the songs. Findings present the interaction among sophisticated, popular and folk societies. Problem of defining industrial folklore.

186 **Griffin**, Kenyon N. and Robert B. Shelton
1978 'Coal severance tax in the Rocky Mountain states'. *Policy Studies* 7 (1): 29–40.
Examines coal-severance tax policies of six coal-exporting states in Rocky Mountain West. Analyses these policies within context of each state's ability to export tax burden to out-of-state consumers.

187 **Grinde**, Jr, Donald A.
    1975 'The Powder Trust and the Pennsylvania anthracite region'. *Pennsylvania History* 42 (3): 207–19.
Formation of Gunpowder Trade Association under leadership of the DuPonts during 1870s. Price competition was generally suppressed for more than four decades.

188 **Gutman**, Herbert G.
    1976 'The negro and the United Mine Workers of America: the career and letters of Richard L. Davis and something of their meaning, 1890–1900'. In Gutman, *Work, Culture and Society in Industrializing America: Essays in American Working-Class and Social History*. New York: Alfred A. Knopf. Pp. 120–208.
Considers certain aspects of early contact between United Mine Workers and negro miners. Attention given mainly to role and ideas of Richard C. Davis, important UMW black leader, who retained and exemplified older traditions of pride, hope and militancy.

189 **Hadley**, Phillips Lance
    1975 'Mining and society in the Santa Eulalia mining complex, Chihuahua, Mexico: 1709–1750'. PhD dissertation, University of Texas at Austin. Ann Arbor: University Microfilms.
Examines structure of silver industry focusing on role of economic elite and the labour force. A comparison with mining communities in other parts of the vice-royalty; places the study in context of eighteenth-century Mexico.

190 **Haines**, Michael R.
    1977 'Fertility, marriage, and occupation in the Pennsylvania anthracite region, 1850–1880'. *Journal of Family History* 2 (1): 28–55.
Demonstrates interaction of mining occupation with ethnicity and other factors in determining normally high fertility of miner families relative to agriculturalists. Setting is four Pennsylvania coal counties. Data are surprisingly detailed.

191 1979 *Fertility and Occupation: Popular Patterns in Industrialization*. Studies in Social Discontinuity Series. New York: Academic Press.
Historical investigation on differential demographic behaviour by an occupational group. Shows how occupational fertility differentials have existed historically, and have remained throughout the modern era. Evidence to support economic interpretation of differentials are from mining and metallurgical populations in Germany, England and Wales and Pennsylvania anthracite region.

192 **Hammond**, Judith Anmarie and Constance W. Mahoney
    1983 'Reward-cost balancing among women coal miners'. *Sex Roles* 9 (1): 17–29.
Answers to two questions are examined in this article: what happens when sub-culture of male coal miners is entered by females as a result legal pressure? What are women's perceptions of rewards and costs resulting from accepting this non-traditional role?

193 **Hanke**, Lewis
    1970 'The social history of Potosi'. In *La Minería Hispana e Iberoamericana*, Ponencias del 1er Coloquio Internacional Sobre Historia de la Minería, 1. A. Vinayo Gonzalez, ed. León, Spain: Cátedra de San Isidoro. Pp. 451–4651.
Historiographical essay on sources available during the colonial epoch. Suggestions and proposals regarding what should be included in such social history research.

194 **Hannah**, Richard L.
    1981 'A case study of underground coal mining productivity in Utah'. PhD dissertation, The University of Utah. Ann Arbor: University Microfilms.

Analyses cluster of Utah underground coal mines showing sizeable differences in productivity. By analysing case by case and using mixture of information sources, concludes that organisational features of work environment are important factors in accounting for productivity differences. Non-union mines are found generally to be more productive than those organised by unions.

195 1982 'The work ethic of coal miners'. *Personnel Journal* 61 (10): 746–51.
Results of a study showed management did not have to motivate coal miners. However, inadequate communications, poor production planning or an inadequate maintenance programme were found to slow production and thus demotivate coal miners.

196 **Harris**, Olivia and Xavier Albo
1975 *Monteras y Guardatojos; Campesinos y Mineros en el Norte de Potosí*. La Paz, Bolivia: Centro de Investigación y Promoción del Campesinado (CIPCA).
Landmark monograph examining peasant and miner syndicalist leadership, cleavages and collaborations in Bolivian mining and agricultural region. Also contains excellent review of various classifications of mine work and of peasant situations which bear on their propensity to participate in collective action.

197 **Harris**, Richard Allan
1981 'The new regulation and the political behavior of individual firms: a case study of the Surface Mining Control and Reclamation Act of 1977'. PhD dissertation, University of Pennsylvania. Ann Arbor: University Microfilms.
Examines political process by which federal government broadened its role as guarantor of corporate environment and social responsibility, using case histories.

198 **Hayduk**, Jeannine Louise
1981 'A retrospective study of hospital primary diagnoses, diseases and injuries, for hospital admissions among underground coal miners and non-coal miners in a rural southern Illinois hospital'. PhD dissertation, Southern Illinois University of Carbondale. Ann Arbor: University Microfilms.
Analysis of 1,766 hospital admissions covering 1960–77. Shows miners have clearly different profile of health afflictions. The author argues health delivery can be better targeted to miners on basis of analyses of hospital admissions records, while unusual profiles for the miners of particular employers can also be discerned.

199 **Helmer**, Marie
1970 'Mineurs Allemands à Potosí: l'expédition Nordenflicht (1788–1798)'. In *La Minería Hispana e Iberoamericana*, Ponencias del 1er Cologuio Internacional Sobre Historia de la Minería, 1. A. Vinzyo Gonzalez, ed. León, Spain: Cátedra de San Isidoro. Pp. 513–28.
The Nordenflicht expedition to Peru and Mexico commissioned by Charles III to improve the technology and management of New World silver mining. Particular attention paid to interpersonal and political problems which undermined original objectives.

200 **Herrera Canales**, Ines, Cuauhtemoc Velasco Avila and Eduardo Flores Clair
1981 *Etnia y Clase, Los Trabajadores Ingleses de la Compañía Real del Monte y Pachuca, 1824–1906*. Cuaderno de Trabajo 38. Mexico City: Departamento de Investigaciones Históricas, Instituto Nacional de Antropología e Historia.
Short monograph describing continuing influence of Cornish miners and managers at great Mexican silver mine both before and after national capitalists bought out the British in 1849. One of a number of works emanating from landmark historical study of Real del Monte being conducted under the auspices of Mexico's Instituto Nacional de Anthropologia y Historia.

201 **Hevener**, John W.
   1978 *Which Side Are You On? The Harlan County Coal Miners, 1931–39.*
       Urbana: University of Illinois Press.
Historical inquiry of Harlan County labour politics examining basic attitudes of miners, bases of resolute opposition by employers, efforts of UMW to organise, means used by operators to resist, effects of public opinion from without, and the New Deal as spur to unionism.

202 **Hickman**, John M. and Jack Brown
   1971 'Adaptation of Aymara and Quechua to the bicultural social context of Bolivian miners'. *Human Organization* 30(4): 359–66.
Interprets the variance in miner behaviour as an assimilation continuum largely managed by the miners' union. Data primarily from the COMIBOL Colquiri mine.

203 **Hill**, Forrest E., Cecil V. Peake and Anthony Sharkey
   1981 *An Analysis of the Applicability and Probable Cost Effectiveness of Advancing Longwall Mining Systems in the United States.* Washington, DC: Department of Energy, Contract no. DE-AC-01-77ET12478.
Department of Energy sponsored feasibility study. Analyses in considerable detail variety of mining systems and specific technologies believed to be capable of improving the mining art. Report assesses applicability, and contains proposal and specifications for a trial of 'advancing longwall' system.

204 **Hitchcock**, Lyman C.
   1973 *Development of Minimum Luminance Requirements for Underground Coal Mining Tasks.* Final report, Contract HO111969. Washington, DC: Department of the Interior, Bureau of Mines.
Although most technical/engineering studies of mining are excluded from this bibliography, readers may find interesting this technical research report on the illumination requirements for efficient and safe work on underground coal faces.

205 **Hodgson**, Bryan
   1972 'Mountain voices, mountain days'. *National Geographic* 142(1): 118–46.
Personal interviews with inhabitants of decaying West Virginia coal towns where unsettled coal market eventually forced most local mine owners to sell out and move. Describes strength, endurance and self-reliance of 'yesterday's people'.

206 **Hoffman**, John N.
   1978 'Pennsylvania's bituminous coal industry: an industry review'. *Pennsylvania History* 45(4): 351–63.
Brief historical overview of state's coal industry, with attention to geology, technology and production.

207 **Hollander**, Neil and Robert MacLean
   1984 'Gold fever hits the Amazonian jungle'. *Smithsonian* 15(1): 88–86.
Although a piece of light journalism, this appears to be one of the few on-site accounts of recent Amazonian gold strikes. Contains useful information on miner organisations that have insisted on excluding labour-saving machinery in order to preserve a labour-intensive and widely shared participation in the wealth recovered.

208 **Hopkins**, George Williams
   1976 'The miners for democracy: insurgency in the United Mine Workers of America, 1970–1972'. PhD dissertation, The University of North Carolina at Chapel Hill. Ann Arbor: University Microfilms.
A study of the insurgent Miners for Democracy. Examines MFD's struggle for union democracy, a new and responsive national leadership, and enforcement of health and safety standards in the mines.

209 **Hornbein**, Marjorie
   1976 'Josephine Roche: social worker and coal operator'. *Colorado Magazine* 53 (3): 243–60.
Short biography of Josephine Roche's career as social worker, politician, coal operator and union official. From 1928 when she gained control of the Rocky Mountain Fuel Company until her death in 1976, she made important contributions to the advance of women and labour.

210 **Hoy**, Don R.
   1974 'Mining (Latin America)'. In *Encyclopedia of Latin America*. Helen Delpar, ed. New York: McGraw Hill.
Overview of mining in Latin America since pre-Columbian times. Tables include (1) principal minerals mined in Latin America, (2) known reserves of minerals and (3) mineral reserves of unknown amount.

211 **Hu-DeHart**, Evelyn
   1981 *Missionaries, Miners, and Indians: Spanish Contact with the Yaqui Nation of Northwestern New Spain, 1533–1820*. Tucson: University of Arizona Press.
Focuses on developing conflict between jesuits and mine entrepreneurs in Sinaloa and Sonora over Yaqui Indians supplying labour to the mines. Yaqui revolt (1740) and expulsion of the jesuits (1767) are major points in author's chronicle.

212 **Hunt**, Richard D.
   1979 *Law and Order vs. the Mines: West Virginia, 1907–1933*. Hamden, Connecticut: Shoe String Press.
Use of the injunction and physical force to establish 'law and order' in the state's coal fields.

213 **Hunt**, T.A.
   1974 'Tissue reaction to pressure stresses in miners'. *Practitioner* 213: 189–94.
Illustrations demonstrate reactions of superficial tissues to the assault of physical forces experienced by miners in the course of their employment.

214 **Hurtado**, Albert Leon
   1981 'Ranchos, gold mines and rancherias: a socioeconomic history of Indians and whites in Northern California, 1821–1860. PhD dissertation, University of California, Santa Barbara. Ann Arbor: University Microfilms.
Compassionately examines massive decline in Native American population in Northern California as the territory population changed from agriculturalist white minority to a booming mining and urban white majority. State policy toward its indigenous population was geared to interests of white entrepreneurs at the tragic expense of the native population.

215 **Hyde**, Doris
   1981 'The mineral industry of Mexico'. In *Bureau of Mines Yearbook*. Washington, DC: Department of the Interior, Bureau of Mines.
Production, exports and imports of mineral commodities, and a commodity review of metals, non-metals and mineral fuels. Text and six definitive and detailed tables; well-researched and thorough report on the industry.

216 International Labour Organisation
   1976 Programme of industrial activities, Coal Mines Committee, tenth session, November 1974. ILO report Session 194. Geneva.
ILO report intended to serve as basis for discussion of: (a) action resulting from conclusions and resolutions adopted at previous committee sessions; (b) proposed committee studies and inquiries in the coal mining industry; and (c) recent events and developments in coal mines.

217 **Iriarte**, Gregorio
   1972 *Galerías de Muerte, Vida de los Mineros Bolivianos.* Montevideo: Tierra
        Nueva.
Broad, angry description of oppressive life in Bolivian tin mines, containing historical
narrative but emphasising present conditions. Focus of book is summed up in author's
final chapter on the life and death of Federico Escobar, hero of Bolivian miner
resistance.

218 1976 *Los Mineros Bolivianos.* Buenos Aires, Argentina: Tierra Nueva-Colección
        Proceso.
Highly readable, sympathetic description of miners' life in Bolivia with some historical
orientation. Mainly for a general audience rather than a contribution to new scholarship.

219 **Jackson**, Jr, John Alexander
   1978 'The Mexican silver schemes: finance and profiteering in the Napoleonic
        era, 1796–1811'. PhD dissertation, University of North Carolina at Chapel
        Hill. Ann Arbor: University Microfilms.
The collusion of France, Spain, Britain and the United States to enrich their treasuries
and further programmes of war through extraction of Mexican silver. The author argues
that international conflicts would have been shortened without Mexican wealth, and
that reaction to Spain's endorsement of schemes strengthened independence movement in Mexico.

220 **Jensen**, Richard Jay
   1974 'Rebellion in the United Mine Workers: the Mines for Democracy, 1970–
        1972'. PhD dissertation, Indiana University. Ann Arbor: University Micro-
        films.
Rhetorical dimensions of the battle fought by the Miners for Democracy to overthrow
incumbent leadership of UWM. Rhetorical procedures are framework used to describe,
analyse and evaluate MFD's battle against Boyle.

221 **Jiminez**, Rosario and Honorio Pinto
   1979 *Minería en Bolivia, 1826–1848 (Documentos).* Fuentes de Historia Social
        Americana VII. Lima: Biblioteca Andina. Mimeo.
Contains eighteen transcribed documents deriving from period of near collapse of silver
and mercury mining in Bolivia.

222 **Johnson**, James P.
   1970 'Reorganizing the United Mine Workers of America in Pennsylvania during
        the New Deal'. *Pennsylvania History* 37 (2): 117–32.
Discusses revolution in organised labour during the 1930s. How a shift in government
attitude, receptive workers, Lewis's daring leadership and demoralised opposition combined to produce UMW resurgence in commercial mines, but failed in 'captive' mines
owned by steel companies.

223 1979 *The Politics of Soft Coal: The Bituminous Industry from World War I
        through the New Deal.* Urbana: University of Illinois Press.
Historical account of political struggles of various interest groups and government in
the war and post-war years. Includes a bibliographical essay as an appendix.

224 **Johnstone**, Bill
   1980 *Coal Dust in My Blood.* Victoria: The British Columbia Provincial
        Museum.
Autobiography of a coal miner. Beginning as a hand miner in England at thirteen,
he rose to district superintendent of mines at Cumberland, Vancouver Island, BC,
before he retired. Provides insight into life of a miner.

225 **Julian**, Louise Chandler
   1980 'Benefit–cost analysis of the Coal Mine Health and Safety Act of 1969'.
      PhD dissertation. Pennsylvania State University. Ann Arbor: University
      Microfilms.
Purely economic-statistical analysis of economic costs entailed by imposition of new
Health and Safety Act, comparing data in Pennsylvania underground bituminous coal
mines, 1950–75. Changes in fatal and non-fatal accident rates are compared to increased
economic costs of implementing act.

226 **Kalisch**, Philip A.
   1972 'Death down below: coal mine disasters in three Illinois counties'. *Journal
      of the Illinois State Historical Society* 65(1): 5–21.
Study of two dozen coal mine disasters in Franklin, Saline and Williamson counties,
Illinois. Discusses incidence of mine disasters in southern Illinois area, and relation
to highly competitive nature of coal industry in region.

227 **Kanarek**, Harold Kenneth
   1972 'Progressivism in crisis: the United Mine Workers and the anthracite coal
      industry during the 1920s'. PhD dissertation, University of Virginia. Ann
      Arbor: University Microfilms.
Examines problems facing industry during 1920s and explains how lack of leadership
in pushing through progressive reform fragmented the movement.

228 Kennecott Copper Corporation
   1971 *Expropriation of El Teniente, The World's Largest Underground Copper
      Mine*. New York: Kennecott Copper Corporation.
Aimed at its stockholders and employees, Kennecott Copper Corp. presents its response
to changes from Chilean government afer Salvador Allende expropriated El Teniente
from Kennecott's subsidiary, Braden Copper Co., in 1971.

229 1971 *Confiscation of El Teniente, 'Expropriation Without Compensation'*.
      Supplement 1. New York: Kennecott Copper Corporation.
Review of amendment in Chile's Constitutional Reform Law of 1971 which enabled
expropriation of El Teniente without compensation. Supplement includes details of
the three deductions, a resumé of financial data, and some of the editorial comment
occasioned by the Chilean government's action.

230 1972 'The Kennecott White Paper on Chile's expropriation of El Teniente copper
      mine'. *Inter-American Economic Affairs* 25: 25–38.
Statement by Kennecott giving its position on expropriation and excess profit claims
of the Allende government, charging violation of international law.

231 **Kerr**, Lorin E.
   1975 'Socioeconomic aspects of coal workers' pneumoconiosis as seen by the
      United Mine Workers of America'. Paper presented at the Coal Workers'
      Pneumoconiosis Meetings (Legal and Social Aspects) held at the National
      Academy of Sciences. Washington, DC, 7 March.
Representative speaker for the United Mine Workers focused attention on sections
of the Safety Act of 1969, as amended, which pertain to black lung benefits.
Recommends drive to make NIOSH directly responsible to Secretary of Health,
Education and Welfare.

232 **King**, Joseph E.
   1977 *A Mine to Make a Mine: Financing the Colorado Mining Industry, 1859–
      1902*. College Station: Texas A&M University Press.
Process and problems of financing a precious metals' mining industry during the nineteenth century. Concerned with explaining mobilisation, migration and impact of
American capital, principally eastern, on a major state in the mining west.

233 **Kline**, Harvey
    1982 *Exxon and Colombian Coal: An Analysis of the North Cerrejon Debates*.
    Occasional Papers Series, no.14, Program in Latin American Studies.
    University of Massachusetts at Amherst.
Role of multinational corporations in Latin American development, using case study
of Cerrejon coal concession (Exxon) in Colombia. Brief background of the coal industry
in the Cerrejon region; concentrates on debate within Colombian political sphere on
proper contract terms for concession.

234 **Knipe**, Edward E.
    1970 'Explaining unexplained variance: technology reconsidered'. Paper presented
    at the Southern Sociological Association, Atlanta, Georgia, April.
Uses *ad hoc* categories of technology to organise data on attitudes and values of coal
miners. Case studies of Appalachian coal miners indicate that knowledge of specific
tasks performed in the mine is better predictor of working preference than type of mine.

235 1974 'Occupational ingress and egress: a survey of Appalachian coal miners'.
    Paper presented at the Southern Sociological Society, Atlanta, Georgia,
    April.
Empirical study of sociological factors related to occupational choice. Results of survey
of 666 coal miners and ex-miners from Appalachia found that characteristics of mines
are best predictors of who stays or leaves mining as occupation.

236 **Knipe**, Edward E. and Helen M. Lewis
    1971 'The impact of coal mining on the traditional mountain sub-culture'. In
    The Not So Solid South. J.K. Morland, ed. Athens, Georgie: University
    of Georgia Press. Pp.25–37.
Commentary and associated documents stemming from interview study of men and
women at several Tennessee coal mines of varying scale.

237 **Kolbash**, Ronald Lee
    1975 'A study of coal mining communities and associated environmental
    problems'. PhD dissertation, Michigan State University. Ann Arbor:
    University Microfilms.
Considers environmental and social problems related to coal mining communities,
particularly those tied to mining operations. Data collected in a community survey
in Grant Town, West Virginia.

238 **Krebs**, Girard
    1975 'Technological and social impact assessment of resource extraction: the case
    of coal'. *Environment and Behavior* 7(3): 307–29.
Research begins with human activity in specified environments and then pursues
systematic effects of coal mining on human social systems and culture. Directed at
discerning different community impacts of two methods of taking coal.

239 **Kruijt**, Dirk
    1982 'Mining and miners in Latin America: an introduction'. Paper presented
    at the forum 'Politics and labor relations in the mining enclaves of Latin
    America in the late seventies', at the tenth national meeting of the Latin
    American Studies Association, Washington, DC, 4–6 March.
Brief overview of miner–company relations, especially in Latin American mineral-
dependent countries of Chile, Bolivia and Peru, where most mining is done in areas
isolated from multi-stranded national life.

240 **Kruijt**, Dirk and Menno Vellinga
    1977 'La politique économique des enclaves minières au Perou'. *Revue Tiers
    Monde* 18(72): 797–832.

Using Peruvian Cerro de Pasco Corporation as a case study, authors analyse the characteristics of 'enclave' industries, extensions of world economy inserted into Third World economies. Presents theoretical model of enclave features and linkages to national economic, judicial and socio-political institutions.

241 1977 'The political economy of mining enclaves in Peru'. *Boletín de Estudios Latino-Americanos 1, del Caribe.* 23: 97–126.
Examines characteristics of the mining enclave in economic, legal and socio-political spheres, and relationship between state and multinational corporation. Based on study of Cerro de Pasco Corporation in Peru.

242 1979 'Bases sociologiques du militantisme chez les ouvriers et les dirigeants syndicaux: une analyse méthodologique de cas'. *Récherches Sociologiques* 10(1): 125–52.
Methodological case analysis of miners in Peru including characterisation of militant agents among workers. Sociological theories are tested against data from employees of Cerro de Pasco Corp. and used to seek reasons for differences in the militant action of different groups of workers.

243 1980 'Las huelgas en la Cerro de Pasco Corporation (1902–1974): los factores internos'. *Revista Mexicana de Sociología* 42(4): 1497–1588.
Monograph-length descriptive piece on the Corporation and its labour history. Readers will find this a very useful source of general information on this enormous mine in its days prior to nationalisation. Concludes that strikes arose out of work issues and grass-roots mobilisation, rather than being personally instigated by union leaders.

244 **Laite**, Julian
1978 'Industrialisation, migration and social stratification at the periphery: a case study of mining in the Peruvian Andes'. *Sociological Review* 26(4): 859–88.
Problem addressed in this paper is that of social stratification in the industrial sectors of underdeveloped nations. Adopts a neo-Weberian approach to problem of class and, in an analysis of industrialisation and migration in highland Peru, analyses market and status aspects of group formation.

245 1980 'Miners and national politics in Peru, 1900–1974'. *Journal of Latin American Studies* 12(2): 317–40.
Organisation of the labour force, strength of trade unionism, and rural sector support have resulted in a cyclical development of miners' politics in Peru, alternating between compromise and confrontation. Analysis contrasts with more standard models predicting increasingly institutionalised conflict as miners extend proletarianisation.

246 1981 *Industrial Development and Migrant Labour*. Manchester, UK: Manchester University Press.
Study of migrant workers comparing Third World industrialisation with the past experiences of developed economies. One analysis investigates history of multinational mining corporation in Peru, structure and role of miners' and metal workers' unions, and industrial policies of the government.

247 1981 'Expansión capitalista, migración y diferención social entre los campesinos de Peru'. *Revista Mexicana de Sociología* 43(1): 193–219.
Interprets increasing elaboration and differentiation in social structure in two Mantaro Valley towns, especially during the twentieth century as response to augmented capitalist expansion in region and elsewhere. Rapid capitalisation of mining at La Oroya and other centres provided major impetus, but parallel expansion in agriculture and in urban opportunities also played a role.

L

248 **Lapp**, N. LeRoy
   1975 'Physiological changes in coal workers' pneumoconiosis'. Paper presented
       at the Coal Workers' Pneumoconiosis Meetings (Medical Aspects) held at
       the National Academy of Sciences, Washington, DC, 31 January.
Discussion of data gathered for an examination of lung capacity of anthracite and
bituminous miners, and radiographic studies of smokers and non-smokers from 9,076
coal miners.

249 **Leamer**, Laurence
   1971 'Eccles no. 6: a West Virginia coal mine, where the job is a privilege'.
       Harpers 243 (12): 100–10.
A journalist's diary-like description of a day's work in a small West Virginia coal mine,
describing the work environment, machinery and social interaction of miner work crews.

250 **Leeper**, Joseph S.
   1977 'The future of Butte, Montana: a special kind of uncertainty'. *Social Science
       Journal* 14(1): 111–17.
Using a Burgess-like concentric zone model of urban structure, Leeper finds that
pressure exerted on city by the growth of the Berkeley pit dramatically altered city's
land-use mosaic and growth pattern.

251 **Legeay**, Stephen Paul
   1980 'The development of advanced capitalism: a case study of retired coal miners
       in southern West Virginia'. PhD dissertation, University of Notre Dame.
       Ann Arbor: University Microfilms.
Analyses changes in work in coal mines between two periods of capitalist development.
Orientation is marxian-dialectical.

252 **Letcher**, Duane Allan
   1975 'Identification and structural analysis of instructional programs for the
       underground coal miner in West Virginia'. PhD dissertation, West Virginia
       University. Ann Arbor: University Microfilms.
Identifies, classifies and analyses instructional system provided for these miners. Intent
was to provide knowledge base necessary for relevant decision-making on federal, state
and local levels.

253 **Levy**, Jerrold E.
   1980 'Who benefits from energy resource development: the special case of Navajo
       Indians'. *Social Science Journal* 17(1): 1–19.
Assesses the benefits derived from coal mining and power production in the western-
most portion of Navajo reservation to determine if tribal as well as local Navajo
expectations have been realised.

254 **Lewis**, Helen Matthews
   1970 'Occupational role and family roles: a study of coal mining families in the
       southern Appalachians'. PhD dissertation, University of Kentucky. Ann
       Arbor: University Microfilms.
Effects of coal mining occupations on family roles and family structure. Typological
models serve as a basis of comparison with data on family organisation.

255 1970 'Fatalism or the coal industry? Contrasting views of Appalachian problems'.
       Mountain Life and Work. 46(11): 4–15.
Two opposing views with implications of each for the solution to Appalachian
problems. First view is Appalachian subculture model (fatalism), second is colonialism-
exploitation model (the coal industry).

256 **Lewis**, Helen Matthews, Linda Johnson, and Donald Askins, eds.
   1978 *Colonialism in Modern America: The Appalachian Case*. Boone, North
       Carolina: The Appalachian Consortium Press.

The causes of poverty and underdevelopment in Appalachia generally, with mining only one of its aspects. Takes internal colonialism as model, proceeds through variety of case materials contributed by various scholars and lay authors, and finally critiques and modifies colonialism into a dependency model around the concept of 'internal periphery'. Specific attention to mining and miners is fairly superficial.

257 **Ley**, Robert David
   1977 'Bolivian tin and Bolivian development'. PhD dissertation, Washington State University. Ann Arbor: University Microfilms.
General analysis of the Bolivian economy focusing on role played by tin production relative to agriculture and other sectors. Data span 1960–71. Assesses success of COMIBOL in achieving improved management and cost–productivity ratio. Includes attention to International Tin Council and international factors underlying tin pricing.

258 **Lingenfelter**, Richard E.
   1974 *The Hardrock Miners: A History of the Mining Labor Movement in the American West, 1863–1893*. Berkeley: University of California Press.
Historical account of growth of early unionism among hardrock miners from founding of first union at Comstock (1863) to organisation of the Western Federation of Miners (1893). Focuses on personalities and conflicts which animated and shaped labour movement. Presented in narrational style.

259 **Lohman Villena**, Guillermo
   1970 'La minería en el Marco del Virreinato Peruano'. In *La Minería Hispana e Iberoamericana*, Ponencias del 1er Coloquio Internacional Sobre Historia de la Minería, 1. A. Vinayo Gonzalez, ed. León, Spain: Cátedra de San Isidoro. Pp. 639–55.
Rather general review of development of Peruvian mining during colonial period, mentioning a number of entrepreneurs and officials of prominence.

260 **Lora**, Guillermo
   1977 *A History of the Bolivian Labour Movement 1948–1971*. Cambridge: Cambridge University Press.
Abridgment and translation of Lora's five-volume history dealing with strengthening and radicalisation of Bolivia's organised labour movement, culminating in profound revolutionary changes of the 1950s. Covers Bolivian history in the century proceding the revolution, major political events following the revolution, renewed repression, labour resistance.

261 **Love**, Frank
   1974 *Mining Corps and Ghost Towns, A History of Mining in Arizona and California Along the Lower Colorado*. Los Angeles: Westernlore Press.
Journalistic history of mine camps in the Arizona–New Mexico area derived from local records, early newspapers and site visits.

262 **Luchembe**, Chilufya Chipasha
   1980 'Finance capital and mine labor: a comparative study of copper miners in Zambia and Peru, 1870–1980'. PhD dissertation, University of California, Los Angeles. Ann Arbor: University Microfilms.
Argues that economic and labour history of mining in these two countries can be better understood if those histories are seen as the product of common strategies and flows of international capital. Interprets both parallels and contrasts in the two national experiences in terms of analogous processes of capital decision-making.

263 **Lunt**, Richard D.
   1979 *Law and Order vs the Miners; West Virginia, 1907–1933*. Hamden, Connecticut: Archon Books.

Struggle to unionise coal miners by the United Mine Workers of America, written in fast-moving narrational style. Author attributes eventual organising success to economic harships of the Great Depression.

264 **Macdowell**, Laurel Sefton
   1979 *Remember Kirkland Lake: The Effects of the Kirkland Lake Gold Miners' Strike, 1941–1942*. Toronto: University of Toronto Press.
Detailed historical study of the Kirkland Lake (north-eastern Ontario) gold miners' strike which provided a focus for the Canadian labour movement's discontent, contributed to its development of common legislative objectives, and eventually influenced Canada's wartime and post-war labour policy.

265 **Macleod**, Donald Eric
   1982 'Mining men, miners, and mining reform: changing the technology of Nova Scotian gold mines and collieries, 1858–1910'. PhD dissertation, University of Toronto. Ann Arbor: University Microfilms.
Traces the quite different histories by which technological and organisational innovation entered the gold and coal mining industries of Nova Scotia during the indicated years. Local 'mining reformers' played major role in the former, while corporations and union insistence predominated in latter.

266 **Maggard**, Sally
   1981 'From farmers to miners: the decline of agriculture in eastern Kentucky'. In *Science and Agricultural Development*. Lawrence Busch, ed. Totowa, New Jersey: Allanheld, Osmun & Co. Pp. 25–66.
Primarily a description of agricultural change occasioned by growth of timbering and mining in eastern Kentucky after 1880. Though mining is the major factor in these changes, the author is primarily concerned with the consequent changes in agricultural enterprise in the area.

267 **Magill**, Jr, John H.
   1974 *Labour Unions and Political Socialization: A Case Study of Bolivian Workers*. New York: Praeger.
Role of four Bolivian labour unions – miners, factory workers, campesinos and petroleum workers – as agents of political socialisation. Object is to explore historical differences and similarities of these unions as basis for explaining observed and measured differences in attitudes, values and participation.

268 **Malone**, Michael P.
   1981 *The Battle for Butte: Mining and Politics on the Northern Frontier, 1864–1906*. Emile and Kathleen Sick Lecture Book Series in Western History and Biography. Seattle: University of Washington Press.
Extensively researched and candid examination of Butte, Montana mining and personalities who dominated it. Focuses on business of copper mining, entrepreneurship and development of state's peculiarly vicious style of factional politics. Excellent introduction to widely analysed political free-for-all that characterised emergence of copper kings of Montana. Readers who want to pursue the political history further will find comprehensive bibliography useful.

269 **Mamalakis**, Markos
   1971 'Contribution of copper to Chilean economic development, 1920–67: profile of a foreign-owned export sector'. In *Foreign Investment in the Petroleum and Mineral Industries*. Raymond F. Mikesell and Associates, eds. Baltimore: The Johns Hopkins Press. Pp. 387–420.
Analyses and measures the various contributions of Chile's large-scale mining sector (the Gran Mineria) to the economic growth of the country. Brief review of copper

industry through three historical stages, and theoretical framework for analysing effects of copper sector.

270 1976 *The Growth and Structure of the Chilean Economy: From Independence to Allende.* New Haven: Yale University Press.
Concentrates on identifying production, distribution and capital formation and interactions among them, to explain Chile's economic morphology. Divided into two parts: 1840–1930 and 1930–73.

271 1977 'Minerals, multinationals, and foreign investment in Latin America'. *Journal of Latin American Studies* 9 (2): 315–36.
Review article covering six books published 1973–5 on multinationals; several contain material on mining (e.g. Moran, Zink, Seidman). Author critiques each against criteria for analysis of corporate decision factors and national penetration.

272 **Mansilla Torres**, Jorge
1978 *Bolivia, Mujeres Mineras, Victoria Popular: Huelga de Hambre.* Lima, Peru.
Journalist's account, published quickly after end of events, of the nationwide hunger strike begun by twenty-five women and children from Llallagua mine to secure free return of husbands and kinsmen imprisoned or exiled for political and unionist activities. Hunger strike, 28 December 1977 to 17 January 1978, forced political releases and repeal of anti-union edicts by Banzer dictatorship. Contains chronicle of events and report from Bolivian Human Rights Assembly.

273 **Margolis**, Eric
1982 'Videoethnography: toward a reflexive paradigm for community studies'. Paper presented at the American Sociological Association, San Francisco, August.
Paradigm designed for use of non-print media in an interactive process of social research is applied to oral-historical study of life of the western coal miner. Interviews with retired coal miners are recorded on videotape, analysed, edited and then revised in a process of 'public editing'.

274 **Martinez Constanzo**, Pedro S.
1970 'La minería en el ultimo tercio del siglo XVII'. In *La Minería Hispana e Iberoamericana*, Ponencias del 1er Coloquio Internacional Sobre Historia de la Minería, 1. A. Vinayo Gonzalez, ed. León, Spain: Cátedra de San Isidoro. Pp. 399–450.
Survey, region by region, of mining activity recorded in archival materials for latter part of eighteenth century throughout Rio La Plata and Alto Peru (Bolivia). Includes accounts of several inspections and visits by foreign engineers of the period, and includes as an appendix a long report (1792) by the German engineer, Juan Daniel Weber, to the Crown.

275 **Martino**, Orlando, Doris Hyde and Pablo Velasco
1981 *Mineral Industries of Latin America.* Washington, DC: Department of the Interior, Bureau of Mines.
Summary review of mineral industries of thirty-four countries and areas in Latin America. Mineral reserves, production, international mineral trade and role of minerals within each country and in terms of world supply.

276 **Martins**, Robert Borges
1980 'Growing in silence: the slave economy of nineteenth-century Minas Gerais, Brazil'. PhD dissertation, Vanderbilt University. Ann Arbor: University Microfilms.
Refutes that slavery in the Minas region developed to supply the gold and diamond mining enterprises. Martins' data show that Minas slavery can be better understood

as a mechanism by which absentee landlords ensured agrarian labour force in an area of abundant free land.

277 **Mayo**, John
 1979 'Before the nitrate era: British commission houses and the Chilean economy, 1851–1880'. *Journal of Latin American Studies* 11 (2): 283–302.
Covers expansion years in Chilean economy when Montt administration attempted to establish infrastructure of modern economy. No self-generating technology was acquired in its mining sectors; British came as shopkeepers and remained as shopkeepers; Chile both gained and lost from their presence.

278 1982 'Commerce, credit and control in Chilean copper mining before 1880'. Paper presented at the 44th International Congress of Americanists, Manchester, UK.
Examination of organisation of copper mining and marketing of its production in Chile. Essentially limited nature of the foreign role in the industry resulted in failure to create an industry whose self-sustaining technology could overcome the challenge of declining ores.

279 **McAteer**, Davitt
 1972 'You can't buy safety at the company store'. *The Washington Monthly* 4 (9): 7–18.
Comparison of safety precautions in US and abroad shows profit and human life are not always compatible. Difference in treatment of miners in US and European mines is reflected in safety precautions routinely carried out in Europe that could be adopted in US.

280 **McCarthy**, Colman
 1976 'Who's who in Appalachia'. *Atlantic Monthly* 238 (5): 68–75.
Descriptive piece on human impact of encroaching strip mine industry on West Virginia mountain communities.

281 **McMahan**, Ronald Loren
 1978 'Visual sociology: a study of the western coal miner'. PhD dissertation, University of Colorado. Ann Arbor: University Microfilms.
Ethnography of coal miners using camera as a research tool for recording interviews. Deals with method of social inquiry aimed at aiding the sociologist and his audience in comprehension of human involvement in coal production.

282 **McNellie**, William C.
 1982 'Research in landholding, business, and mining in northern Mexico; the Registro de Propiedades of Saltillo, Coahuila'. *The Americas* 39 (2): 246–9.
Short description of the Saltillo archive containing wealth of useful pointers. Breadth of ownership and property characteristics given in the Libro de Minas (1894 to present) will intrigue historians of Saltillo mining history.

283 **Melzer**, Richard
 1976 *Madrid Revisited, Life and Labor in a New Mexican Mining Camp in the Years of the Great Depression*. Santa Fe, New Mexico: The Lightning Tree.
This sixty-three-page book will be of some interest to those seeking descriptive texture of mining town life. Recounts the harsh realities of a company-owned coal town which enjoyed an overlay of gloss in a Christmas pageant during the Depression years.

284 **Mendez Beltrán**, Luz María
 1979 *Instituciones y Problemas de la Minería en Chile 1787–1826*. Santiago: Universidad de Chile, Vicerectoria de Extensión y Comunicaciones.
Excellent, scholarly study of early mining in Chile covering exploration, legal setting, technology and finance as these factors guided development of Chilean mining during period.

285 **Mikesell**, Raymond F.
  1971 'Iron ore in Brazil: the experience of the Hanna Mining Company'. In
       Foreign Investment in the Petroleum and Mineral Industries. Raymond
       Mikesell, ed. Baltimore: Johns Hopkins University Press. Pp. 345–64.
Case analysis of the serious political difficulties encountered by Hanna Mining Co.
of Cleveland, Ohio, during the period 1958–70. Object lesson in corporate insensi-
tivities, political consequences and advisability of joint ownership with lost country
nationals.

286 **Mikesell**, Raymond F. *et al.*
  1971 *Foreign Investment in the Petroleum and Mineral Industries.* Baltimore:
       Johns Hopkins University Press.
Combined efforts of eleven American and Latin American economists to understand
relationship between governments of host countries and foreign investors in the resource
industries. Most case studies from field research in Latin America, including Venezuela,
Argentina, Mexico, Brazil and Chile.

287 1975 *Foreign Investment in Copper Mining: Case Studies of Mines in Peru and
       Papua New Guinea.* Baltimore: Johns Hopkins University Press.
Study constitutes portion of long-range investigation of world copper industry with
emphasis on contribution of developed countries. Case studies of two of the world's
largest open-pit copper mines: Southern Peru Copper Corporation's Toquepala mine
and the Bougainville mine in Papua, New Guinea.

288 **Miller**, Arnold
  1975 Statement of Arnold Miller, President of the United Mine Workers of
       America, Before the Subcommittee on Labor Standards, Committee on
       Education and Labor on 1975 Black Lung Amendments, 26 February
       1975. Supplementary paper provided by Dr Kerr at the Coal Workers'
       Pneumoconiosis Meetings (Legal and Social Aspects) held at the National
       Academy of Sciences, Washington, DC, 7 March.
Testimony on form legislation should take. Summation reminds the committee that
in the disagreement about fine points of black lung: 'Coal mine work produces dust,
exposure to coal dust over time causes pneumoconiosis, pneumoconiosis disables and
kills human beings'.

289 1975 'The wages of neglect: death and disease in the American workplace'.
       American Journal of Public Health 65 (11): 1217–20.
Address by president of the United Mine Workers dealing primarily with shortcomings
of present laws and institutions dealing with occupational disease, with substantial
attention to black lung and related mine illnesses.

290 **Miller**, Frances
  1975 'Department of labor administration: its responsibility for case development
       and treatment'. Paper presented at the Coal Workers' Pneumoconiosis
       Meetings (Legal and Social Aspects) held at the National Academy of
       Sciences, Washington, DC, 7 March.
Describes the method of administration of the Black Lung Benefits Act of the Department of Labor which assumued the administration of the programme from the Social
Security Administration in July 1973.

291 **Mills**, Nicolaus
  1977 'Harlan County, U.S.A.' *Dissent* 24 (3): 307–10.
Review of Barbard Kopple's film of the Brookside mine workers (from July 1973 to
August 1974) in struggle to win contract from Eastern Mine Co. Comment by Irving
Howe, reply by Mills.

292 **Mitchell**, Roger S.
    1975 'Prevalence of coal workers' pneumoconiosis (CWP) and related matters'.
    Paper presented at the Coal Workers' Pneumoconiosis Meetings (Medical
    Aspects) held at the National Academy of Sciences, Washington, DC,
    31 January.
Medical definitions and identifying of pathological lesion of CWP. Recommends that
disability and death from CWP and research in CWP be merged, within NIOSH, with
research on all other occupational inhalants.

293 **Mitre**, Antonio
    1977 'The economic and social structure of silver mining in XIX century Bolivia'.
    PhD dissertation, Columbia University.
Historical study of Bolivian silver mining from 1810 to the first years of twentieth
century, examining both internal and external factors shaping mining industry and
its social and economic position within national economy. Huanchaca Mine is given
particular attention, as is silver boom, 1873–95.

294 1978 'La minería Boliviana de la plata en el siglo XIX. In *Estudios Bolivianos
    en homenaje a Gunnar Menfoza L*. Martha Urioste Aguirre, ed. La Paz:
    manuscript. Pp. 143–68.
Good detailed account of silver mining activity in Bolivia during nineteenth century,
responding to international and national economics, and changes in technology.
Importance of Antofagasta–Uyuni railway is shown in continued production of silver
and mercury in Bolivia despite low world prices.

295 **Mittelman**, Eugene
    1975 'A legislative history'. Paper presented at the Coal Workers' Pneumoconiosis
    Meetings (Legal and Social Aspects) held at the National Academy of
    Sciences, Washington, DC, 7 March.
Legislative history of the 1969 Federal Coal Mine Health and Safety Act which began
with the 1968 explosion in the Farmington, West Virginia mine in which seventy-eight
were killed. Passage of the 1969 law was first time Congress had mandated and
recognised an occupational disorder.

296 **Mohiuddin**, Kazi Golam
    1981 'A cross-country prediction model for political risk assessment by American
    mining corporations operating in developing countries'. PhD dissertation,
    University of California, Los Angeles. Ann Arbor: University Microfilms.
Offers complex statistical model, using data from ninety-five multinational mining
corporations in sixty-eight countries, to test linkage between general political instability
and risk to foreign-held mining properties, and also link between political instability
and lowered investment by these companies. Both correlations found to be high. Author
further investigates likelihood of property loss among three types of mining enterprises.

297 **Molina Martínez**, Miguel
    1981 'El real tribunal de minería de Lima (1785–1821)'. Doctoral thesis,
    Universidad de Granada, Granada, Spain.
Detailed analytic history of Bourbon-initiated agency established to manage mining in
the Peruvian vice-royalty. Author concerned with forces which led to agency's establishment, persons and policies that animated it, and its organisational disintegration as the
independence movement neared. Based on extensive research in Spanish archives.

298 **Monroe**, Douglas Keith
    1977 'A decade of turmoil: John L. Lewis and the anthracite miners 1926–1936'.
    PhD dissertation, Georgetown University. Ann Arbor: University Micro-
    films.

Attempts to place saga of anthracite mine workers of north-east Pennsylvania, their union, the government and John L. Lewis in perspective: conflict surrounding the hard coal fields from 1926 until the collapse of their industry in 1936.

299 **Moran**, Theodore H.
  1971–2 Dependencia y futuro del nacionalismo económico en la industria cuprifera Chilena'. *Estudios Andinos* 2 (3): 29–45.
Written during the Allende period, argues that Chile's copper nationalisation strategy can only succeed if international, oligopolistic, vertically integrated nature of the industry is fully understood. Predicts Chile's risks are high and failure would bring a newer, terrifying form of national dependency.

300 1972 'The alliance for progress and "The foreign copper companies and their local conservative allies" in Chile, 1955–1970'. *Inter-American Economic Affairs* 25 (4): 3–24.
Argues that frequent alliance between foreign companies and local conservatives opposing reformist, nationalist and expropriation moves is not necessary constant. Uses the Frei government nationalisation as an historical case.

301 1973 'Paradigma de la dialectica de las inversiones extranjeras en las concesiones de importantes recursos naturales: el caso del cobre en Chile'. *Estudios Andinos* 3 (2): 5–15.
Moran posits model linking cycles in natural resources profits, fluctuations in foreign investment levels, increases in country participation in the industry, and a growing feasibility of some form of national control, to account for nationalisation policies in Latin America; Chile is central case.

302 1974 *Multinational Corporations and the Politics of Dependence: Copper in Chile.* Princeton: Princeton University Press.
Argues that *dependencia* has had both an economic and a political character, and that the two essences do not always coincide. Reviews Chilean effort to capture some national autonomy over its participation in the copper market.

303 **Morgan**, W. Keith C.
  1975 'Definition of terms'. Paper presented at the Coal Workers' Pneumoconiosis Meetings (Medical Aspects) held at the National Academy of Sciences, Washington, DC, 31 January.
Estimation of size of problem of coal workers' pneumoconiosis, what might be done to control black lung, its cost and areas where both human and dollar cost might be reduced.

304 **Morgan**, W. Keith C., *et al.*
  1973 'The prevalence of coal workers' pneumoconiosis in U.S. coal mines'. *Archives of Environmental Health* 27: 221–6.
Results of National Institute of Occupational Safety and Health field study to determine prevalence of coal miners' pneumoconiosis in working US coal miners. A total of 9,076 miners from twenty-nine bituminous and two anthracite mines were examined between 1967 and 1971.

305 1974 'Ventilatory capacity and lung volumes of U.S. coal miners'. *Archives of Environmental Health* 28 (4): 182–9.
Lung volumes and ventilatory capacities of 9,076 US coal miners were determined. Results showed that reductions of ventilatory capacity were minimal in the absence of complicated pneumoconiosis and would not be associated with respiratory disability.

306 **Mulchansingh**, Vernon C.
  1971 'The bauxite/alumina industry of Jamaica'. *Journal of Tropical Geography* 33 (12): 20–30.

Explains the four essential stages in the bauxite/alumina industry and why, although Jamaica is locus of production of over 25 per cent of the world's bauxite, benefits to country are disproportionately small.

307 **Munn**, Robert F.

1977 *The Coal Industry in America: A Bibliography and Guide to Studies*, 2nd edition. Morgantown, West Virginia: West Virginia University Library.

Voluminous bibliography of publications on coal industry in America. Multidisciplinary approach with principal emphasis on social sciences. Includes titles of real significance and works whose bibliographies will serve as guide to those interested in a specific area or problem.

308 **Munoz Pérez**, José, ed.

1974 *Documentos Existentes en el Archivo General de Indias, Sección de Lima.* VI Congreso Internacional de Minería. *La Minería Hispana e Iberoamericana, Contribución a su Investigacion Histórica*, 8, parts 1 and 2. León, Spain: Cátedra de San Isidoro.

Monumental annotated bibliography of some 7,000 documents found in Spain's Archives of the Indies, incorporated into 1,628 legajos (binders). Documents stem from early conquest to end of colonial period. Order follows the enumeration of the legajos, which itself manifests no order of year or topical category. Thus the annotation, though not indexed, is a critical contribution to further use of the archive's contents.

309 1974 *Documentos Existentes en el Archivo General de Indias, Sección de Guatemala.* VI Congreso Internacional de Minería. *La Minería Hispana e Iberoamericana, Contribución a su Investigación Historica*, 4. León, Spain: Cátedra de San Isidoro.

Annotated listing of Gobierno Section (Section 5) of Saville's Archives of the Indies pertaining to mining in Audiencia of Guatemala (including modern Guatemala, Honduras, Nicaragua etc.). Over 1,100 documents are described; majority date from eighteenth century. Munoz' introduction constitutes valuable brief description of internal organisation of Saville archive, giving the reader some sense of what may be obtained in sections of archive not included.

310 **Murilo de Carvalho**, José

1978 *A Escola de Minas de Ouro Preto, O Peso da Gloria.* São Paulo: Editora Nacional, Financiadora de Estudos e Projetos.

History of a Brazilian school of mines from establishment in 1870s to present. Argues that school suffered a decline in quality beginning in 1930s which has not been reversed. Various remedies proposed.

311 **Nash**, June C.

1970 'Mitos y costumbres en las minas nacionalizadas de Bolivia'. *Estudios Andinos* 1 (3): 69–82.

Argues that MNR's effort to infuse Bolivian miners with revolutionary ideology has thwarted its own success by ignoring and rejecting indigenous beliefs and customs of miners, such as the *ch'alla* and tios. Suggests accommodation to these strong beliefs would yield greater success.

312 1972 'The devil in Bolivia's nationalized tin mines'. *Science and Society* 36 (2): 221–33.

Recombination of themes to prove different postulates about beliefs in society is illustrated with the changing significance of the devil (Tio) in Bolivian tin mines. Discusses change in beliefs and rituals in honor of Tio from time of the tin barons to the military dictatorship of Barrientos.

313 1972 'Devils, witches and sudden death'. *Natural History* 81 (3): 52–8, 82–3. The only readily available source containing photographs of grotesque icons of *tios* (deities of mining underworld) honoured in all Bolivian mines, and constructed as large icons in some. Good description and photos of ch'alla llama sacrifice.

314 1974–6 'Conflicto industrial en los Andes: los mineros Bolivianos del estaño'. *Estudios Andinos* 4 (2): 219–57. Analyses patterns of Bolivian mining labour conflict since the 1930s within fairly straight-forward framework of proletarianisation and class conflict, modified to accommodate specifics of Bolivian miner background and disadvantaged economic and political position.

315 1979 'Ethnology in a revolutionary setting'. In *The Politics of Anthropology.* Gerrit Huizer and Bruce Manheim, eds. The Hague: Mouton Publishers. Pp.353–70. Anthropological study of ideology and social change requires the participant-observer to adopt a perspective more adequate to the demands revolutionary stress puts on their role in the field. Case setting is the tin-mining community of Oruro, Bolivia.

316 1979 *We Eat the Mines and the Mines Eat Us: Dependency and Exploitation in Bolivian Tin Mines.* New York: Columbia University Press. Good example of case study material within a dependency theoretical framework. Based on extensive first-hand ethnography in major mine in Oruro, Bolivia, with some comparative data from other Bolivian mines. Includes useful discussion of dependency issues, extending precision with which assumptions and terms can be applied to actual work settings.

317 1979 'Anthropology and the multinational corporation'. In *Politics of Anthropology.* Gerrit Huizer and Bruce Manheim, eds. The Hague: Mouton Press. Pp.421–46. Although only marginally related to mining, this is a particularly candid source of observational data on how decisions are made in distant corporate headquarters, forming the other end of chain which links miners and other workers and their communities in the production process. Valuable piece of writing and field work.

318 1979 'Ethnology in a revolutionary setting'. In *Politics of Anthropology.* Gerrit Huizer and Bruce Manheim, eds. The Hague: Mouton Press. Pp.353–70. Describes field work process of Nash and her assistant within context of armed conflict and aggressive ideological mobilisation among Bolivian miners and their union leaders. Vivid observations about forces governing anthropologist–informant relationships in settings of turmoil, both improving and reducing data quality.

319 **Nash**, June C., Jorge Dandler, and Nicholas S. Hopkins, eds. 1976 *Popular Participation in Social Change: Cooperatives, Collectives, and Nationalized Industry.* The Hague: Mouton Press. Distributed in the United States by Aldine, Chicago, Illinois. Aimed at understanding the social form of cooperatives. Included are case studies of cooperatives in peasant societies, the relationship between socio-political movements and a communalistic ideology, and the role of co-ops in a commercial setting. Relates to co-op forms of Andean mining.

320 **Nash**, June C. and Manuel María Rocca 1976 'Dos mujeres indígenas (Basilia, by June Nash and Facundina, by Manuel María Rocca)'. *Antropología Social*, 14. Mexico City: Instituto Indigenista Interamericano. 'Basilia' is a self-account of the life of a woman miner who worked both underground and above ground in Bolivia's largest mines, 1930s–60s through labour conflict,

revolution and technology changes. Recounts special struggles faced by a strong woman in world where both allies and enemies presume special domination over women. The second account, unrelated to mining, deals with Chiriguano native.

321 **Nash**, Michael Harold
  1975 'Conflict and accommodation: some aspects of the political behavior of America's coal miners and steel workers, 1890–1920'. PhD dissertation, State University of New York at Binghamton. Ann Arbor: University Microfilms.
Major theme is failure of socialism and why violent class conflicts did not permanently radicalise American coal miners and steel workers.

322 1982 *Conflict and Accommodation: Coal Miners, Steel Workers and Socialism, 1890–1920*. Westport, Connecticut: Greenwood Press.
Case study of workers who theoretically should have embraced socialism, but who accommodated themselves to capitalism following bitter struggles. Interesting insights on labour movement as part of a general culture rather than as a separate problem.

323 **Neff**, James Alan and Robert J. Constantine
  1979 'Community dissatisfaction and perceived residential alternatives: an interactive model of the formulation of migration plans'. *Journal of Population* 2 (1): 18–32.
'Push–pull' model of migration with respect to the migration plans of Appalachian coal miners. Data obtained via mailed questionnaire from sample of 113 West Virginia coal miners.

324 **Nellis**, Lee
  1974 'What does energy development mean for Wyoming?' *Human Organization* 33 (3): 229–38.
Impact of energy resource development on small western United States communities, designed to illumine changes in population, lifestyle and land-use patterns. Study based on experience of a coal town, Hanna, Wyoming.

325 **Nelson**, Joel I. and Robert Grams
  1978 'Union militancy and occupational communities'. *Industrial Relations* 17 (3): 342–6.
Attempts to specify the pre-conditions and confirm hypothesis that for dissatisfied workers, access to occupational communities is associated with structural blame, which is associated with militancy. Minnesota miners are sample.

326 **Newcomb**, Richard
  1978 'The American coal industry'. *Current History* 74 (437): 206–9, 228.
Brief review of factors affecting growth and current status of US coal industry with projections of future to 2050.

327 **Newell**, Dianne C. E.
  1981 'Technological change in a new and developing country: a study of mining technology in Canada West-Ontario, 1841–1891'. PhD dissertation, University of Western Ontario. Ann Arbor: University Microfilms.
Comparative analysis of mining in three ecologically different zones of Canada's mining frontier over fifty-year period, focusing on specific, often personalistic, path by which technological innovations entered mine operations. Identifies contextual variables which appear to govern speed and direction of diffusion.

328 **Newson**, Linda
  1982 'Labor in the colonial mining industry of Honduras'. *The Americas* 39 (2): 185–203.

Gold and silver mining in Honduras relied principally on various forms of coercive labour until end of colonial period, much later than in neighbouring Mexico. Traces cause to low and fluctuating profitability, discouraging payment of wages sufficient to attract a permanent force of free workers.

329 **Noonan, J. Robert**
  1976 **'Analysis of contingencies in the Appalachian coalfield'.** *Community Mental Health Journal* 12 (1): 99–105.
Argues that affluence–deprivation alternate according to demand for coal in national economy. Author analyses region in terms of contingencies, selecting responses in order to regard population as operating according to variable interval schedule.

330 **Northrop**, Stuart A.
  1975 *Turquois and Spanish Mines in New Mexico*. Albuquerque: University of New Mexico Press.
New Mexico mineralogy and mining, tracing back to the prehistoric utilisation of minerals. Includes description of outstanding minerals of New Mexico and where they occur.

331 **Novelo**, Victoria
  1980 'De huelgas, movilizaciones y otras acciones de los mineros del carbon de Coahuila'. *Revista Mexicana de Sociología* 42 (4): 1355–77.
Investigates history of union of coal and zinc miners at the ASARCO coal mines in Nueva Rosita, Coahuila. How national union, controlled by conservatives, refused local's support in the 1950 strike. Government troops moved in, resulting in failure of miners' petitions for redress of grievances, automation of mines.

332 **Nyden**, Linda
  1977 'Black miners in western Pennsylvania, 1925–1931: the National Miners Union and the United Mine Workers of America'. *Science and Society* 41 (1): 69–101.
How National Mine Union helped change UMW and mould the CIO approach to black workers. Examines militant unionism of the IWW and the CIO, and role of Communist Party in coal fields.

333 **Nyden**, Paul J.
  1970 'Coal miners, "their" union and capital'. *Science and Society* 34: 194–223.
Free-ranging piece on American Appalachian coal mining, labour force, federal and state legislation, and internal divisions within miners' union.

334 1972 'The coal miner's struggle in eastern Kentucky'. *Mountain Life and Work*, February: 4–11.
Journalistic report on miner oppression and response in region.

335 1974 *Black Coal Miners in the United States*. Occasional paper 15. New York: American Institute for Marxist Studies.
Describes diversity of races and nationalities among US coal miners, and contribution made by black men throughout history of miners' unions. Explores steps toward racial solidarity during struggles, and progress under new UMW leadership since 1972.

336 1977 'As the coal business booms, turmoil hits the mineworkers'. *Working Papers for a New Society* 4 (4): 46–53, 108.
Discusses period when Miners for Democracy under Arthur Miller took over UMW, beginning long process of housecleaning. Union membership burgeoned, factions reappeared, and miners struck in record numbers. What UMW's ambivalent membership faces in future.

337 1978 'Rank-and-file organizations in the United Mine Workers of America'. The Insurgent Sociologist 8 (2–3): 25–39.

Interprets 1977–8 national mining strike as victory of rank-and-file miners against weak national union leadership seeking ends contrary to miner needs. Offers brief historical perspective and useful quantitative data.

338 1979 'An internal colony: labor conflict and capitalism in Appalachian coal'. *The Insurgent Sociologist* 8 (4): 33–43.
Applies internal colony model within marxist perspective to account for the economic improverishment of Appalachia.

339 **O'Brien**, Jr, Thomas F.
1979 'Chilean elites and foreign investigators: Chilean nitrate policy, 1880–82'. *Journal of Latin American Studies* 11 (1): 101–21.
Causes leading to Chile's decision to turn over to private hands the nitrate industry captured in War of the Pacific. Decision eventually led to entirely foreign ownership of Chile's key industry. O'Brien finds chief motives were to increase tax revenues and remove government as target of indemnity suits.

340 **Oihus**, Collen A.
1978 'Lignite: North Dakota's fledging [sic] coal industry, 1878–1900. *North Dakota Quarterly* 46 (4): 51–67.
Recounts early entrepreneurial efforts to initiate lignite exploitation in North Dakota, subsequently to be displaced by large corporations, including railways.

341 **O'Leary**, John
1975 'An executive perspective of the black lung compensation program'. Paper presented at the Coal Workers' Pneumoconiosis Meetings (Legal and Social Aspects) held at the National Academy of Sciences, Washington, DC, 7 March.
Combination of liberal group of congressional staff people, individuals within the Bureau of Mines interested in the black lung problem, general disgust regarding whole pneumoconiosis issue among congressmen, and presidential resignation resulted in the passage of the 1969 law.

342 **Orlove**, Benjamin
1975 'Relations of production in industrial capitalism: a mine in Central Peru'. Manuscript.
Focuses on one unit of production, a copper mine in southern Peru. Name of mine and location are fictitious, demonstrating how relations of production generate class conflict.

343 **Ortega**, Lisandro
1982 'The first four decades of the Chilean coal mining industry, 1840–1879'. *Latin American Studies* 14 (1): 1–32.
Recounts rapid and sizeable development of Chilean coal industry in competition with British coal brought cheaply to Chile as ship ballast. Chilean producers' success seen to raise doubts about tenets of dependency theory as it might apply to this mineral in these decades.

344 **Ortmeyer**, Carl E., *et al.*
1974 'Mortality of Appalachian coal miners, 1963–1971'. *Archives of Environmental Health* 29 (2): 67–72.
Results of prevalence study of coal workers' pneumoconiosis in which random samples consisting of (*a*) 2,549 employed miners and (*b*) 1,177 ex-miners were examined. Conducted by Public Health Service.

345 **Palomba**, Catherine A. and Ronald J. Althouse
1975 'West Virginia miners view mine safety'. *Labor Law Journal* 26 (3): 139–45.
Describes results of questionnaire survey of 132 miners voluntarily enrolled in university

job advancement extension programme. Questions deal with perceived hazards and safety knowledge.

346 **Palowitch**, Eugene Robert
  1982 'The social efficiency of the coal industry'. PhD dissertation, University of Pittsburgh. Ann Arbor: University Microfilms.
Social costs to health, environmental damage and lost natural resources of the US coal industry are assessed, as well as industry and government-induced paybacks. The latter, particularly to black lung disability and abandoned mine reclamation funds, are found inadequate. Various policies to encourage lowered initial costs by industry are proposed.

347 **Parker**, Morris B.
  1979 *Mules, Mines, and Me in Mexico 1895–1932*. James M. Day, ed. Tucson: The University of Arizona Press.
Memoir dictated by Morris Parker to his daughter in 1947, covering his experiences as mining engineer in Mexico during Diaz period, the war period and early years of the post-war era (only a few pages devoted to latter). Mainly narrative, covering personal adventures, mine personalities and technical problems in several mines which hired Parker to increase profit.

348 **Parlow**, Anita
  1976 'Millionaires and mobile homes'. *Southern Exposure* (winter): 25–30.
Social power in Pikeville, Kentucky continues to consolidate, leaving little room for economic or political alternatives. With inflated price of coal, Pikeville is now technocratic economy and boomtown but employment is declining, welfare increasing.

349 **Paul**, John Edward
  1983 'Clinic utilization and organizational response: results from changes in the United Mine Workers of America Health and Welfare Fund Payment'. PhD dissertation, University of North Carolina, Chapel Hill. Ann Arbor: University Microfilms.
Analyses effect on miners' health clinic visits of introduction of cost-sharing feature in union health coverage. Usage declined markedly, entailing differing rates of service reduction and staff adjustments in clinics.

350 **Pearson**, Jessica
  1980 'Hazard visibility and occupational health problem solving: the case of the uranium industry'. *Journal of Community Health* 6 (2): 136–47.
Although one might question whether study design conclusively establishes cause and effect, plausible case is made for necessity of visible hazards (e.g. deaths from lung cancer) before administrative and legislative remedy occurs. Uranium mines in southwestern US, 1947–1970, provide basis for study.

351 **Pellet**, Gail
  1977 'The Making of *Harlan County, U.S.A.*: an interview with Barbara Kopple'. *Radical America* 11 (2): 33–42.
Interview with Barbara Kopple, filmmaker (*Harlan County, U.S.A.*) who documented strike in Harlan County, Kentucky, after the mine owners refused to sign a contract with newly formed local of the United Mine Workers.

352 **Perez**, Lisandro
  1982 'Iron mining and socio-demographic change in eastern Cuba, 1884–1940'. *Journal of Latin American Studies* 14 (2): 381–405.
Argues that great demographic expansion in Cuba's Oriente was driven not only by rapid capitalisation of sugar industry, as is widely held, but also by substantial investments in iron mining by Cuban and American companies.

353 **Perez de Tudela y Bueso**, Juan
1970 'El problema moral en el trabajo del Indio (siglos XVI and XVII)'. In *La Mineria Hispana e Iberoamericana*, Ponencias del 1ᵉʳ Coloquio Internacional Sobre Historia de la Mineria, 1. A. Vinayo Gonzalez, ed. Leon, Spain: Catedra de San Isidoro. Pp. 355–71.
Philosophical concerns in Spanish administration regarding appropriateness of forced labour, especially in mines, in New World colonies.

354 **Perkins**, Hon. Carl
1975 'Summary of House Bill 3333'. Paper presented at the Coal Workers' Pneumoconiosis Meetings (Legal and Social Aspects) held at the National Academy of Sciences. Washington, DC, 7 March.
Summary of United Workers/Black Lung Association sponsored 1975 Black Lung Amendments.

355 **Petersen G.**, Georg
1970 *Minería y Metalurgia en el Antiguo Peru*. Publicaciones del Instituto de Investigaciones Antropológicas 12. Lima, Peru: Museo Nacional de Antropología y Arqueología.
United States, examining personal characteristics, business strategies, labour practices and other features from their biographical records. The three decades corresponded to the period when great fortunes were made, but before the ascendency of large corporate mine management.

356 **Petrik**, Paula Evans
1982 'The bonanza town: women and family on the Rocky Mountain frontier, Helena, Montana, 1865–1900'. PhD dissertation, State University of New York at Binghamton. Ann Arbor: University Microfilms.
Anchored in quantitative analyses of demographic, economic, inheritance, court and property records, author seeks to explain early prominence and success of women's suffrage in the frontier environment of a mining town. Analysis suggests coping with rigours and affronts of a male-centred frontier led, one and two generations later, to women's substantial autonomy and political involvement.

357 **Phipps**, Stanley Stewart
1983 'From bull pen to bargaining table: the tumultuous struggle of the Coeur d'Alene miners for the right to organize, 1887–1942'. PhD dissertation, University of Idaho. Ann Arbor: University Microfilms.
Historical account of growth of Idaho mine unionism through a period of ideological surges, violent confrontations and concerted repression.

358 Pittsburgh Research Center, Staff
1980 Mine Safety Education and Training Seminar. Proceedings: Bureau of Mines Technology Transfer Seminars, Pittsburgh, Pennsylvania, 9 December 1980, Springfield, Illinois, 12 December 1980, and Reno, Nevada, 16 December 1980. Information circular 8858. Washington DC: US Department of the Interior, Bureau of Mines.
Proceedings discuss findings and products of research on mine safety training. Research emphasis directed toward development of baseline materials and instructional models; methods for structuring and evaluating health and safety, supervisory and occupational training.

359 1981 'Ergonomics – human factors in mining'. Bureau of Mines Technology Transfer Seminars, Pittsburgh, Pennsylvania, 3 December 1980, St Louis, Missouri, 10 December 1980, and Denver, Colorado, 15 December 1980. Information circular 8866. Washington, DC: US Department of the Interior, Bureau of Mines.

Papers on ergonomics in the mining industry. Covers topics relating to ergonomics and mining in man–machine interface, and design of work environment. Directed toward reducing accidents associated with human error.

360 **Poppe**, Rene
    1973 *Koya Loco, Cuentos de Rene Poppe*. La Paz: Ediciones ISLA.
Eight short stories dealing with Bolivian mining life. Story from which collection takes name deals with first encounter of new recruit with the figure of the *Tio*, a deity of Bolivian mines.

361 1975 *La Khola (Relato Minero)*. La Paz: Ediciones ISLA.
Searching, poetic novel involving a violent, brutal conflict between miners and soldiers in the precincts of a Bolivian mine.

362 **Powell**, H. Benjamin
    1974 'Establishing the anthracite boomtown of Mauch Chunk, 1814–1825: selected documents'. *Pennsylvania History* 41 (3): 249–61.
Documents illuminating activities of Jacob Cist and Isaac Chapman in history of one of Pennsylvania's oldest and most famous boom towns.

363 1980 'The Pennsylvania anthracite industry, 1769–1976'. *Pennsylvania History* 46 (1): 3–28.
General review with attention to impact of changes in transportation, markets and technology on mining of anthracite coal.

364 President's Commission on Coal
    1980 *The American Coal Miner, A Report on Community and Living Conditions in the Coal Fields*, Washington, DC: The President's Commission on Coal.
Documentary report concentrating on employed miners and how lives have changed over past thirty years. Data, personal interviews, government reports, historical and current literature and formal surveys depict contemporary living conditions and daily life in coal mining communities from Appalachia to the west.

365 Price Waterhouse & Co.
    1972 *The British Columbia Mining Industry in 1971*. Report by Price, Waterhouse and Co. New York.
Presents comprehensive information on metal, asbestos and coal mining in British Columbia. Covers 1971 exploration, development and production activities of fifty-six companies. Supplements similar reports for 1967–70, providing continuing basic information about the industry.

366 **Prieto**, Carlos
    1973 *Mining in the New World*. Publication of the Spanish Institute, Inc. New York: McGraw-Hill Book Company. (Original: *La Mineria en el Nuevo Mundo*.)
Central thesis is that mining was creator of peoples and nations of Ibero-America as they exist today. Discovery of New World, and rapid expansion and occupation of an entire continent in less than sixty years, was due to the search for an exploitation of precious metals. Emphasis is mainly on conquest and ensuing colonial period in Mexico, Peru and Brazil. Strengths are its discussion of early techniques of refining ores, a fifty-four page, well organised general bibliography and the plates.

367 **Probert**, Alan
    1975 'Silver quest: episodes of mining in New Spain'. *Journal of the West* 14 (2): 5–142.
An entire book is included in this one issue. The author is mining engineer who has been associated with mining companies from Alaska to Guatemala, and from Utah to California's Mother Lode. Interesting, readable history.

M

368 **Przeworski**, Joanne Fox
    1980 'The decline of the copper industry in Chile and the entrance of North
        American capital, 1870–1916'. PhD dissertation, Washington University
        (1978). New York: Arno Press.
Apparently unrevised doctoral dissertation acquired for the 'Multinational corporations: operations and finance' series of Arno Press. Principal concern is interlock
of capital investment, transport and national and international events as they affected
the composition, growth and decline of mining, 1870–1916. Information from various
archives in US and Chile, though resolution of conflicting information is sometimes
weak. Little on the mining communities themselves beyond output figures, investment
and progress made in railway and plant construction, but does illuminate machinations
of capital and market manipulation characteristic of Chilean copper interests.

369 **Purser**, W.F.C.
    1971 *Metal-Mining in Peru: Past and Present*. Special Studies in International
        Economics and Development. New York: Praeger.
Although survey of history of Peruvian mining is included, main emphasis is general
description of contemporary mining in various regions in 1970. Consciously adopts
Peruvian perspective. Particularly valuable regional surveys.

370 **Querejazu Calvo**, Roberto
    1977 *Llallagua: Historia de una Montana*. La Paz: Editorial Los Amigos del
        Libro.
History of first great tin strike of Bolivia, La Salvadora, and fortune it provided to
Simon Patiño. Mine history is well contextualised within Bolivian politics, international
price structures and Patiño's business acumen. Heavy use of Patiño's personal
correspondence. Sticks to the narrative without heavy value judgements.

371 **Ragsdale**, Kenneth Baxter
    1976 *Quicksilver, Terlingua and the Chisos Mining Company*. College Station:
        Texas A & M University.
Social and economic study of industrialist Howard E. Perry, and influence he wielded;
his company and the region, the Terlingua Quicksilver District. Terlingua is now ghost
town in Texas's Big Bend.

372 **Rama**, Carlos M.
    1974 'La Lucha de Chile con las empresas multinacionales in 1970–1973'.
        Cuadernos Americanos 194 (3): 15–33.
Spirited discussion of tactics used by Kennecott, ITT, Dow and Pfizer to complicate
the socialist policies of Allende government.

373 **Ramos**, Demetrio
    1970 'Ordenación de la minería en Hispanoamérica durante la época provincial
        (siglos XVI, XVII y XVIII)'. In *La Minería Hispana e Iberoamericana*,
        Ponencias del 1er Coloquio Internacional Sobre Historia de la Minería, 1.
        A. Vinayo Gonzalez, ed. León, Spain: Cátedra de San Isidoro. Pp. 373–97.
Excellent review of legislation pertaining to mining and mine labour not only at level
of Crown, but also at provincial level within New World colonies.

374 1970 *Minería y Comercio Interprovincial en Hispanoamérica (Siglos XVI, XVII,
        y XVII)*. Estudios y Documentos 31. Valladolid, Spain: Universidad de
        Valladolid, Departamento de Historia Moderna.
Argues Crown philosophies of resource management through three centuries nourished
substantial diversity in mining and commercial establishments in various regions of
the New World colonies. Provides separate analyses of Antilles, Argentine and Pacific
Coast regions. Useful index of names included.

375 **Randall**, Alan, *et al.*
    1978 'Reclaiming coal surface mines in central Appalachia: a case study of the benefits and costs'. *Land Economics* 54 (4): 472–89.
Applies sophisticated economic analysis to 1,600 square mile coal strip mining area of Kentucky (USA) to determine ratio of cost to benefit of 1977 federal law requiring land reclamation for strip mining. Finding: benefits far outweighed costs.

376 **Randall**, Laura
    1979 'Política energética de Mexico'. *Revista Mexicana de Sociología* 41 (4): 1123–66.
Energy policy and the extent of energy deposits. Considers government subsidies, structure of decision making, impact of 1973 embargo, conservation efforts and specific policies for oil, gas, electricity, geothermal energy, coal and nuclear and solar energy.

377 **Randall**, Robert W.
    1972 *Real del Monte, a British Mining Venture in Mexico*. Latin American Monographs, no. 26, Institute of Latin American Studies. Austin: University of Texas Press.
Detailed account of Company of Adventurers in the mines of Real del Monte, from formation in 1824 through its dissolution in 1849. Analyses technology built up which formed foundation for financial bonanza that fell to later owners. Microcosmic view of foreign penetration into Mexican mining industry in first quarter-century after independence.

378 **Represa**, Amando, ed.
    1970 *Archivo General de Simancas; Indice de Documentación sobre Minas (1316–1832)*. VI Congreso Internacional de Minería. *La Minería Hispana e Iberoamericana*, V. León, Spain: Cátedra de San Isidoro.
885-item list of documents and document groupings in Spain's Simancas Archive relating to mining. While nearly all documents deal with mining in Spain, a small portion relate to New World mining. An advantage is its index, assisting user to locate items relating to specific New World dominions, personalities and mines. Very brief annotations.

379 **Rhinehart**, John Paul
    1973 'The UMWA in a Kentucky county, 1931–1941: the unionization of "bloody Harlan".' Masters thesis, University of Texas at Austin.
Account of decade-long struggle in Harlan County, which, accompanied by violence, led to UMW unionisation by 1941; by native of Harlan County.

380 **Rittenberg**, Libby Tamara
    1980 'The coal mining industry: a case study of decline in labor productivity'. PhD dissertation, Rutgers University. Ann Arbor: University Microfilms.
Of all major United States industrial sectors, productivity of bituminous coal mining has declined more steeply than any other. Declines are found to be rooted in rising industry costs from factors other than wages.

381 **Roberts**, Donald Frank
    1974 'Mining and modernization: the Mexican border states during the Porfiriato, 1876–1911'. PhD dissertation, University of Pittsburgh. Ann Arbor: University Microfilms.
Attempts to disprove validity of Dias's strategy by showing that direct foreign mining investment in Mexico's northern border states did not result in economic development and modernisation in areas where mining investments were greatest.

382 **Rodriquez-Rivas**, Daniel Alonso
    1970 'La legislación minera Hispano-colonial y la intrusión de labores'. In *La Minería e Iberoamericana*, Ponencias del 1er Coloquio Internacional Sobre Historia de la Minería, 1. A. Vinayo Gonzalez, ed. León, Spain: Cátedra de San Isidoro. Pp. 657–68.
Provides general comparative survey of mining law in Peru, Mexico and Spain during colonial period. By a lawyer, study indicates major codifications for law code in these regions and offers short, useful bibliography.

383 **Rojas**, Juan and June C. Nash
    1976 *He Agotado mi Vida en la Mina, Una Historia de Vida*. Cuadernos de Investigación Social. Buenos Aires, Argentina: Ediciones Nueva Visión.
Important autobiography by miner in COMIBOL's San Jose tin mine of Oruro, Bolivia. Incapacitated by silicosis at forty-four, Rojas was retired from mine work and composed this autobiography with Nash's assistance. Rojas's view of miner's life and work is not that of union leader or sage, but that of working man, keenly aware of hopes and futility of mining life.

384 **Romero Pittari**, Salvador
    c. 1972 'Notas sobre la estratificación social en Bolivia'. *Revista Paraguaya de Sociología* 13 (35): 7–23.
Commentary on changes in class structure and intra-class social and economic distance that have developed since 1952 revolution.

385 **Rose**, Dan
    1981 *Energy Transition and the Local Community; A Theory of Society Applied to Hazelton, Pennsylvania*. Philadelphia: University of Pennsylvania Press.
Anthropological discourse on region named. Author sets up explanatory model of social forces to show how they developed, and then explains region's economics and the specific, emergent social organisation as it responded.

386 **Rowe**, John
    1974 *The Hard-Rock Men; Cornish Immigrants and the North American Mining Frontier*. New York: Barnes and Noble.
Sizeable study of Cornish contribution to mining in 1800s in Great Lakes, Rockies, California, south-west and British Columbia. Though extensively researched, book is easily read, conveying detailed understanding of spread of mining technology, skills and labour along with Cornish ethnicity.

387 **Samame Boggio**, Mario
    1974 *Peruvian Mining*. Lima: Labor.
Covers three basic aspects of mining activity in Peru: history of how mining was done, present situation, future. For generalist.

388 **Sanders**, Mark S., Terry V. Patterson, and James M. Peay
    1976 *The Effect of Organizational Climate and Policy on Coal Mine Safety*. Final report, Contract HO242039. Washington, DC: Department of the Interior, Bureau of Mines.
Develops homeostatic model for factors affecting incidence of job injury. Data derive from questionnaire study of miners in twenty-two coal mines; posits decentralised decision making, managerial flexibility and innovation and high morale lead to lowered incidence of injury.

389 **Santistevan**, Jorge and Angel Delgado
    1980 *La Huelga en el Peru, Historia y Derecho*. Lima: Centro de Estudios de Derecho y Sociedad.

Miners and mine strikes play only incidental part in this work which analyses political and legal history of Peruvian strikes from early in twentieth century to present. Seeks to show that in capitalist environment strikes more often initiate process of cooptation and consensus building rather than struggle between management and labour. In state-controlled environment, right to strike becomes more problematic and responds to radically different legal context.

390 **Sariego Rodriguez**, Juan Luís
   1980 'Los mineros de la Real del Monte: un Proletariado en formación y transición'. *Revista Mexicana de Sociología* 42(4): 1379–1404.
Important analysis of effect heavy labour turn-over and close peasant–miner cyclical migration with respect to delayed proletarianisation of the labour force, covering 1920 to late 1970s at major silver mine at Pachuca, Mexico.

391 **Schweider**, Dorothy Ann
   1981 'A social and economic study of Iowa's coal mining population, 1895–1925'. PhD dissertation, University of Iowa. Ann Arbor: University Microfilms.
History of Iowa mining sector with attention to role mines played in inducting European immigrants and eventually American blacks into American economic life during this period of strong growth.

392 **Seager**, Charles Allen
   1982 'Proletariat in Wild Rose Country: the Alberta coal miners, 1905–1945'. PhD dissertation, York University. Ann Arbor: University Microfilms.
As Alberta coal industry rapidly expanded during decades before first world war, very strong sense of class consciousness developed among unionised miners. In early 1920s strong sentiments existed to expropriate mine owners and bring about economic revolution. After 1925 fervour receded and more durable relationship between capital and labour developed.

393 **Seltzer**, Curtis Ian
   1974 'The unions – how much can a good man do?' *Washington Monthly* 6(4): 7–24.
Questions whether union can demystify energy bamboozle in time to prevent mine operators from busting union under guise of solving energy crisis.

394 1977 'The United Mine Workers of America and the coal operators: the political economy of coal in Appalachia, 1950–1973'. 2 vols. PhD dissertation, Columbia University, Ann Arbor: University Microfilms.
Analyses intersection of politics and economics in Appalachian bituminous coal industry, 1950–73. Impact of coal development on people and resources of coal fields described and assessed.

395 1977 'Health care by the ton: crisis in the mine workers' health and welfare programs'. Health/PAC (Health Policy Advisory Center) Bulletin no. 79: 25–33.
Establishment, growth, turmoil and decline of UMW welfare and retirement funds, pre-paid industrial health care system which, at peak, ran network of clinics and hospitals for health needs of two million workers. Recounts politics and corruption which plagued system in early years. Substantial attention to problems emerging under guidance of managers in the Miller reformist administration.

396 **Seskin**, Eugene P.
   1975 'Letter commenting on proceedings of March 7, 1975 meeting of National Academy of Sciences on Coal Workers' Pneumoconiosis'. In *Mineral Resources and the Environment*. Washington, DC: National Academy of Sciences.

Expresses concern over Federal Coal Mine and Safety Act of 1969 and possibility that programme will serve as model for workers in other occupations: little economic justification. Economic burden should be shifted from federal government to mine operators.

397 **Short**, John *et al.*
> 1979 *A Study to Determine the Manpower and Training Needs of the Coal Mining Industry.* Final report, Contract JO395038. Washington, DC: Department of the Interior, Bureau of Mines.

National study of training requirements entailed in anticipated growth of coal mine labour force as US moves toward greater reliance on coal-based energy. Findings: most mine labour is trained by mine companies; governmental, private and school training is mainly ineffective. Advocates programme to enhance company-based efforts.

398 **Sigmund**, Paul
> 1981 *Multinationals in Latin America: the Politics of Nationalization.* Madison: University of Wisconsin Press.

Implications of nationalisation of US corporate enterprises in Latin America for future of foreign investment and for relations with the Third World. Nationalisation in five major Latin American countries, identifying conditions underlying demands for nationalisation.

399 **Simon**, Richard M.
> 1980 'Mill village and coal town compared, a review essay'. *Appalachian Journal* 8(1): 67–71.

Reviews two books (including Williams 1976); both offer explanations for transition from pre-capitalist to capitalist economy in North Carolina and less so in West Virginia. Focuses heavily on inadequacy of 'revolution from above' model, citing inability to handle realities of class relations in these regions.

400 1981 'Uneven development and the case of West Virginia: going beyond the colonialism model'. *Appalachian Journal* 8(3): 165–86.

Excellent analysis of state's coal economy, arguing post-1880 entrance of the coal operators reinforced patterns of regional underdevelopment. Useful description of institutionalised exploitation and control of miners, and corporate strategies of capital investment, which led to region's economic underdevelopment.

401 1982 'The development of working class culture: a review essay'. *Appalachian Journal* 9(4): 311–15.

Review of Corbin 1981; makes important contribution to miner history, showing how working-class culture developed out of repression in southern West Virginia. Argues Corbin downplays contributions to militancy of coal mine work process and coal industry economics and understates extent of pro-UMW feeling during attempts at unionisation.

402 1983 'Hard times for organized labor in Appalachia. Review of *Radical Political Economy*'. In press.

How energy crisis intensified labour–management relations in coal fields. Future relations depend on relative weight of post-1978 trend to miner cooperation/apathy and counter forces: need and desire to resist contract take-away, workplace conflict, and the need to organise the unorganised.

403 **Sinclair**, Ward E.
> 1974 'Building a hillbillies' union: the UMW'. *Race Relations Reporter* (March): 14–16.

Perceptive, brief account of Black Lung Movement of 1969 which resulted in replacement of Boyle regime of United Mine Workers with populist Miller leadership, and passage of state and federal legislation and compensation programme for victims.

404 **Singer**, Alan Jay
   1982 '"Which side are you on?"': ideological conflict in the United Mine Workers of America, 1919–1928'. PhD dissertation, Rutgers University, the State University of New Jersey.
Struggles of class-conscious, working-class movement among bituminous coal miners following first world war to win control of UMW from John L. Lewis. After defeat, remaining members founded communist-supported National Miners Union leaving Lewis and business unionists in undisputed control of UMW.

405 **Sisselman**, Robert
   1975 'Rosario Dominicana launches Latin America's largest gold mine'. *Engineering and Mining Journal* 176(10): 71–8.
Dominican Republic's $45 million open-cut complex at Pueblo Viejo; second largest gold producer in western hemisphere and fifth largest in free world. Dominicans expected to benefit most, realising 65 per cent of anticipated $20–30m annual profits.

406 1979 'Peru'. *Engineering and Mining Journal* 180(11): 114–42.
Review of major copper enterprises, including Cuajone, Toquepala, the Ilo smelter/refinery, Cobriza, Morococha, Tomomocho, La Oroya smelter, Cerro Verde and several new deposits. Summarises deposits, installations, investment plans and future reserves for each.

407 **Smith**, Barbara Ellen
   1981 'Black lung: the social production of disease'. *International Journal of Health Services* 11(3): 343–59.
Historical background and medical politics of bitter controversy over control of definition of black lung. Considers black lung as medical construct, occupational disease and as object of mass movement.

408 1981 'Digging our own graves, coal miners and the struggle over black lung disease'. PhD dissertation, Brandeis University. Ann Arbor: University Microfilms.
Black lung rebellion within United Mine Workers Union, 1968–9, which lead to defeat of older union leadership establishment and election of reformist president, widening of federally mandated recognition, and treatment of black lung. Author has flair for dramatic phrasing and interprets movement as part of historic changes in power relations in Kentucky mining.

409 **Sonnichsen**, C. L.
   1974 *Colonel Greene and the Copper Skyrocket*. Tucson: The University of Arizona Press.
Spectacular rise and fall of William Connell Greene, copper king, cattle baron, and promoter extraordinary in Mexico, American south-west and New York financial district. Covers Greene before, during and after 'Copper Skyrocket'.

410 **Spence**, Clark C.
   1970 *Mining Engineers & the American West, The Lace-Boot Brigade, 1849–1910*. Yale Western Americana Series, 22. New Haven and London: Yale University Press.
One of few scholarly studies of development of a single mining profession, the mine engineer. Covers transition from 'on-the-job' training to professional schooling organised through various schools of mines in US. The profession, with particular episodes and individuals used as descriptive illustrations of its development. Extensive bibliographic listing included, emphasising original sources.

411 **Sperry**, J. R.
   1973 'Rebellion within the ranks: Pennsylvania anthracite, John L. Lewis, and the coal strikes of 1943'. *Pennsylvania History* 40(3): 293–312.

Clashing personalities and threatened violence of 1943 coal strikes and fundamental issues raised in walkout. Argues unfortunate myth of Lewis as beloved champion of nation's hard coal miners long distorted Lewis–anthracite relationship.

412 **Stapleton**, Darwin H.
  1978 'The diffusion of anthracite iron technology: the case of Lancaster county'. *Pennsylvania History* 15 (2): 147.
Local history piece describing the spread of anthracite-fired coking in 1840s, displacing prior bituminous coal process.

413 **Steenrod**, Susan A.
  1979 'The Navajo and Peabody Coal Company: a case study of communication in intercultural labor relationships'. Master's thesis, University of Denver.
Field study of frictions and conflict between Navajo miners and Euro-American managers traceable to differences of culture. Participant-observation and interview data regarding inter-ethnic contacts, including bargaining sessions and two wild-cat strikes. Absence of cultural tolerance, empathy were two variables characterising the most hostile contacts.

414 **Stein**, Jane
  1973 'Coal is cheap, hated, abundant, filthy, needed'. *Smithsonian* 3 (11): 18–27.
Concern over new wave of coal promotion; energy planners are turning again to polluting, but ever abundant, coal. Government geologist sees US on brink of generations of coal strip mining.

415 **Stickell**, Arthur Lawrence
  1979 'Migration and mining: labor in southern Chile in the nitrate era, 1880– 1930'. PhD dissertation, Indiana University. Ann Arbor: University Microfilms.
Chilean nitrate mining industry relied on domestic sources of labour to fulfil fluctuating requirements, drawing mainly from Norte Chico area. Documents the ebb and flow of labour supply and adaptations of both labourers and companies to changing work availability.

416 **Stillman**, Don
  1975 'Company stores thrive'. *Mountain Life and Work* 51 (12): 36–44.
Examines control of company stores on lives of coal miners and families. In isolated areas of Virginia, Kentucky and West Virginia, coal companies often recoup in profits at company store a sizeable portion of miners' pay.

417 **Strebig**, Kelly C., and H. William Zeller
  1975 *The Effect of Depth of Cut and Bit Type on the Generation of Respirable Dust*. Bureau of Mines Report of Investigations 8042. Washington, DC: Department of the Interior, Bureau of Mines.
Results of laboratory tests showed drill bit type did not significantly influence dust production, but deep-cutting was more efficient and generated substantially less respirable dust per pound of coal broke than shallow-cutting.

418 **Straw**, Richard Alan
  1980 'This is not a strike, it is simply a revolution: Birmingham miners struggle for power, 1894–1908'. PhD dissertation, University of Missouri-Columbia. Ann Arbor: University Microfilms.
UMW coal miners' strike of 1908, characterised by tumultuous violence and determined intervention by industry, national union and state and federal government to quell it. Strategies on both sides analysed to understand destruction of coal unions. Special attention given to racially mixed union locals and implications for nature of southern racism of the period.

419 **Stucki**, Larry
   1970 'The entropy theory of human behavior: Indian miners in search of the ultrastable state during a prolonged copper strike'. PhD dissertation, University of Colorado. Ann Arbor: University Microfilms.
Maximisation hypothesis using language and concepts partially derived from cybernetics. Model is developed to study dynamic systems through time. Analyses stress on systems present at Ajo, Arizona, due to prolonged copper strike.

420 **Suggs**, Jr, George G.
   1972 *Colorado's War on Militant Unionism: James H. Peabody and the Western Federation of Miners*. Detroit: Wayne State University Press.
Important study of decisive period in brief history of first national miners' union when governor waged vigorous assault on WFM. Covers earlier context of western mining and labour developments, then concentrates on 1903–5 Peabody governorship.

421 **Sweeney**, John Vincent
   1977 'An economic analysis of the nationalization of the Gran Mineria of copper in Chile'. PhD dissertation, the Catholic University of America. Ann Arbor: University Microfilms.
Issues involving nationalisation of basic foreign-owned industries in developing countries through analysis of Chilean copper's Gran Mineria. Distinguishes impact on public from private sectors of economy.

422 **Swift**, Bert, Robert Decker and Mike McKeown
   1975 'Mental health in Appalachia: an emerging problem'. *Appalachia* (October–November): 36–44.
More about regional setting than with miners *per se*. Mental health problems appearing to stem from poverty, unemployment, class distinctions and other sources are described, together with care delivery strategies of Appalachian Regional Commission (regional coordinating agency for public programmes).

423 **Tabershaw**, Irving R.
   1972 'The health of the coal miner, an expendable resource?' *Journal of Occupational Medicine* 12: 453–7.
Keynote address to 1970 National Conference on Medicine and Federal Coal Mine Health and Safety Act of 1969. Mutual interdependence of worker health, ecological health and economic and social legislation in US may well depend upon how this legislation is administered and how it may be modified.

424 **Takamiya**, Makoto
   1978 *Union Organization and Militancy: Conclusions From a Study of the United Mine Workers of America, 1940–1974*. Konigstein: Verlag Anton Hain. (Also, 'The impact of organizational factors on union militancy: a historical study of the United Mine Workers of America'. PhD dissertation, Massachusetts Institute of Technology. Ann Arbor: University Microfilms, 1977.)
Organisational characteristics modify influence of economic conditions on propensity of unions to strike. From analysis of interviews with former union officials and inspection of union documents, elaborates new theory of 'official' union militancy.

425 **Tall**, Edward
   1975 'Black lung revisited'. Paper presented at the Coal Workers' Pneumoconiosis Meetings (Legal and Social Aspects) held at the National Academy of Sciences, Washington, DC, 7 March.
Gives number of applications for black lung benefits under 1969 law, under 1972 amendments and funds paid by state to January 1975. With each federal employee increase in salary, black lung beneficiaries also get proportionate incremental increase:

426 **Taussig**, Michael T.
     1980 *The Devil and Commodity Fetishism in South America*. Chapel Hill:
           University of North Carolina Press.
Relationship between image of the devil and capitalist development explored in sugar
plantations of western Colombia and Bolivian tin mines. Interprets why proletarianised
peasants assign importance to devil while peasants working land in accord with pre-
capitalist principles of production do not.

427 **Temple**, John
     1972 *Mining, An International History*. New York: Praeger.
Although areas outside western hemisphere are covered, bulk of text and illustrations
deal with examples from Americas. Treatment is general and simple, with emphasis
on history of mining technology.

428 **Thomas**, Jerry Bruce
     1971 'Coal country: the rise of the southern smokeless coal industry and its effect
           on area development, 1872–1910'. PhD dissertation, University of North
           Carolina at Chapel Hill. Ann Arbor: University Microfilms.
Reviews development of coal mining industry in southern West Virginia, arguing that
singularity of coal industry and infrastructural supports lead to serious underdevelopment in other economic and social sectors of the region, which continues today.

429 **Thurman**, James B.
     1978 'The foreman's role in autonomous work groups: a description and
           exploratory study in an underground coal mine'. PhD dissertation, The
           Pennsylvania State University. Ann Arbor: University Microfilms.
Study conducted to determine role and functions of first-level supervisor of autonomous work group in Pennsylvania underground coal mine. Combines open-ended
observation and systematic recording structure. Recommendations based on findings.

430 1982 'The supervision of autonomous work groups'. Paper presented at the 44th
           International Congress of Americanists, Manchester, UK, 5–10 September.
The results of study of role and functions of first-level supervisors of autonomous work
groups in American coal mine. Intent was to provide empirical evidence to support
or refute current theory.

431 **Tomb**, Thomas
     1975 'Review of respirable dust data from underground coal mines'. Paper
           presented at the Coal Workers' Pneumoconiosis Meetings (Medical Aspects)
           held at the National Academy of Sciences, Washington, DC, 31 January.
Effects of Federal Coal Mine Health and Safety Act of 1969 on underground environmental conditions; coal mine operator data reliability; dust exposure by occupation;
and current status of industry compliance with present dust standard.

432 **Tudela de la Orden**, José
     1970 'La minería y la metalurgía de la América Española en los manuscritos de
           las bibliotecas de España'. In *La Minería Hispana e Iberoamericana*,
           Ponencias del 1ᵉʳ Coloquio Internacional Sobre Historia de la Minería, 1.
           A. Vinayo Gonzalez, ed. León, Spain: Cátedra de San Isidoro.
           Pp. 679–89.
Lists significant manuscripts or documents contained in various lesser-known libraries
of Spain pertaining to mining, dating mostly from colonial period.

433 **Twinam**, Ann
     1976 'Miners, merchants, and farmers: the roots of entrepreneurship in Antioquia,
           1763–1810'. PhD dissertation, Yale University. Ann Arbor: University
           Microfilms.

Entrepreneurship in colonial Antioquia, Colombia, identifying those elements from seventeenth and eighteenth centuries which contributed to conspicuous economic success of later years. Archival research focuses on mining, commerce and agriculture.

434 1982 *Miners, Merchants, and Farmers in Colonial Colombia.* Austin: University
       of Texas Press.
History of Antioquia, Colombia, provides one case in which, because of certain human and natural resources, a colonial heritage proved a positive source. How gold was mined, how much and who mined it are fundamental to understanding eighteenth-century Antioquia.

435 **Ury**, William Langer
       1982 'Talk out or walk out: the role and control of conflict in Kentucky coal mine'.
       PhD dissertation, Harvard University. Ann Arbor: University Microfilms.
Investigates wild-cat strikes which, Ury concludes, served to reassert miners' control of features of their jobs and, through costliness of these strikes to both sides, eventually reinforced bargaining process for dispute mediation.

436 **Van Winkle**, Mary Louise
       1982 'Education and ethnicity in the 1930s in a Minnesota mining community'.
       PhD dissertation, Harvard University. Ann Arbor: University Microfilms.
Variable use by immigrant ethnic groups of public education for their offsprings' social mobility. Use by all ethnic groups was high, leading to greater level of achievement for schools of this community than national average.

437 **Vaught**, Charles
       1983 'A framework for analyzing militancy in two isolated masses'. Paper pre-
       sented at the annual meeting of the North Central Sociological Association.
Framework for analysing differential patterns of militancy among textile workers and underground coal miners. Kerr's and Siegel's isolated mass hypothesis defines several indicators of internal status of an industry.

438 **Vecsey**, George
       1974 'Day in the life of a coal miner'. *Business and Society Review/Innovation*
       10: 65–70.
'Life in the pits is backbreaking and tedious. And if a collapsing wall doesn't get you, the black lung will.' Article is based on author's book, *One Sunset a Week*, autobiographical story of unusual miner.

439 1974 *One Sunset a Week: The Story of a Coal Miner.* New York: Saturday Review
       Press.
Sympathetic life history of West Virginia miner, Dan Sizemoɪe, intermingled with moment-by-moment account of working days of typical week.

440 **Veliz**, Claudio
       1975 'Engaña, Lambert, and the Chilean mining associations of 1825'. *Hispanic*
       American Historical Review 55(4): 637–63.
Describes establishment of four mining companies to work Chilean copper mines, financed by London stock sales just prior to sudden slump on London exchange. All four companies quickly collapsed, but still managed to introduce lasting innovations in Chile's smelting, laying groundwork for Charles Lambert's fortune, and reinforced Chile's conservative policies through influence on Mariano Engana.

441 **Vietor**, Richard Henry Kingsbury
       1975 'Environmental politics of the coal industry'. PhD dissertation, University
       of Pittsburgh. Ann Arbor: University Microfilms.
Formation and implementation of regulatory policy applicable to coal-related environmental problems of surface mining and stationary source air pollution.

442 1978 'Environmental politics in Pennsylvania: the regulation of surface mining, 1961–1973'. *Pennsylvania History* 45(1): 19–46.
Account of conservationist industry and regulatory forces that shaped implementation of first major state strip-mining legislation. Undercut conservationist momentum while yielding some level of environmental restoration.

443 **Viezzer**, Moema, compiler; Domitila Barrios de Chungara
1978 *'Si Me Permiten Hablar ...'; Testimonio de Domitila, una Mujer de las Minas de Bolivia.* Mexico City: Editores Siglo XXI.
Highly regarded and eloquent personal narrative by a woman of the mines, wife to a miner, indigenous organiser and target of various political repressions.

444 **Villalobos R.**, Sergio
1979 *La Economía de un Desierto; Tarapaca durante la Colonia.* Santiago, Chile: Ediciones Nueva Universidad, Pontificia Universidad Católica de Chile.
Unusual and careful ecological interpretation of mining and agricultural economy of the northern province of Tarapaca: labour flows, foodstuff support, mineral extraction technology and colonial social structure.

445 **Volk**, Steve
1974 'Tin and imperialism'. *NACLA's Latin American and Empire Report* 8(2): 12–18.
Examines Bolivian miners' struggle against imperialism: tin barons, mechanisms of imperialist control, US and refining of Bolivian tin, investment codes and reintroduction of direct foreign investment in mining, and miners' response to the military dictatorship.

446 **Volk**, Steven Saul
1983 'Merchants, miners, moneylenders: the *habilitacion* system in the Norte Chico, Chile: 1780–1850'. PhD dissertation, Columbia University. University Microfilms.
Examines mineral production with two objectives: (1) development of commodity and credit markets prior to formal structuring by state; (2) information and perspective on development and underdevelopment.

447 **Watkins**, T.H.
1976 'Idle and industrious miner – tale of California life'. *American West* 13(3): 39–45.
California life during the gold rush. Booklet of verse first published in Sacramento in 1854 and reproduced in this article. Woodcut illustrations by Charles Christian Nahl; introduction by T.H. Watkins.

448 1976 'The centennial homestake'. *The American West* 13(6): 10–17.
Visit to Homestake mine in Black Hills of South Dakota; west's largest and oldest gold mine. Only major gold mine still operating in the west, and deeper than any other mine in US, shafts reaching 8,000 feet.

449 **Whalen**, Eileen and Ken Lawrence
1975 'American workers and liberation struggles in Southern Africa: the boycott of coal and chrome'. *Radical America* 9(3): 1–16.
Discusses broad multinational movement in the south. Strength has been ability to link the struggle for liberation in South Africa with struggle against racism and oppression in the US. Boycott activity is combined with black coal miners' struggle for equality.

450 **Wheeler**, Hoyt N.
1976 'Mountaineer mine wars: an analysis of the West Virginia mine wars of 1912–1913 and 1920–1921'. *Business History Review* 50(1): 69–91.

Analysis of the causes precipitating armed violence between miners and coal operators in Cabin Creek/Paint Creek and Mingo/Logan counties. Useful overview of these two historic confrontations in American labour history.

451 **Whitehead**, Lawrence
  1980 'Sobre el radicalismo de los trabajadores mineros de Bolivia'. *Revista Mexicana de Sociología* 42(4): 1465–96.
Bolivian miners in their national role as worker vanguard. Notorious for their labour and political militancy, they have suffered under military dictatorship, particularly from 1971–8, but maintain their sectional strength through solidarity of union locals.

452 1981 'Miners as voters; the electoral process in Bolivia's mining camps'. *Journal of Latin American Studies* 14(1): 1–32.
Careful examination of voting power of miners from beginning of the twentieth century to revolution of 1952. Particular attention to elections of 1923, 1940, 1944 and 1951. Bolivia's miners come to vote increasingly as bloc and exercise political power on sympathetic candidates and parties.

453 **Wice**, Marsha Nye
  1973 'Revolution in the mines: an analysis of the miners' revolt of 1969–1970'. PhD dissertation, University of Illinois. Ann Arbor: University Microfilms.
Union control of its membership and cause/effects of attempted overthrow of that control. Two major movements controlled by the membership to change union policies are considered: black lung and the Yablonski campaign in West Virginia.

454 **Widerkehr**, Doris E.
  1975 'Bolivia's nationalized mines: a comparison of a cooperative and a state-managed community'. PhD dissertation, New York University. Ann Arbor: University Microfilms.
Two nationalised mines in Bolivia, one organised as corporative, other as state-managed enterprise. Focus is on ways they relate to rank-and-file, to each other and to national power structure.

455 **Wilkes**, Elizabeth
  1981 'Accumulation and crisis in the Peruvian mining industry, 1900–1977'. PhD dissertation, Columbia University. Ann Arbor: University Microfilms.
Analyses Peruvian mining as a test of dependency theory. Data mainly from Cerro de Pasco and Marcona. Internal factors, such as wage cost relative to changing mining productivity and technology, account for most of variance, while external factors cited in dependency theory had only small effects.

456 **Windham**, Joey Samuel
  1981 'Grand Encampment mining district: a case study of the life cycle of a typical western frontier mining district'. EdD dissertation, Ball State University. Ann Arbor: University Microfilms.
Posits six-stage cycle for typical mining towns (discovery, boom, transition, mature, decline, ghost town) and illustrates it through history of Rocky Mountain copper district.

457 **Wionczek**, Miguel
  1971 'A foreign-owned export-oriented enclave in a rapidly industrializing economy: sulphur mining in Mexico'. In *Foreign Investment in the Petroleum and Mineral Industries*. Raymond F. Mikesell and Associates, eds. Baltimore: The Johns Hopkins Press. Pp. 264–311.
Historic prominence of precious metals mining in Mexico diverts attention from its recently achieved status as one of world's great producers of sulphur. Detailed political history. Goes well beyond sulphur itself to analyse in detail the evolution of recent Mexican mining policy to assert control of a major enclave multinational. Recency of sulphur's rise (since 1954) makes case unusual.

458 **Wolfe**, Lee Morrill
   1976 'Radical third-party voting among coal miners, 1896–1940'. PhD disser-
      tation, The University of Michigan. Ann Arbor: University Microfilms.
Why organised coal miners in Great Britain were able to support an independent labour
party; and why comparable elements in US labour did not.
459 1983 'Socialist voting among the coal miners, 1900–1940'. *Sociological Forum*
      16(1): 37–47.
Identifies economic factors that can explain fluctuation of socialist voting among coal
miners. Of several models tested, most plausible relates level of wage to level of vote.
Suggests that support for mass socialist parties moves in tandem with economic cycles.

460 **Wyman**, Mark
   1979 *Hard Rock Epic, Western Miners and the Industrial Revolution, 1860–1910.*
      Berkeley: University of California Press.
Well-written and well-researched book on metal mining in western US, especially gold
and silver. Accounts for growth of unions and collective action by miners on basis
of hardship, danger and impact of newer technology leading to diminishment of
significance of individual miners and skills. Photographs particularly good; extensive
footnotes and bibliography.

461 **Yarrow**, Michael Norton
   1982 'How good strong union men line it out: explorations of the structure and
      dynamics of coal miners' class consciousness. PhD dissertation, Rutgers
      University. Ann Arbor: University Microfilms.
From extensive interviews with West Virginia active, retired and incapacitated miners,
applies various extant models of class consciousness, not only as expressed in retrospective interviews, but also during extended strike. Existing models fit data poorly;
suggests aspects of an improved formulation.

462 **Young**, Jr, Otis E., with Robert Lenon
   1976 *Black Powder and Hand Steel; Miners and Machines on the Old Western
      Frontier.* Norman: University of Oklahoma Press.
Main strength of this generally informal description of mining in western America
during the last half of nineteenth century is discussion of techniques and equipment
used underground. The drawings and discussion of the Cornish pump, for instance,
are usefully detailed.

463 1970 *Western Mining; an Informal Account of Precious-Metals Prospecting,
      Placering, Lode Mining, and Milling on the American Frontier from Spanish
      Times to 1893.* Norman: University of Oklahoma Press.
Thrust of this book, which author defines as neither history nor engineering manual,
is to present account and explanation of surface and underground mining methods
in use from Spanish times to the end of the Sherman Silver Purchase Act in 1893 in
the western American states. Extensive glossary is handy for deciphering the many
terms that ornament the mining literature.

464 **Zapata**, Francisco S.
   1975 *Los Mineros de Chuquicamata: Productores o Proletarios?* Cuadernos del
      Centro de Estudios Sociológicos 13. Mexico City: El Colegio de Mexico.
Detailed analysis of labour politics in Chile's nationalised copper mine during the
Allende period (1970–3). The struggle to define orderly political relations between
miners and state representatives is recounted, and then subjected to interpretations
which may be useful in other instances of nationalisation and efforts at political
maturation.

465 1975 'Action syndicale et comportement politique des mineurs Chiliens de
      Chuquicamata'. *Sociologie du Travail* 17(3): 225–42.

Argues copper miners of Chile's Chuquicamata, isolated in country's north, identified not so much with class and national interests during the Allende period, as with those supporting local, pragmatic interests.

466 1977 'Enclaves y sistemas de relaciones industriales en América Latina'. *Revista Mexicana de Sociología* 39: 719–31.
Applies concept of enclave economies to production settings where a society is arranged around a principal enterprise. Comments on differences among types of ownership (including state ownership), and among production types (mine, plantation, factory), and analyses process of relationship between labour, management, conflict and power associated with enclaves.

467 1980 'Mineros y militares en le coyuntura actual de Bolivia, Chile, Peru (1976–1978)'. *Revista Mexicana de Sociología* 42(4): 1443–64.
Considers three labour actions: the Bolivian miners' strike; Peruvian mining conflict; and labour problems at Chuquicamata, Chile, to illustrate how military regimes have had to deal differently with miners in their efforts to undermine syndical organisations because of disproportionate power of the miners.

468 **Zavaleta Mercado**, René
1978 'El proletariado minero en Bolivia'. *Revista Mexicana de Sociología* (April–June): 517–59.
Complex treatise on development of Bolivian proletariat, especially among miners, focusing on 1952 revolution to Torres coup in 1970. Argues that development of proletarian classes can become very advanced in a very backward national setting, as in Bolivia in recent decades.

469 c. 1982 'La acumulación de clase del proletariado minero en Bolivia'. Manuscript.
Vigorously disputes Bonilla's characterisation of Andean miners as 'incipient proletarians'. Analyses Bolivian miners, developing interpretation of their actions within class conflict framework.

470 **Zulawski**, Ann
1982 'Labor and migration in an Andean mining center'. Paper presented at the 44th International Congress of Americanists, Manchester, UK, 5–10 September.
Analysis of 1683 census of industrial mining centre of Oruro, Bolivia, concentrating particularly on measures of labour origin and labour stability. Concludes mine labour in Oruro came from nearby and exhibited greater stability when compared to Potosi, suggesting that the lack of labour drafts at Oruro stimulated more costly, but more stable labour supply.

Addendum

471 **Espinosa Uriarte**, Humberto and Jorge Osorio Torres
1971 'Dependencia y poder económico, caso minero y pesquería'. In *Dependencia Económica y Tecnológica, Caso Peruano*. H. Uriarte *et al.*, eds. Lima: Universidad Nacional Federico Villarreal, Centro de Investigaciones Económicas y Sociales. Pp. 69–229.
A detailed study of the interlinkage of capital and power in two major Peruvian industries, minerals and fishmeal. Conclusion: these two industries are strongly dominated by intricate networks of power and capital, to the detriment of the Peruvian populace. Provides a large number of data charts for each industry.

472 1972 'La concentración del poder económico en el sector minero'. *Cuadernos Polémicos de la Izquierda* 1(2): 9–37.

Remarkable study of the interlocking directorates and patterns of cross-ownership with the Peruvian mining sector, with extensive charts naming corporations and individuals. Cerro Corporation is shown to be the center of the largest block, with links to various levels of Peruvian mining activity. Includes a sizeable portion of the text of 1971 citation above, but expended with a broadened theoretical context.

473 **Fuentes Royo**, Julio
   1981 *Reservas de Minerales en Bolivia* (2nd edition). La Paz: 'SIC' Ltda.
Author is one of the principal geologists, mine managers and academic authorities in Bolivia. Though mostly a handbook of mineralogy, it would be useful to readers untrained in mineralogy in order to portray the geological and geochemical resources of Andean mine enterprises. Contains listing of Bolivian mines indicating location and minerals produced. Maps.

474 **Fuller**, Carol Andrews
   1973 'An analysis of fluctuations in strike activity in the bituminous coal industry'. MA thesis, West Virginia University. Morgantown, West Virginia.
Statistical analysis of wildcat strikes between 1952 and 1970 in US bituminous coal, as reflected in Bureau of Labor Statistics figures. Correlations found with degree of unemployment in alternative work, and the level of mining accidents.

475 **García-Barceña**, Joaquín
   1975 'Las minas de obsidiana de la Sierra de las Navajas, Hgo. Mexico', in *Actas del XLI Congreso Internacional de Americanistas*, Vol. 1. José Carlos Chiaramonte, compiler. Mexico City: Instituto Nacional de Antropología y Historia.
Describes three late pre-Columbian obsidian sources. Extraction was by means of pits 1 to 1½ meters deep and several wide.

476 **Greaves**, Thomas C.
   1984 'When the price of livelihood is life'. Paper presented at the workshop, Medieval and Contemporary Aspects of Risk and Pollution. New York Academy of Sciences, New York City.
Questions applicability of model advanced by W. K. Viscusi of worker health risk to miners. Using Bolivian case, finds that demographic, political and economic constraints appear to foreclose worker options to decline health risks or to receive commensurately greater wages.

477 **Helm**, Michael, ed.
   1982 *City Country Miners*. Berkeley: City Miner Books.
Anthology of urban and rural stories, poems, articles, letters and graphic work organized around the four categories of relationships, places, work and politics in northern California. Presents a diversity of indigenous voices.

478 **Holingsworth**, Sandra
   1978 *The Atlantic: Copper and Community South of Portage Lake*. John H. Forster Press.
Account of Atlantic Mine, a small mining community in northern Michigan. Depicts the opening, operations and closing of one of the successful mining ventures in the Copper Country. Based on materials collected by Wilfred Erickson, local historian.

479 **Judkins**, Bennett M.
   1979 'The black lung movement: social movements and social structure'. In *Research in Social Movements, Conflicts and Change*. Louis Kriesberg, ed. Connecticut: JAI Press, Inc. Pp. 105–129.
Findings from a study of the Black Lung movement in Appalachia. Includes a brief historical picture of the Social Movement Organization.

480 **Kapsoli**, E. Wilfredo
   1975 *Los Movimientos Campesinos de Cerro de Pasco, 1800–1963*. Huancayo, Perú: Instituto de Estudios Andinos. Also published as *Anales Científicos* No. 4 of La Universidad Nacional del Centro del Perú.
Excellent and ambitious study of the conditions of life and work for hacienda workers in the Cerro de Pasco region, showing strong linkages between the needs and policies of the mine company and the severity of aggression against hacienda workers and adjacent free communities. Organized around various hypotheses, tested against historical and case data.

481 **Levy**, Elizabeth and Tad Richards
   1977 *Struggle and Lose, Struggle and Win: The United Mine Workers*. New York: Four Winds Press.
Describes for general readers the rise of organized labor in the American coal industry; the leadership and internal politics which sometimes improved, sometimes exploited, union members.

482 **Marinovic Olivos**, Hector
   1978 'The small mining industry in Chile'. In *Small Scale Mining of the World*. Conference Proceedings, Jurica, Queretaro, Mexico. New York: United Nations Institute for Training and Research. Pp. 529–542.
A critique of ENAMI, the Chilean agency charged with promoting small scale mining (mainly copper, gold and silver) and purchasing its output. Concludes ENAMI's success has been limited; the plan to reallocate its promotional mission holds promise of better results.

483 **Molina**, Daniel
   1978 *La Caravana del Hambre*. Colección Fragua Mexicana. Mexico City: Ediciones El Caballito.
Describes a bitter strike by mine workers against two Coahuila coal mines, Nueva Rosita and Cloete (American Smelting & Refining Co.) in 1951. Strikers eventually marched in Mexico City where they faced various confrontations and harrassments from the national government, and were eventually forced to return to work.

484 **Muñoz Rioseco**, Liliana
   1971 *Estudio Ocupacional de la Minería del Cobre*. 2 vols. Santiago de Chile: Servicio Nacional del Empleo, Departamento de Estudio y Programación.
Provides a data base on occupational structures in Chilean copper mining to support national labor planning following industry nationalization. Author visited 15 mines ranging from the Chile's largest to mines of less than a dozen workers. Job structure is described and characteristics of principal jobs are listed. Maps, tables, plates.

485 **Nash**, June C.
   1974 'Paralelos revolucionarios en una historia de vida'. In *Las Historias de Vida en Ciencias Sociales; Teoría y Técnica*. Jorge Balán, ed. *Cuadernos de Investigación Social*. Buenos Aires, Argentina: Ediciones Nueva Vision. Pp. 193–213.
Offers detailed account of the life of a Bolivian tin miner, Juan Rojas, and analyzes the cultural and political setting which makes the life events and human responses of Rojas understandable.

486 1975 'Resistance as protest: women in the struggle of Bolivian tin-mining communities. In *Women Cross-Culturally; Change and Challenge*. Ruby Rohrlich-Leavitt, ed. The Hague: Mouton Publishers. Pp. 261–271.
Seeks to define resistance as a political act typically associated with women, used when normal channels of political protest are thwarted. The pattern is illustrated through a recounting of various 20th century episodes of organized resistance by women, principally in the Siglo XX tin mine. Quotations from eye witnesses are often dramatic.

487 **Pelaez** R. Segundino and Marina Vargas S.
   1980 *Estaño, Sangre y Sudor (Tragedia del Minero Locatario)*. Oruro, Bolivia: Editorial Universitaria, Universidad Técnica de Oruro.
Probably the best single study of the Bolivian locatarios, piece work miners working in small teams in nationalized mines after salaried workers have been withdrawn as ore grade falls. Centered at the Cataví mine, book was produced with the cooperation and endorsement of the locatario union, and contains substantial statistical data, plates and a brief glossary.

488 **Poppé**, René
   1979 *El Paraje del Tio y Otros Relatos Mineros*. La Paz, Bolivia: Ediciones 'Piedra Libre', Imprenta y Libreria 'Renovación', Ltda.
Fourteen short stories dealing with human striving within the context of the miner's underground world of belief and reality. A sensitive interpreter of this world, Poppé's short stories are often powerful.

489 **Probert**, Alan
   1971 'A lost art rediscovered: primitive smelting in Central America'. *Journal of the West*, 10(1): 116–128.
Exceptional description of simple lead smelting furnaces called 'chimbos' in Guatemala. Construction and operation apparently resembled counterparts in the early colonial period.

490 1972 'Three Landesio landscapes found in an English attic'. *Journal of the West* 11(3): 437–459.
Describes three paintings by Eugenio Landesio completed for Real del Monte manager John Buchan depicting scenes of the Real del Monte mine and related areas. Describes life of Buchan and the mine of the time of his employment. Plates of paintings included.

491 **Quiroga Santa Cruz**, Marcelo
   1973 *El Saqueo de Bolivia*. Buenos Aires, Argentina: Colección Política, Ediciones de Crisis, Gráfica Devoto.
A political tract written by founder of the Bolivian Socialist party, also minister of mines and petroleum during nationalization of Gulf Oil. Demonstrates an enduring pattern of extraction of national wealth by foreign and domestic capitalists, the frustration of popular interests, and a reassertion of this pattern following Banzer's coup.

492 **Salas**, Guillermo P.
   1978 'Small scale mining in Mexico'. In *Small Scale Mining of the World*. Conference Proceedings, Jurica, Queretaro, Mexico, November, 1978. New York (?): United Nations Institute for Training and Research. Pp. 11–36.
Summarizes Mexican government policies and programs promoting small scale mining, accompanied by tables and charts covering current levels of production and finance.

493 **Samame Boggio**, Mario
   1972 *Minería Peruana, Biografía y Estrategía de una Actividad Decisiva*. Lima, Peru: Editorial 'La Confianza'.
Lengthly general description of Peruvian mining, written by mining engineer holding managerial and professional posts dealing with mines, engineering education. Proposes national plan; reflects author's campaign for Peruvian presidency.

494 **Schweider**, Dorothy Ann
   1983 *Black Diamonds: Life and Work in Iowa's Coal Mining Communities, 1895–1925*. Ames, Iowa: Iowa State University Press.
Surprisingly, coal was a major Iowa industry during the early 20th century. Using archival, census and interview data, author examines labor sources among European immigrants, the often seasonal nature of mining jobs, and union strength among Iowa coal miners during these years.

495 **Stucki**, Larry
  1973 'Who controls the Indians? Social manipulations in an ethnic enclave'.
  American Indian Urbanization. J. Waddell and O. Watson, eds. Institute
  for the Study of Social Change, Monograph 4. Lafayette, Indiana: Department of Sociology and Anthropology, Purdue University. Pp. 28–50.
Local politics in a copper company town, Ajo, using both Papago and other labor.
Company interests, local social stratification, a Tribal Council, and federal pressures
in the 1950s and 1960s are analyzed. Proposes abstract systems model as explanatory
framework.

496 **Sulmont**, Denis
  1975 *El Movimiento Obrero en el Perú, 1900–1956*. Two vols. Lima, Peru:
  Pontificia Universidad Católica del Perú.
Thorough analytical history of social and political development of Peruvian organized
labor, written by former student of Alain Touraine. Large data appendices and
bibliography. Limited material specific to miners, but larger context of Peruvian labor
movement is provided.

497 **Townley**, John M.
  1971 'El Placer: A New Mexico mining boom before 1946'. *Journal of the West*
  10(1): 102–115.
Short account of gold placer mining in two central New Mexico localities beginning
in the early 1830s, and how these efforts were affected by political changes leading
to the Mexican war.

498 **Villavicencio Chumacero**, Ismael
  1972 *Catavi: Analysis Histórico, Económico y Social de la Empresa Minera
  Catavi en el Vigésimo Aniversario de la Nacionalización de las minas.*
  Cochabamba: Serrano Hnos., Ltda.
Small descriptive handbook of Catavi, one of the largest state-owned Bolivian tin mines.

499 **Williams**, John Alexander
  1976 *West Virginia and the Captains of Industry*. Morgantown: West Virginia
  University Library.
Focuses on Johnson N. Camden, Henry G. Davis, Stephen B. Elkins and Nathan
B. Scott. Bituminous coal has been the leading industrial product of the state since
the 1890s; these four industrial magnates shaped the economy of contemporary West
Virginia.

General index

Index to bibliography

A glance at the index to follow will quickly reveal its plan. We have catalogued almost every source first geographically (usually by nation), and then mainly by whether it dealt with miners or mining. Those looking for all sources on, say, labour movement history will have to look under each country and, for some, sub-region. That will be a frustration, no doubt, but it is our perception that the descriptive youthfulness of mining studies continues to lead us more often to seek information first by place and then by topic.

Numbers identify *sources*, not pages.

Kentucky 43, 70, 82, 99, 127, 137, 201, 266,
291, 334, 335, 348, 351, 375, 379,
408, 416, 435
Michigan 478
miners (*see also entries by state or region*)
attitudes, work culture 7, 179, 192, 195,
234, 252, 255, 345, 436, 477
blacks, negroes 26, 82, 148, 188, 332, 335,
391, 418, 449
Cornish influence, history 386
convict labour 104
fiction and art 9, 447, 477
folklore and folk life 61, 167, 185
gold mining 29
health, safety and disabilities 7, 38, 50, 99,
163, 345, 422
pneumoconiosis (black lung) 37, 60, 86,
87, 92, 148, 149, 231, 248, 303, 304,
305, 344, 350, 403, 407, 408, 453, 479
labour's health concerns 55, 113, 171,
208, 288, 289, 349, 395, 453
labour movement
Brookside strike 43, 137, 291, 351
future 402
history 18, 25, 26, 38, 55, 61, 62, 63, 68,
74, 82, 84, 95, 96, 104, 120, 133, 136,
140, 141, 142, 148, 152, 161, 162,
176, 190, 191, 201, 212, 222, 227,
258, 263, 298, 321, 322, 332, 335,
357, 379, 391, 401, 411, 413, 416,
420, 450, 460, 494
immigrants and ethnicity 25, 436, 494
Miners for Democracy 36, 68, 88, 89,
208, 220, 336, 453
Mother Jones 140
National Miners' Union 124
ownership patterns, political economy 12,
162, 394
proletarianisation, class consciousness
401, 402, 461
political parties, participation, ideology
12, 84, 95, 124, 136, 159, 321, 322,
393, 404, 458, 459
strikes 2, 43, 62, 104, 120, 137, 141, 152,
161, 201, 291, 325, 351, 357, 419,
424, 435, 437, 450, 474
United Mine Workers of America
(UMW) 42, 89, 105, 136, 139, 142,
152, 176, 201, 222, 263, 298, 332,
337, 379, 394, 403, 404, 408, 411,
424, 481
Western Federation of Miners 64, 420
life narratives 1, 9, 13, 49, 82, 87, 114,
141, 205, 249, 281, 283, 438, 439
living and working conditions 7, 13, 61,
170, 171, 175, 179, 181, 205, 235,
251, 252, 254, 323, 329, 334, 338,
364, 416, 429, 430
native American miners 178, 214, 253,
413, 419, 495
psychology, physiology of mining 3, 50,

128, 163, 205, 213, 255, 256, 422
race relations 18, 26, 82, 148, 178, 188,
214, 253, 332, 335, 391, 413, 418,
449, 495
women miners, women of miners 87, 98,
192, 209, 356
mining (*see also entries by state or region*)
bibliography 65, 96, 223, 268, 307
capitalisation 225, 232
coal 2, 3, 7, 13, 20, 25, 29, 34, 37, 38, 43,
46, 49, 60, 61, 62, 70, 86, 98, 99, 105,
114, 119, 121, 127, 133, 134, 135,
137, 139, 161, 187, 190, 191, 194,
203, 205, 206, 209, 212, 225, 226,
227, 248, 249, 283, 298, 304, 324,
334, 340, 346, 348, 362, 363, 380,
385, 391, 397, 400, 404, 411, 412,
413, 414, 418, 423, 428, 461, 474,
494, 499
community management 1, 70, 134, 181,
416
copper, history 44, 268, 419, 456, 478,
495
entrepreneurs, owners 127, 187, 209, 268,
340, 499
gold, history 79, 118, 121, 214, 447, 448,
460, 497
governmental policy
environmental impact 44, 93, 121, 135,
178, 197, 329, 346, 375, 414, 441, 442
health, safety, accidents 20, 60, 149, 166,
171, 225, 231, 279, 288, 289, 290,
292, 295, 341, 354, 396, 417, 423,
425, 431
history 22, 38, 55, 119, 139, 214, 222,
223, 227, 386, 404
labour control 2, 104, 161, 212, 334, 393,
420
taxation 186
health, safety, accidents 33, 34, 37, 60, 86,
92, 98, 99, 101, 103, 105, 145, 146,
171, 198, 204, 225, 226, 248, 279,
304, 344, 346, 350, 358, 359, 388,
history 1, 18, 25, 35, 38, 46, 49, 61, 65, 70,
74, 79, 86, 90, 118, 119, 134, 139,
161, 166, 170, 187, 190, 191, 206,
209, 212, 214, 222, 225, 227, 232,
258, 261, 266, 268, 283, 298, 330,
340, 356, 362, 363, 371, 404, 411,
412, 416, 420, 428, 436, 447, 456,
460, 462, 463, 478, 494, 497, 499
homestake mine 79, 448
impact on local area, economy 1, 44, 134,
178, 190, 191, 205, 236, 237, 238, 249,
250, 253, 266, 280, 324, 385, 416,
456, 460
labour supply and management, labour
relations 2, 62, 104, 175, 194, 386,
397, 429, 430, 474, 494, 495
Mercury, history 371
mineral deposits and investment 107, 330